Impedances and Wakes in High-Energy Particle Accelerators

Impedances and Wakes in High-Energy Particle Accelerators

Bruno W Zotter
CERN, Geneva. Switzerland

Semyon A Kheifets
SLAC, Stanford, USA

World Scientific
Singapore • New Jersey • London • Hong Kong

Published by

World Scientific Publishing Co. Pte. Ltd.

P O Box 128, Farrer Road, Singapore 912805

USA office: Suite 1B, 1060 Main Street, River Edge, NJ 07661

UK office: 57 Shelton Street, Covent Garden, London WC2H 9HE

Library of Congress Cataloging-in-Publication Data
Zotter, Bruno W.
Impedances and wakes in high-energy particle accelerators / Bruno W. Zotter, Semyon A. Kheifets.
p. cm.
Includes bibliographical references and index.
ISBN 9810226268
1. Particle accelerators -- Design and construction. 2. Storage rings -- Design and construction. 3. Particle beams -- Design and construction. 4. Impedance (Electricity) 5. Wakes (Fluid dynamics) I. Kheifets, Semyon A. II. Title.
QC787.P3Z68 1997
539.7'3--dc21 97-44022
CIP

British Library Cataloguing-in-Publication Data
A catalogue record for this book is available from the British Library.

First published 1998
Reprinted 2000

This book is printed on acid-free paper.

Printed in Singapore by Uto-Print

Foreword

Modern particle accelerators and storage rings, whether used for high-energy physics, as synchrotron light sources, or for any other purpose, require particle beams with the highest possible intensity. In order to achieve optimum performance, a good understanding is required of the interaction of the charged particle beams with the surrounding structures.

This interaction can be described by impedances in the frequency domain, or equivalently by wake fields in the time domain. These quantities need to be known at least approximately in order to estimate the thresholds of coherent instabilities, or other collective effects, which may limit the achievable beam current. Such considerations have to be taken into account early during the design of accelerators, as they limit the choice of materials and shapes of components required for their operation.

The present book attempts to explain the basic concepts of impedances and wakes, and to describe a number of methods which have been used to calculate them, without any pretence of completeness. The emphasis is on cyclic or *circular particle accelerators* and storage rings, with which the authors are more familiar, but these concepts are equally useful for *linear accelerators* or colliders. Most likely the book will have a number of omissions of topics of special interest to particular readers. The collection of all the subjects presented here was a major task, and their descriptions will probably contain a number of misprints and misinterpretations for which the authors ask the readers' indulgence.

The calculation of impedances and wakes for a complete accelerator is practically impossible due to the enormous complexity of such a task. The usual approach is to calculate impedances of separate components and to add their individual contributions. This is permitted as long as the structures are far enough apart that their interaction can be neglected. Other limitations and restrictions of the impedance and wake field concept are discussed.

The book is mainly written for physicists and engineers working with or on particle accelerators or storage rings, and who want to understand some of the more common methods which have been used for the calculation of impedances. The book is *not* an introductory text, and a reasonable familiarity with basic particle accelerator concepts, as well as some knowledge of higher mathematics, such as special functions and calculus, is expected from the reader. A number of more detailed derivations are grouped in chapter appendices which can be skipped in a first reading.

We are grateful to our home laboratories *CERN - the European Laboratory for High-Energy Physics* and *SLAC - the Stanford Linear Accelerator Center* for technical support of this project, without which our task would have been substantially more difficult. We also want to thank our wives, Helgard and Julia, for their enduring patience and unmitigated support during the writing of the book.

Many people have helped in the preparation of this book, and we would like to express our gratitude in particular to B. M. Montague (*CERN*), P. Wilson and T. Fieguth (*SLAC*) for proof-reading the manuscript, K. Bane and S. Heifets (*SLAC*) for providing some material, and J. S. Berg, O. Meincke and A. Wagner (*CERN*) for helping with the intricacies of modern computer systems. We further express our appreciation to the *CERN* management, in particular to J. Gareyte for suggesting and supporting the writing of this book as a means of transferring our accumulated experience to the next generation.

Contents

Glossary: List of Symbols

Small Roman Letters

a .... beam radius
b .... tube radius
c .... light velocity $= 2.997925 \times 10^8$ m/s
d .... cavity radius
e .... electron charge $= 1.602 \times 10^{-19}$ C
f .... frequency
g .... length of cavity gap
h .... harmonic number, hole radius
j .... imaginary unit
k .... wave number $= \omega/v$
k_0 .... wave number $= \omega/c$
l .... length, summation index
m .... axial mode number
m_e .... electron rest mass $= 511.0$ KeV/c^2
m_p .... proton rest mass $= 938.2$ MeV/c^2
m, n ..... summation index (0/1 to ∞)
n .... longitudinal mode number $= \omega/\omega_0$
$p_{\parallel}$ .... longitudinal momentum
$p_{\perp}$ .... transverse momentum
p, q ...... summation index ($-\infty$ to ∞)
q .... charge
r .... radial coordinate
r_e .... classical electron radius $= e^2/4\pi\varepsilon_0 m_e c^2 = 2.818 \times 10^{-15}$ m
r_p .... classical proton radius $= e^2/4\pi\varepsilon_0 m_p c^2 = 1.535 \times 10^{-18}$ m
r_L .... loss factor ratio $= (\sigma\kappa_{\perp})_{max}/(\sigma\kappa_{\perp})_{\infty}$
s .... longitudinal distance from bunch center
s, t .... summation index
t .... wall thickness, time
u, v ...... transverse or radial coordinates
v .... velocity $= \beta c$
w .... complex error function
x .... horizontal coordinate
y .... vertical coordinate
z .... axial coordinate

Large Roman Letters

A .... vector potential, Amperes
B .... magnetic induction $= \mu H$
C .... machine circumference $= 2\pi$R, capacitance
$\mathcal{C}$ .... integration path
D .... electric displacement $= \varepsilon E$
E .... electric field strength (time domain)
$\widetilde{E}$ .... electric field strength (frequency domain)
$\mathcal{E}$ .... particle energy
$\mathcal{E}_0$ .... rest energy $= mc^2$
F .... force
$F_{0,1}$ ...... Bessel function combination
G .... conductance
$G_{0,1}$ ...... Bessel function combination
$G_{\|}$ .... longitudinal wake function
$G_{\perp}$ .... transverse wake function
H .... magnetic field strength
I .... current, unity matrix
$I_{0,1}$ ...... modified Bessel function first kind
J .... current density
$J_{0,1}$ ...... (regular) Bessel function first kind
K .... surface current density
$K_{0,1}$ ...... modified Bessel function second kind
L .... inductance, length, $|\mathbf{r} - \mathbf{r}'|$
O .... order of expansion term
P .... period length, power
Q_r .... quality factor
Q'_r .... reduced quality factor $= \sqrt{Q_r^2 - 1/4}$
Q''_r .... reduced quality factor $= \sqrt{1/4 - Q_r^2}$
R .... average machine radius $=$ C$/2\pi$
R_s .... longitudinal shunt impedance, resistance,
$R_{\perp}$ .... transverse shunt impedance (resistance)
$R_{0,1}$ ...... modified Bessel function combination
$\mathcal{R}(\omega)$ .... response function
S .... energy flow
$\mathcal{S}$ .... surface of integration

$S_{0,1}$ ...... modified Bessel function combination
T_0 .... revolution time
U .... field energy
$V_{(ind)}$ ..... (induced) voltage
$W(\tau)$ .... wake potential
$W_{\parallel}$ ...... longitudinal wake potential
$W_{\perp}$ ...... transverse wake potential
$X(\omega)$ .... reactance
$Y(\omega)$ ..... admittance
$Y_{0,1}$ ...... Neumann functions (Bessel functions 2-nd kind)
$Z(\omega)$ ..... impedance
$Z_{\parallel}$ .... longitudinal impedance
$Z_{\perp}$ .... transverse impedance
Z_0 .... impedance of free space $= \sqrt{\mu_0/\varepsilon_0} = 120\ \pi\Omega = 377\Omega$
Z_{eff} ..... effective impedance
Z_{hom} ..... higher-order mode impedance
Z_{sc} ...... space-charge impedance

Small Greek Letters

α .... circumference factor $= g/2\pi R$
α_L .... resonant frequency factor $= 2f_r/\pi\hat{\sigma}$
α_p .... momentum compaction factor $=\mathcal{E}\partial C/C\partial\mathcal{E} = pdR/\beta^2 Rdp$
α_s .... (axial) propagation constant
α_E .... electric polarisibility
α_M ...... magnetic polarisibility
β .... beta function, velocity factor $= v/c$,
γ .... energy factor $= \mathcal{E}/mc^2$
γ_r .... damping decrement
δ_s .... skin depth $= \sqrt{2\rho_c/\mu\omega}$
δ_E .... electric loss angle (in $\tan\delta_E$)
δ_M .... magnetic loss angle (in $\tan\delta_M$)
ε .... permittivity (dielectric constant)
ε_0 .... permittivity of free space $= 1/(36\ \pi 10^9)$ C/Vm
ε_c .... complex permittivity$= \varepsilon(1 - j\tan\delta_E) + \sigma_c/j\omega$
ε_r .... relative permittivity $= \varepsilon/\varepsilon_0$
ϵ_{0n} .... Neumann symbol

ζ .... surface impedance
η .... surface charge density, cavity-step transition parameter
θ .... azimuthal coordinate
$\kappa_{\parallel}$ .... longitudinal loss factor
$\kappa_{\perp}$ .... transverse loss (kick) factor
λ .... line density, wave length
λ_s .... longitudinal propagation constant $= \sqrt{k^2 - \mu_s^2}$
μ .... permeability
μ_0 .... permeability of free space $= 4\pi 10^{-7}$ Vs/Am
μ_c .... complex permeability $= \mu(1 - j\tan\delta_M)$
μ_r .... relative permeability $= \mu/\mu_0$
μ_s .... propagation constant $= \pi s/g$
ν_p .... radial propagation constant $= \sqrt{k^2 - p^2}$
ν_s .... synchrotron tune $= \omega_s/\omega_0$
$\nu_{(\beta)}$ ...... betatron tune $= \omega_\beta/\omega_0$
ξ .... parameter for cavity-array transition
ξ_β .... chromaticity $= pd\nu_\beta/\nu_\beta dp$
ρ .... bending radius, charge density
ρ_c .... resistivity
σ .... rms bunch length (in time) $= \sigma_s/v$
σ_c .... conductivity $= 1/\rho_c$
σ_s .... rms bunch length (in distance)
τ .... distance behind bunch center (in time)$= s/v$
$\tau_{(k)}$ ...... propagation constant $= k/\gamma$
ϕ .... azimuthal angle
ϕ_{st} .... matrix element Eq.(8.34)
φ .... phase angle
φ_s .... synchronous phase angle
χ .... propagation constant $= \sqrt{k^2 - \mu\varepsilon\omega^2}$
ψ .... phase advance per cell
ω .... (circular) frequency $= 2\pi f$
ω_0 .... revolution frequency $= 2\pi f_0$
$\omega_{1,2}$ ...... complex frequency poles $= \omega_r/Q_r(j/2 \pm Q_r')$
ω_r .... resonant frequency
ω_s .... synchrotron frequency $= \mathrm{Q}_s\omega_0$
ω_β .... betatron frequency$= \mathrm{Q}_\beta\omega_0$

Large Greek Letters

Γ .... Gamma function, damping decrement
Δ .... difference, Laplacian
Λ .... ratio of time to spatial mesh size
Π .... Hertz potential
Σ .... summation symbol
Σ_s .... surface charge
Φ .... scalar potential, magnetic flux
Ψ .... Debye potential
Ω .... Ohm

Abbreviations and Acronyms

EM ...... electro-magnetic
HOM .... higher-order mode
rf .... radio-frequency
rms ..... root mean square
e.g. ...... for example
i.e. .... namely
l.h.s. ..... left hand side
r.h.s. .... right hand side
w.r.t. .... with respect to

General Conventions

vectors are in bold characters, e.g. ($\mathbf{F}$),

matrices are in script characters, e.g. ($\mathcal{F}$),

Fourier transforms are indicated by a tilde, e.g. ($\widetilde{F}$).

Chapter 1

Introduction

1.1 Particle Accelerators and Storage Rings

We begin this book with a description of the basic characteristics of *high-energy particle accelerators* and *storage rings.* We concentrate mainly on accelerators which close on themselves - usually called cyclic or *circular*, although their real shape may be more complicated, e.g. they may contain a number of straight sections or even inverse bends. Particle beams in such machines perform a periodic motion around the circumference with the *revolution time* T_0. More often we will make use of the *circular revolution frequency* $\omega_0 \equiv 2\pi f_0 = 2\pi/T_0$. However, the concepts of impedances and wakes discussed below are equally useful for *linear accelerators*, colloquially called *linacs.* For such single-pass machines, the discrete harmonics of the revolution frequency have to be replaced by a continuous frequency spectrum.

In general, *electromagnetic* (*EM*) fields are used to store and accelerate beams of charged particles. These beams must move in high vacuum in order to avoid particle loss by scattering on the residual gas molecules. The vacuum chamber usually consists of a metallic pipe which is a good conductor of electric current. However, it may also be made - at least partially - of insulating materials such as ceramics, e.g. for fast cycling synchrotrons where *eddy currents* induced in conducting chamber walls by the varying magnetic field could be destructive for the beam. But even in that case the inner surface of the chamber is often covered by a thin metal layer, or protected by metallic shields, in order to prevent distortions of the beam motion by electric charges accumulating on the insulating wall.

In many accelerators - in particular those operating with light particles such as electrons and positrons - the particles are grouped into one or more *bunches* such that all particles can pass the applied rf fields in the correct, accelerating phase, needed to gain energy or to overcome the energy loss due to emission of synchrotron radiation. On the other hand, storage rings for heavy particles, such as protons or ions, may operate with *continuous beams* as long as the loss of energy by synchrotron radiation is negligible. A coasting, uniform beam suffers no energy loss due to interaction with the walls since it does not induce any currents there.This is strictly correct only as long as the beam and the vacuum chamber have the same cross section at the entrance and the exit of a structure (this is obviously always the case for an arbitrary point on the circumference of a circular machine) and if the fluctuations of the beam current are negligible.

In most cases, the acceleration is provided by a set of *resonant cavities* which are fed by high-frequency rf power source. Usually the cavities are part of the vacuum system of the accelerator, but they may also be separated from it by insulating *windows*. The cavities are designed either to provide standing *EM* waves, or travelling ones with a phase velocity reduced to be approximately equal to the instantaneous particle velocity. The reduction of the phase velocity of a *EM* wave is achieved by *loading* the vacuum pipe with obstacles. An example of this is the disc-loaded accelerating structure used mainly in linacs.

1.2 Interaction of Beams with Surroundings

The core of an accelerator, the part which is in closest proximity to the beam of accelerated particles, is the vacuum vessel. Its design and materials may be quite different in various accelerators. But the main components are usually the same: beam pipes with expansion bellows, connection flanges, and pumping ports; accelerating rf cavities; monitors, collimators, kickers, separators, and similar beam handling equipment. Such complexity of the vacuum vessel, as well as the diversity of materials out of which its elements may be built, constitutes an unavoidable non-uniformity of the beam surroundings.

A charged particle beam generates *EM* fields at any cross-section variation of the vacuum vessel, whether it is abrupt or a gradual (or *tapered*) transition, only the strength of the interaction will be different. These fields may be trapped or propagate in the vacuum chamber, but in any case they react back on the beam and interact with the elements of the vacuum chamber.

Generally the *EM* fields generated by the beam contain three spatial components. These are usually classified as longitudinal, i.e. directed along the average particle motion, and transverse (horizontal and vertical, or radial and azimuthal). The longitudinal component of the electric field may change the effective amplitude and phase of the applied accelerating field, and thereby the rate of acceleration, as well as the energy distribution in the bunch and its effective length. This in turn may influence the optics of the accelerator, in particular by changing the focusing, and the *closed orbit* around which particles perform transverse *betatron* oscillations. Similarly, transverse field components may increase the cross section of a bunch, change its closed orbit, its frequency of oscillation around it, and shorten the life time of the beam in storage rings due to increased loss of particles.

These effects become more severe at higher beam currents. Depending on the design of the vacuum chamber and other accelerator elements, longitudinal and transverse instabilities of a beam will occur above some *threshold current.* Then the beam may start to oscillate, develop tails and halos, and be lost partially or even completely.

The interaction of *EM* field with charged particles is most effective when one of the field components has a phase velocity v_{ph} close to the particle velocity. Such a field component is called *synchronous.* When the phase φ_s of a particle with respect to that of the longitudinal electric field $E_z = E_{zo} \sin \varphi_s$ is such that $\sin \varphi_s > 0$, the particle gains energy from the wave and is accelerated. In the opposite case it is decelerated and the amplitude of the wave increases.

The strength of the interaction can be characterised by the *coupling impedance* which is defined as the integral over the normalised Fourier transform of the *EM* force along the particle trajectory. Thus the longitudinal impedance is the integral over the synchronous component of the corresponding electric field strength normalised (divided) by the charge. The transverse impedance is the averaged component of the transverse *Lorentz force*, normalised by the dipole moment of the charge in the direction considered.

Other important considerations are the effects of the induced *EM* fields on elements of the vacuum chamber. High intensity *EM* fields can excessively heat sensitive elements close to the beam, which may damage or even destroy them in some cases or produce undesirable increase in vacuum pressure. The generation of *EM* fields by the beam also increases the required power output of the rf generators.

Particles which are lost from the beam due to induced *EM* fields may hit the walls of the vacuum chamber and thereby produce excessive radiation. The characteristics of this radiation are of importance for designing personnel protection systems for an accelerator. Furthermore, stray particles can degrade electronic equipment or cable insulation, and even damage the vacuum chamber in severe cases.

For all these reasons, the elements of the vacuum chamber should be designed to *minimise* their coupling impedances. For example, chambers with different cross sections should be connected with tapered transitions (tapers); bellows need to be separated from the beam by shielding; plates should be grounded or terminated to avoid reflections, etc. In general, it requires that vacuum chambers should be as smooth as possible.

1.3 Wake Fields

When traversed by charged particles, cross section variations of the vacuum chamber wall generate *EM* fields which are called *wake fields* since they remain usually behind the (ultra-relativistic) exciting particles. These wake fields can influence the motion of trailing particles, in the longitudinal and in one or both transverse directions. This may lead to *energy loss*, to *beam instabilities*, or produce undesirable secondary effects such as excessive heating of sensitive components at or near the chamber wall. Without a good knowledge of these wake fields and of their interactions, an accelerator can therefore hardly be operated at the desired top performance. For high-energy beams below the threshold of instability one can usually assume that the average particle motion remains undisturbed by the induced fields. This assumption is well justified when particles move with nearly light velocity, where their relativistic mass becomes very large. It is often sufficient to study the integrated effect of all fields of which those synchronous with the beam motion comprise the wake fields. The normalised integral over the *EM* force due to fields excited by a point charge or *delta function distribution* is called the *wake function.* It is a function of the distance (or time delay) between the exciting and test charges. Since it is the response to a delta function excitation, it is also a *Green function* for the structure investigated. In the limit $v \to c$, the *wake function* depends only on the surrounding structure.

On the other hand, the integrated effect over a finite distribution of charged particles is described by the *wake potential* (actually per unit charge or - in the transverse case - per unit moment). Using the superposition property of the (linear) Maxwell equations, the wake potential can be determined by the convolution of the wake function with the charge distribution in the bunch. Consequently, the wake potential depends not only on the structure, but also on the charge distribution which excites it. Wake functions and wake potentials will be discussed in Chapter 3.

In order to determine a wake function one needs solutions for the propagation, diffraction and interference of *EM* fields radiated by the beam, which requires solving the inhomogeneous Maxwell equations including source terms with the appropriate initial and boundary conditions. Analytically this is possible only for a very limited number of idealised geometries. More realistic cases need to be solved numerically, but this cannot be done for a delta-function excitation with the commonly used mesh codes, since the bunch length must be multiple of the mesh size. The wake function is therefore often approximated by the wake potentials of very short bunches, but the bunches can rarely be chosen as short as would be desired, since both computing time and required memory size increase drastically with the smaller mesh sizes needed. Hence the wake function close to the origin - where it is most important for short bunches - cannot be determined correctly.

It is often suggested to obtain the wake function from the wake potential of a finite bunch by a *deconvolution*. However, this is not possible without severe loss of accuracy, due to the enormous increase of the weighting function proportional to the inverse of the distribution function at the tails. This fact is easier to understand in the frequency domain, since the effective impedance of a long bunch simply does not contain the high-frequency components excited by a short bunch, and in particular by a delta function pulse.

By expanding an arbitrary particle distribution into a (finite) set of basis functions, its wake potential can be obtained by superposition of the wake potentials of these bases. This can be done for example with *Hermite polynomials* multiplied with a Gaussian distribution [Siemann83, Nys86]. Since these functions are orthogonal, the expansion coefficients can be obtained rapidly. However, the convergence of such series is not very good except for distributions which do not differ significantly from a Gaussian.

The approximation of a distribution function with a number of shifted, simple basis functions is more promising. In particular, piecewise linear functions (e.g. rectangles or equilateral triangles [Sabbi95]) have given good results. A distribution can be divided into a rather large number of slices, and the wake potential of the basis functions needs to be obtained only once. It can be stored in a lookup table, which is particularly well suited for computer simulation.

Triangles are better suited than rectangles since they do not have infinitely steep edges, and thus their wake potentials converge faster to the desired accuracy. However, for structures with a very large high-frequency content, it is better to use higher-order polynomial basis functions or even shifted Gaussians to reduce the numerical noise appearing in the wake potential computation.

In many cases it is sufficient to consider effects averaged over the particle distribution, e.g. for the total energy loss, or for changes of the transverse momenta ('kicks'), which a bunch experiences as a whole when passing through a structure. These averaged effects are described by the loss factors and effective impedances which are discussed in Chapter 5.

1.4 Impedances

Analytic calculations are usually easier in the frequency domain, where one uses the concept of *coupling impedances* which are the *Fourier transforms* of the wake functions. From this definition it follows that they are integrated quantities which can be used only under the assumption that the average beam motion remains essentially undisturbed. Then they are a property of the surrounding structure only, and do not depend on beam parameters except on its velocity v. The particle velocity in high-energy accelerators is generally *ultra-relativistic*, i.e. close to the speed of light, and most of the time we will therefore assume $v \to c$. Impedances will be described in more detail in Chapter 4. We will denote the Fourier transform of any time-dependent function $F(t)$ by $\widetilde{F}(\omega)$. The beam in the time domain is usually taken to be a line charge, moving on a straight trajectory parallel to the axis, and which in the frequency domain can be analysed at any frequency. The impedance is in general a complex function, the real and imaginary parts referring to field components which are either in phase with, or in quadrature to the oscillation of the source. The real, i.e. *resistive*, part of the longitudinal impedance describes the energy loss, while the imaginary, i.e. *reactive*, part shifts the frequency of the beam oscillation. The transverse impedance is conventionally defined per unit displacement and with an additional factor j (the imaginary unit). This makes its dimension *Ohm/m* and takes into account that the transverse *EM* forces do not act on the displacement itself, but rather on its derivative. The symmetry properties of real and imaginary parts of the transverse impedance are inverted in respect to the longitudinal one. If the frequency shifts of any two low-order oscillation modes lead to their degeneracy, a *mode-coupling instability* may occur.

Multi-bunch instabilities depend mainly on narrow resonances at frequencies below or near the cut-off of the beam pipe. Above this frequency, the *EM* fields can propagate out of a structure and thus will not build up resonantly, unless they are reflected back by adjacent cross-section variations. In the time domain, such narrow-band resonances correspond to long-range oscillations caused by resonant modes with a *long memory*.

Single bunch instabilities depend on the short range wake fields, corresponding to *wide-band impedances* which occur mostly at higher frequencies. For a cavity the number of resonant modes increases rapidly with frequency, their widths tend to grow and their amplitudes tend to decrease. Hence the peaks overlap more and more at high frequencies and produce a smoothly falling frequency dependence of the impedance. Another contribution to these *high-frequency tails* comes from the unavoidably large number of very small cross-section changes of beam pipes. These may become important for short bunches which may be influenced by very high frequencies since their spectrum reaches to frequencies of the order of $\omega \approx 1/\sigma$.

The *EM* fields usually decay sufficiently fast - due to the finite resistivity of the vacuum chamber walls, etc. - that fields left over from previous turns can be neglected for single bunch effects. Then wake fields behind a bunch have no effect on its motion. In the frequency domain, this corresponds to the fact that the detailed frequency behaviour of the impedance cannot be resolved during the short time of a bunch passage. Therefore the actual impedance consisting of many narrow resonances can be replaced by a smooth, broad-band one. Because of its simplicity, the *broad-band resonator model* [Hofmann79b] turned out to be particularly useful as it permits analytic solution of many problems.

The design of a new generation of accelerators, such as high-energy colliders, synchrotron light sources, storage rings for the production of various mesons (ϕ, τ, and $B - factories$) etc. is often based on the use of very short bunches. In this case, beam stability and energy loss are largely determined by the behaviour of the impedance at high-frequencies, which led to a number of studies [Bisognano88, Heifets90d] and the development of more elaborate models as described in Chapter 11 [Hofmann88, Palumbo94].

1.5 Coordinates, Dimensions and Units

In order to make practical applications easier we shall use the MKSA system of units, although the use of the CGS system is still quite common, in particular in the USA. However, the MKSA system is better adapted to electrical quantities and is replacing the CGS system more and more. A conversion table for the more common electrical quantities can be found e.g. in ref.[Jackson75]. Usually a positive charge is assigned to a proton, and a negative one to an electron. Hence the direction of the beam current in an electron accelerator is opposite to the direction of particle motion.

In the frequency domain, we shall generally describe oscillations with complex variables. Throughout the book, they are understood to be proportional to $\exp(j\omega t)$, which is sometimes called the *electrical engineering convention.*

The equally valid convention $\exp(-i\omega t)$ is often preferred in physics, but for a subject dealing mainly with *EM* quantities the first choice seems more appropriate. The two conventions can be easily converted into each other simply by replacing the imaginary unit j by $-i$ or vice versa.

Throughout this book we use a right-handed system of coordinates. The choice of suitable coordinates is dictated by the geometry of the system, in particular, its symmetry properties. The most common geometry encountered in accelerators is the circular cylindrical geometry with axial (rotational) symmetry (often not quite rigorously called *cylindrical* - which really refers to any translationally uniform cross section rather than to the implied circular cylindrical one). Cylindrical coordinates r, θ, z are therefore used most often to describe the geometry of vacuum chamber elements. In most cases it is convenient to direct the z axis along the average beam velocity. The choice of the origin of the longitudinal axis is usually arbitrary, while the radial coordinate is most often centered on the beam axis.

In general, the *rational* MKSA system of units (*Meter* m, *Kilogram* kg, *Second* s, *Ampere* A) is used, but we often replace the current by the charge measured in Coulombs C, and measure potential differences in Volts V. A detailed discussion of this and other systems of units can be found, e.g. in ref.[Jackson75]. Table 1.1 contains dimensions and numerical values of relevant electromagnetic quantities.

Since a charge of 1 *Coulomb* contains $0.625 \cdot 10^{19}$ electrons, it is rather inconvenient to use MKSA units for single particles. Hence in the accelerator literature more practical units are used, e.g. the particle energy is measured in *electron-Volt* (eV)(or its multiples KeV $= 10^3$ eV, MeV $= 10^6$ eV, GeV $= 10^9$ eV, TeV $= 10^{12}$ eV). 1 eV is the energy which a particle of charge e acquires by passing through a potential difference of 1 V. In MKSA units this energy is approximately $1.6 \cdot 10^{-19}$ Joule.

The energy of the electron at rest is $\mathcal{E}_e = m_e c^2 = 0.511$ MeV, which corresponds to a rest mass $m_e = 9.11 \cdot 10^{-31}$ kg. The classical electron radius is $r_e = e^2/4\pi\varepsilon_0 m_e c^2 = 2.81810^{-15}$m. For a proton, the values are $\mathcal{E}_p = 938.24$ MeV, $m_p = 1.67 \cdot 10^{-27}$ kg and $r_p = 1.53510^{-18}$m.

The numerical values of several useful constants are given below:

the *permeability of free space* $\mu_0 = 4\pi 10^{-7} \approx 1.257 \cdot 10^{-6}$ Vs/Am;

the *permittivity of free space* $\varepsilon_0 = 1/\mu_0 c^2 \approx 8.854 \cdot 10^{-12}$ As/Vm;

the *velocity of light* in vacuum $c = 1/\sqrt{\varepsilon_0 \mu_0} \approx 2.998 \cdot 10^8$ m/s;

the *impedance of free space* $Z_0 = \sqrt{\mu_0/\varepsilon_0} \approx 120\pi\ \Omega \approx 376.6\ \Omega$.

Materials are characterised by their permittivity or *dielectric constant* $\varepsilon = \varepsilon_0 \varepsilon_r$, permeability $\mu = \mu_0 \mu_r$, and conductivity σ_c. The quantity ε_r is called *relative permittivity*, μ_r *relative permeability*, and $\rho_c = 1/\sigma_c$ *resistivity*. All these quantities depend in general on frequency.

The conductivity[1] of metals is usually quite high, e.g. for high purity copper, from which rf cavities are often made, conductivity is about $5.8 \cdot 10^7$ S/m, for aluminium $3 \cdot 10^7$ S/m, and for stainless steel only $1 \cdot 10^6$ S/m.

The *complex permittivity* is defined by $\varepsilon_c = \varepsilon(1 - j \tan \delta_E) - j\sigma_c/\omega$, where $\tan \delta_E$ is the *electric loss tangent*. For dielectrics, the low frequency relative permittivity is usually larger than one, e.g. typically 8–10 for some common ceramic materials. On the other hand, the conductivity of such materials is generally very small, e.g. for glass it is of the order of 10^{-12} S/m. Furthermore it may depend on external factors such as temperature, humidity and, for anisotropic materials, also on the direction of the field.

For magnetic materials the resistive losses are usually specified by the *magnetic loss tangent* $\tan \delta_M$. Then the *complex permeability* is given by $\mu_c = \mu(1 - j \tan \delta_M)$.

In homogeneous, isotropic, continuous media, the parameters ε and μ are constants and can be represented as the products $\varepsilon_0 \varepsilon_r$ and $\mu_0 \mu_r$, where ε_r is the *relative permittivity*, and μ_r the *relative permeability*.

[1] We use the unit *Siemens* = 1/*Ohm* for conductance, while in the USA it is more common to use *mho* for an inverse Ohm.

Table 1.1: **Dimensions of Electromagnetic Quantities**

Quantity	**Symbol**	**Dimensions**	**MKSA unit**
Force	F	m kg s^{-2}	Newton N
Energy	$\mathcal{E}$	m^2kg s^{-2}	Joule J
Power	P	m^2kg s^{-3}	Watt W = J/s
Current	I	C s^{-1}	Ampere A
Current Density	J	C $s^{-1}m^{-2}$	A/m^2
Charge	q	C	Coulomb C = A s
Charge Density	ρ_c	C m^{-3}	C/m^3
Linear Density	λ	m^{-1}	1/m
Potential	V	m^2kg $s^{-2}C^{-1}$	Volt V = J/C
Electric Field Strength	E	m kg $s^{-2}C^{-1}$	V/m
Resistance	R	m^2kg $s^{-1}C^{-2}$	Ohm Ω = V/A
Conductance	G	$m^{-2}kg^{-1}$s C^2	Siemens S = $1/\Omega$
Conductivity	σ_c	$m^{-3}kg^{-1}$s C^2	S/m
Capacitance	C	$m^{-2}kg^{-1}s^2C^2$	Farad F = A s/V
Permittivity	ε	$m^{-3}kg^{-1}s^2C^2$	A s/V.m
Magnetic Field Strength	H	$m^{-1}s^{-1}C$	A/m
Magnetic Flux	Φ	m^2kg $s^{-1}C^{-1}$	Weber W = V s
Magnetic Induction	B	kg $s^{-1}C^{-1}$	Tesla T = V s/m^2
Inductance	L	m^2kg C^{-2}	Henry H = V s/A
Permeability	μ	m kg C^{-2}	V s/A.m
Longitudinal Wake Potential	$W_{\parallel}$	m^2kg $s^{-2}C^{-2}$	V/C
Transverse Wake Potential	$W_{\perp}$	m kg $s^{-2}C^{-2}$	V/C m
Longitudinal Coupling Impedance	$Z_{\parallel}$	m^2kg $s^{-1}C^{-2}$	Ω
Transverse Coupling Impedance	$Z_{\perp}$	m kg $s^{-1}C^{-2}$	Ω/m
Longitudinal Loss Factor	$\kappa_{\parallel}$	m^2kg $s^{-2}C^{-2}$	V/C
Transverse Loss Factor	$\kappa_{\perp}$	m kg $s^{-2}C^{-2}$	V/C m

Appendix 1.A Generalised Concept of Impedance

The concept of an *impedance* is useful for studying any *linear system*. In particular, it is especially suitable for *EM* problems due to the fact that the Maxwell equations are linear. The idea behind this concept is to study a system by analysing its response to a monochromatic excitation. The concept of an impedance is intimately connected to the Fourier analysis. The analogous concept in the time domain is the *wake function*, which describes the response to a short pulse or *delta-function*. The description in the frequency domain is usually simpler and more adequate for analytic treatment than in the time domain. We recall that the Fourier transform of any time-dependent function $F(t)$ is denoted by $\widetilde{F}(\omega)$.

1.A.1 Impedance of Mechanical Systems

The concept of an impedance can be applied in general to any linear mechanical system [Morse53]. This is often described by pairs of canonical variables q_m, p_m where the index m refers to the m-th independent coordinate.

The dynamics of such a system can be characterised by two quantities. One of them is the work performed by the external driving force on the system in the direction of the m-th coordinate. This is given by the product of the m-th component of this force F_m and the corresponding component of the response velocity $\dot{q}_m$:

$$W_m = \dot{q}_m F_m \ . \tag{1.1}$$

The other quantity is the ratio of the external driving force to the velocity. This quantity is called the *impedance* in the m-th direction. The impedance can be written as a coefficient in the expression:

$$\widetilde{F}_m = \sum_n Z_{mn} \dot{q}_n \ . \tag{1.2}$$

While the work W_m is a purely real quantity, the impedance is in general complex. However, as will be shown later, its imaginary part is the Hilbert transform of the real part, and in general does not carry independent information. On the other hand, if the real part is known only over a limited frequency range, the imaginary part does provide additional information.

The concept of an impedance is particularly useful when Z_{mn} does not depend on the strength (or amplitude) of the force. In most cases that happens in the vicinity of a local minimum of the potential energy of the system. There the potential energy is a quadratic form in q which can be written $\frac{1}{2}\sum b_{mn}q_m q_n$. The kinetic energy is another quadratic form in p which can be written $\frac{1}{2}\sum a_{mn}p_m p_n$. The system thus is linear in this region. If the coefficients a_{mn} and b_{mn} do not depend on time, we get:

$$F_m = \sum_n \left[a_{mn}\ddot{q}_n + b_{mn}q_n\right] . \tag{1.3}$$

For a dissipative system, the expression for the force acquires an additional term $\sum_n r_{mn}\dot{q}_n$, where the coefficients r_{mn} describe the *friction* in the system.

If the external force $F_m = \hat{F}\exp(j\omega t)$ is a single harmonic with amplitude $\hat{F}$ at frequency ω, and the velocity is $\hat{v}\exp(j\omega t)$, then the impedance Z_{mn} defined by Eq. (1.2) becomes a matrix of the form:

$$Z_{mn}(\omega) = j\omega a_{mn} + \frac{b_{mn}}{j\omega} + r_{mn} . \tag{1.4}$$

The inverse matrix Y_{mn} is defined in such a way that $\sum_i Y_{mi}Z_{in} = E_{mn}$, where E_{mn} is the unit matrix. It is called the *admittance* of the system. The impedance Z_{mn} or the admittance Y_{mn} describe a system in the frequency domain. The behaviour of the system close to its equilibrium can be understood from the properties of these functions.

The elements of these matrices depend on the choice of canonical variables q_m and p_m. In particular, by a suitable choice, the matrix can be brought into diagonal form. The elements of such a matrix are called its *eigenvalues* and are invariants of the system. The trace (i.e. the sum of the diagonal elements) and the determinant (i.e. the product of the eigenvalues) of the impedance (or admittance) matrix are also invariants. These invariants fully define the properties of the impedance of the system.

Whenever one of the eigenvalues of the impedance matrix equals zero at some frequency ω_r, the reciprocal matrix can not be defined. Such a frequency is called a *resonant frequency* of the system. In the vicinity of these resonances, this eigenvalue Z_r^0 can be written in the following form:

$$Z_r^0 = \frac{1}{Y_r^0} = j\omega M_r + \frac{K_r}{j\omega} + R_r , \tag{1.5}$$

where M_r, K_r and R_r are (real) constants which are determined by the matrices a_{mn}, b_{mn} and r_{mn}. One can see that the eigenvalues of the impedance of a conservative system (i.e. a system without dissipation for which $R_r = 0$) are purely imaginary. For such a non-dissipative system the resonant frequency is then given by

$$\omega_r = \pm\sqrt{\frac{K_r}{M_r}} \ . \tag{1.6}$$

For a dissipative system ω_r acquires an additional imaginary part.

The expression Eq. (1.5) in the vicinity of the resonant frequency ω_r can be rewritten in the form:

$$Z_r = \frac{jM_r}{\omega}(\omega - j\gamma_r - \omega_r)(\omega - j\gamma_r + \omega_r) \ , \tag{1.7}$$

where the constant $\gamma_r = R_r/2M_r$ is always positive. From this expression it is clear that the roots of the impedance are located in the upper half of the complex frequency plane. The time dependence of the oscillation is proportional to $\exp(-\gamma_r t + j\omega_r t)$. The oscillation is damped, as required for a dissipative system, and γ_r is called the *damping decrement.* The expression Eq. (1.5) shows that Re $Z_r(\omega)$ is an *even*, and Im $Z_r(\omega)$ an *odd* function of frequency.

Many properties of the impedance found here appear to be true for the impedances, which will be studied in this book.

1.A.2 Impedance of Equivalent Circuits

The concept of an impedance is also quite useful for a system of electric circuits. The reason for that is simple: the relation between the instantaneous current $I(t)$ and the electromotive force $\mathcal{E}(t)$ is linear for linear circuits with constant resistance R (or conductance $G = 1/R$), inductance L and capacitance C. The current plays the role of a velocity, and the electromotive force corresponds to a driving force considered in the previous section.

The assumed linearity makes the concept of an impedance applicable to the whole range of circuit parameters. The most general linear relation between $\mathcal{E}$ and I can be written [Landau60]:

$$\mathcal{E}(t) = \mathcal{Z}I(t). \tag{1.8}$$

In this expression $\mathcal{Z}$ is an arbitrary linear operator. Expanding the functions $\mathcal{E}(t)$ and $I(t)$ into Fourier integrals, one gets for each monochromatic oscillation, with complex amplitudes $\tilde{\mathcal{E}}$ and $\tilde{I}$:

$$\tilde{\mathcal{E}} = Z(\omega)\tilde{I} \ , \tag{1.9}$$

where the complex function $Z(\omega)$ is called *circuit impedance.* The dependence of this impedance on frequency can be expressed in terms of the circuit parameters L, C and R, e.g. by using the law of the energy conservation.

We consider first a *series* R, L, C resonator circuit. By the definition of the electromotive force $\mathcal{E}$, the work per unit time, done by the electric field on a charges moving in the circuit, is the product $\mathcal{E}I$. This work is spent partly to change the energy stored in the magnetic field of the current $LI^2/2$, partly to change the energy stored in the electric field of the charges $q^2/2C^2$ and partly to cover the Joule losses in its resistance RI^2:

$$\mathcal{E}I = LI\frac{dI}{dt} + \frac{q}{C}\frac{dq}{dt} + RI^2 \ . \tag{1.10}$$

Since the current $I = dq/dt$, this can be rewritten in the form:

$$\mathcal{E}(t) = L\frac{dI}{dt} + RI + \frac{1}{C}\int Idt \ . \tag{1.11}$$

Each term on the right hand side of this equation represents the partial voltage drop over the corresponding element of the circuit. Hence the sum of these voltages equals the electromotive force applied to the circuit.

For an electromotive force and a current which are periodic in time with the circular frequency ω, this equation gives the linear relation between the amplitude of the Fourier component of the electromotive force $\tilde{\mathcal{E}}$ and the amplitude of the current $\tilde{I}$. The coefficient of proportionality in this relation is the impedance:

$$Z_{ser}(\omega) = R + j\left(\omega L - \frac{1}{\omega C}\right) \ . \tag{1.12}$$

The zeros of the impedance of this circuit are given by

$$\omega_{1,2} = \pm\sqrt{\frac{1}{LC} - \frac{R^2}{4L^2}} - j\frac{R}{2L} \ . \tag{1.13}$$

The *quality factor* Q of the circuit can be expressed as

$$Q_r = \frac{1}{R}\sqrt{\frac{L}{C}} \ . \tag{1.14}$$

When the resistivity R is so small that $Q_r \gg 1$, the resonant frequency can be approximated by the expression:

$$\omega_{1,2} \simeq \pm\omega_r - j\gamma_r \; , \tag{1.15}$$

where

$$\omega_r = \frac{1}{\sqrt{LC}}, \qquad \gamma_r = \frac{R}{2L} \; . \tag{1.16}$$

The expression for ω_r is called the *Thomson formula*, cf. Eq. (1.6). The quantity γ_r is proportional to the width of the resonance.

For a more complex circuit which consists of several inductively coupled simple circuits, Eq. (1.11) can be easily generalised. The currents $I_n = dq_n/dt$ in the nth circuit are now influenced not only by the inductances of the circuit L_{nn} but also by the mutual inductances with all the other coupled circuits $L_{nm} = M_{nm}$:

$$\mathcal{E}_n = \sum_{nm} L_{nm}\frac{dI_m}{dt} + R_n I_n + \frac{1}{C_n}\int I_n dt \; . \tag{1.17}$$

The corresponding impedance in this case is:

$$Z_{nm} = \delta_{nm}\left(R_m + \frac{1}{j\omega C_m}\right) + j\omega L_{nm} \; . \tag{1.18}$$

The eigenfrequencies of the system are defined by the zeros of the determinant of the impedance $|Z_{nm}| = 0$.

We turn now to the *parallel* R, L, C resonator. For this case the total current in the circuit should be equal to the sum of the partial currents flowing in each of the circuit elements. In analogy with Eq. (1.11) we can write:

$$I(t) = C\frac{dV}{dt} + GV + \frac{1}{L}\int V dt \; . \tag{1.19}$$

Each term in this equation represents a partial current flowing in one of the circuit elements: the capacitance C, the conductance G, and the inductance L. For monochromatic current $I(t) = \tilde{I}\exp(j\omega t)$ and voltage $V(t) = \tilde{V}\exp(j\omega t)$, this equation yields $\tilde{I} = Y_{par}(\omega)\tilde{V}$, with the *admittance*:

$$Y_{par}(\omega) = G + \frac{1}{j\omega L} + j\omega C \; . \tag{1.20}$$

Note that the expression for the admittance of the parallel circuit is similar to that for the impedance of a series circuit, cf. Eq. (1.12).

The impedance of the parallel circuit $Z_{par}(\omega) = 1/Y_{par}(\omega)$ is then:

$$Z_{par}(\omega) = \frac{R_s}{1 + jQ_r(\omega/\omega_r - \omega_r/\omega)} , \tag{1.21}$$

where we introduced the following notations:
the *resonant frequency* $\omega_r = 1/\sqrt{LC}$,
the *quality factor* $Q_r = R_s\sqrt{C/L}$,
and the *shunt impedance* $R_s = 1/G$.

It is interesting to note that the ratio $R_s/Q_r = \sqrt{L/C}$ does not depend on the shunt impedance R_s of the circuit.

1.A.3 Impedance of a Transmission Line

Another important application of the concept of impedance appears in the study of transmission lines. We assume that all parameters of the line are uniformly distributed along it, and that their values are constant. We denote these values as the *resistance* R_1 or the *conductance* G_1, the *inductance* L_1 and the *capacitance* C_1, all defined per unit length.

The voltage drop $V(z,t)$ over any small portion of the line of length dz is linearly dependent on the current at this position:

$$\frac{\partial V}{\partial z} = -R_1 I - L_1 \frac{\partial I}{\partial t} , \tag{1.22}$$

Similarly, the decrease of current $I(z,t)$ along the line is is linearly dependent on the voltage:

$$\frac{\partial I}{\partial z} = -G_1 V - C_1 \frac{\partial V}{\partial t} . \tag{1.23}$$

The minus signs on the right hand sides of these equations are chosen because both voltage and current are decreasing along the line.

These two first order differential equations are equivalent to a single, second order equation called *telegraph equation*:

$$\frac{\partial^2 V}{\partial z^2} - L_1 C_1 \frac{\partial^2 V}{\partial t^2} - \left[L_1 G_1 + R_1 C_1\right] \frac{\partial V}{\partial t} - R_1 G_1 V = 0 . \tag{1.24}$$

A similar equation holds for the current I.

Hence both the voltage V and the current I in the line are oscillating with the circular frequency ω: $V = \widetilde{V}(z)\exp(j\omega t)$ and $I = \tilde{I}(z)\exp(j\omega t)$. Their amplitudes $\tilde{V}$ and $\tilde{I}$ are functions of the coordinate z and satisfy the following equations:

$$\frac{\partial \tilde{V}}{\partial z} = -Z_1 \tilde{I} \ , \quad \frac{\partial \tilde{I}}{\partial z} = -Y_1 \tilde{V} \ , \tag{1.25}$$

where we introduced the *impedance* $Z_1(\omega) = R_1 + j\omega L_1$ and the *admittance* $Y_1(\omega) = G_1 + j\omega C_1$.

Eqs. (1.22) and (1.23) have two sets of independent solutions:

$$I_1(z,t) = I_1 \, \mathrm{e}^{-jkz + j\omega t} \ , \quad V_1(z,t) = Z_c I_1(z,t) \ , \tag{1.26}$$

and

$$I_2(z,t) = I_2 \, \mathrm{e}^{jkz + j\omega t} \ , \quad V_2(z,t) = Z_c I_2(z,t) \ , \tag{1.27}$$

where the *propagation constant* k and the *characteristic impedance* Z_c of the transmission line are defined by:

$$\begin{aligned} k = -j\sqrt{Z_1 Y_1} &= -j\sqrt{(R_1 + j\omega L_1)(G_1 + j\omega C_1)} \ , \\ Z_c &= \sqrt{\frac{Z_1}{Y_1}} = \sqrt{\frac{R_1 + j\omega L_1}{G_1 + j\omega C_1}} \ . \end{aligned} \tag{1.28}$$

The first solution describes the wave propagating in the positive z-direction. Since the values of the current and the voltage at $z \to +\infty$ should remain finite, the imaginary part of the propagation constant k must be negative. Asimilar condition holds for the second solution which describes the wave propagating in the negative z-direction. There the current and the voltage must be damped exponentially for $z \to -\infty$.

1.A.4 Impedance of a Plane Wave

We now consider a plane *EM* wave propagating in an isotropic and homogeneous medium, which is characterised by three constants: *permittivity* ε, *permeability* μ and *conductivity* σ_c. The homogeneous Maxwell equations for the *EM* fields which are not driven by currents or charges are [Stratton41]:

$$\nabla \times \mathbf{E} + \mu \frac{\partial \mathbf{H}}{\partial t} = 0, \quad \nabla \cdot \mathbf{H} = 0 \ , \tag{1.29}$$

and

$$\nabla \times \mathbf{H} - \varepsilon \frac{\partial \mathbf{E}}{\partial t} - \sigma_c \mathbf{E} = 0, \quad \nabla \cdot \mathbf{E} = 0 \ . \tag{1.30}$$

In a *plane wave*, the fields do not depend on the transverse coordinates x and y. Such a wave can have two polarisations. For one of them we can assume that the vector $\mathbf{E}$ is in the direction of the x axis, and the vector $\mathbf{H}$ in the direction of the y axis. In the frequency domain, the Maxwell equations for the transverse field components of a monochromatic wave are given by:

$$\frac{\partial \widetilde{E}_x}{\partial z} = -j\omega\mu\widetilde{H}_y, \quad \frac{\partial \widetilde{H}_y}{\partial z} = -(j\omega\varepsilon + \sigma_c)\widetilde{E}_x \ . \tag{1.31}$$

The solutions of these equations can be found analogously to those of Eqs. (1.25). The impedance per unit length is replaced by $j\omega\mu$, and the corresponding admittance by $j\omega\varepsilon + \sigma_c$. Hence the propagation constant k_p and the impedance Z_p in the direction of propagation of a plane wave are:

$$k_p = \sqrt{\varepsilon\mu\omega^2 - j\mu\omega\sigma_c} \ , \quad Z_p = \sqrt{\frac{\mu\omega}{\varepsilon\omega - j\sigma_c}} = \frac{\omega\mu}{k} \ . \tag{1.32}$$

Then the expressions for the *EM* fields of two plane waves, propagating in opposite directions, can be written:

$$E_{x1}(z,t) = E_1 \exp(-jk_p z + j\omega t) \ , \quad H_{y1}(z,t) = Z_p E_{x1}(z,t) \tag{1.33}$$

and

$$E_{x2}(z,t) = E_2 \exp(jk_p z + j\omega t) \ , \quad H_{y2}(z,t) = Z_p E_{x2}(z,t). \tag{1.34}$$

The density of energy flow is given by the Poynting vector $\mathbf{S} = \mathbf{E} \times \mathbf{H}^\star$. For a plane wave this vector has only one non-zero component in the direction of wave propagation. For example, for the wave propagating in the positive z direction: $S_z = Z_p E_1^2 \exp[-j(k_p - k_p^\star)z]$. The energy flow is constant and independent of z only when the propagation constant k_p is real. In a conductive or dispersive medium k_p is always complex, with the sign of the imaginary part defined such that the wave always decays.

Consider, for example, the propagation of a plane wave into a metal wall. The second term in the expression for the propagation constant, Eq. (1.32), is much larger than the first one up to extremely high frequencies. For all metals used for vacuum chambers of accelerators the conductivity is actually high enough that this condition is fulfilled over the whole range of frequencies considered in this book. Then $k_p \approx (1-j)\sqrt{\mu\omega\sigma_c/2}$ and the *EM* fields decay in the direction z over the characteristic distance

$$\delta_s \equiv \frac{1}{|\mathrm{Im}\, k_p|} = \sqrt{\frac{2}{\mu\omega\sigma_c}} \ . \tag{1.35}$$

The parameter δ_s, which characterised the penetration of *EM* fields of frequency ω into the metal, is called the *skin depth.* For example, at a frequency of $\omega/2\pi = 500$ MHz, the skin depth is $\approx 3.10^{-6}$ m in copper with a conductivity of $\sigma_c \approx 5.8 \cdot 10^7$ S/m. The field is effectively confined in a very thin layer near its surface. The effective wave length of the field inside the metal is $2\pi/|\mathrm{Re}\, k_p|$ and is also much smaller than in vacuum.

When the conductivity $\sigma_c = 0$, the propagation constant becomes $k = \sqrt{\varepsilon\mu}\omega$. Then the expressions given in Eq. (1.33) represent an undamped plane wave traveling in the positive z-direction. Its phase is $\varphi = kz - \omega t$. The condition $\varphi =$ constant defines the *phase velocity* of the wave:

$$v_{ph} = \frac{dz}{dt} = \frac{\omega}{k} \ . \tag{1.36}$$

A monochromatic wave is only an idealisation - in reality, every smooth (continuous and differentiable) pulse of finite duration contains harmonics over a finite frequency range. The frequency range $\Delta\omega$, over which the amplitudes of harmonics are not negligible, is inversely proportional to the time duration of the pulse. Since the wave numbers k depend on frequency, they have also a finite spread Δk. Different harmonics move with slightly different phase velocities, producing the effect that the envelope of the oscillations moves with a velocity which is, in general, different from these phase velocities. The velocity of the envelope is called the *group velocity.* If the range of frequencies in a pulse is narrow, the spread of the wave numbers is also small. Then all amplitudes in a pulse are approximately constant if the condition $z\Delta k - t\Delta\omega =$ constant is fulfilled. That means that the group velocity is:

$$v_{gr} = \frac{dz}{dt} = \frac{d\omega}{dk} \ . \tag{1.37}$$

From here we see that the group and the phase velocity are always different when the dependence $k = k(\omega)$ is non-linear. It is the case for the propagation of a plane wave in conducting (or more general in dispersive) media.

In vacuum the propagation constant and the impedance become:

$$k = \omega\sqrt{\varepsilon_0\mu_0} = \frac{\omega}{c}, \quad Z_0 = \sqrt{\frac{\mu_0}{\varepsilon_0}} \ . \tag{1.38}$$

The first of these equations shows that the group velocity $v_{gr} = d\omega/dk$ and the phase velocity $v_{ph} = \omega/k$ of a wave in free space are both equal to the velocity of light $c = 1/\sqrt{\varepsilon_0\mu_0}$. The free space impedance $Z_0 = 376.6$ Ohms provides a scale for all impedances.

Appendix 1.B Bunch-Cavity Interaction

Here we give an somewhat intuitive description of the interaction of a bunch of charged particles with a cavity [Morton93]. We consider a bunch passing with (nearly) light velocity $v \approx c$ through a cavity as shown in Fig. 1.1. In order to simplify the considerations we neglect the beam pipes and at first also assume that the cavity has only one mode. It will be not very difficult to generalise later to the case of many modes being simultaneously present.

When a bunch with current I_b traverses the cavity, beam induces a current flow I_L in cavity walls. The direction of flow of the induced current is such that its magnetic field opposes that of the beam current. The current I_L causes also accumulation of electric charges on the cavity gap. The signs of the induced charges are such as to produce an electric field directed opposite to the direction of the beam current I_b.

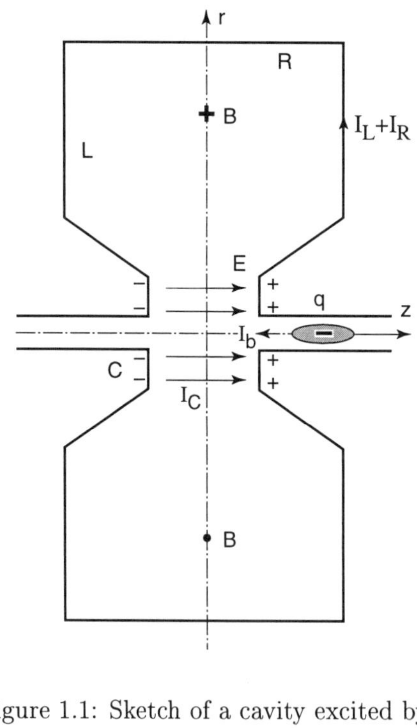

Figure 1.1: Sketch of a cavity excited by beam current.

The displacement current I_C is the integral of the time derivative of the electric field over the cavity cross section. If the walls of the cavity have a finite conductivity, there is an additional current I_R flowing through the resistivity.

Applying Kirchhoff's sum rule for currents we find:

$$I_b + I_L + I_R + I_C = 0 \; . \tag{1.39}$$

The impedance of the cavity can be represented by the equivalent circuit of a parallel resonator with inductance L, resistance R, and capacitance C.

Expressing the voltage V excited by the beam in each element of this equivalent circuit, we have:

$$\frac{1}{L}\int V dt + \frac{1}{R}V + C\frac{dV}{dt} = -I_b \; . \tag{1.40}$$

We can now use the result Eq. (1.21), found in a previous Section, to obtain the *coupling impedance* of the cavity in terms of its equivalent parameters:

$$Z_c(\omega) = \frac{R}{1 + jQ(\omega/\omega_r - \omega_r/\omega)} \; , \tag{1.41}$$

The dependence of this coupling impedance Z_c on the geometrical parameters of the cavity will be discussed in the following Chapters.

For a cavity with several resonant modes, the voltages V_n of the n-th mode excited by the beam current I_b are given by:

$$\sum_n \Big(\frac{1}{L_n}\int V_n dt + \frac{1}{R_n}V_n + C_n\frac{dV_n}{dt}\Big) = -I_b \; , \tag{1.42}$$

similar to Eq. (1.41). In this case the total voltage is the sum over the partial voltages of all cavity modes: $V_{tot} = \sum_n V_n$. The coupling impedance is represented by the sum over all resonant terms:

$$Z_c(\omega) = \sum_n \frac{R_n}{1 + jQ_n(\omega/\omega_{rn} - \omega_{rn}/\omega)} \; . \tag{1.43}$$

The real part of such an impedance is given by a series of narrow resonant peaks with high quality factors $Q_n \gg 1$. This behaviour is characteristic for frequencies below the (lowest) cutoff frequency of the beam pipe ω_{co}: Above it, the field oscillations are damped due to the fact that energy can escape into the beam pipes, and hence the resonant peaks broaden and can overlap, in particular since their number per frequency interval increases. The impedance becomes a smooth function, which is often called *broad-band impedance*.

Chapter 2

Calculation of EM Fields

Calculation of the EM fields generated by a beam, and diffracted or reflected on wall discontinuities, is *in principle* quite simple. One only needs to find the solution of a system of differential equations which satisfies the proper boundary conditions, depending on the shape and material of the vacuum chamber walls. In some cases, initial conditions may also have to be fulfilled, but are not required if only the equilibrium solution is desired.

In practice, such a problem is quite formidable because of complex chamber geometries. In most cases, the only workable method is a purely numerical calculation of the fields with the help of a computer. Depending on the size and shape of the structure, in particular for 3-dimensional problems without simplifying symmetries, such calculations may exceed the capabilities of todays most advanced computers.

Several numerical codes exist which perform direct calculations of the EM fields by solving the discretised Maxwell equations on a large, but finite, number of mesh points in the time domain, cf. Appendix to Chapter 11. However, to achieve reasonable accuracy, the bunch length must exceed some multiple of the mesh size and cannot be chosen arbitrarily small. Numerical calculation of the wake function of a point charge (i.e. a delta-function distribution) is not possible due to its singularity.

An alternative method of finding the EM fields is to first determine the impedance in the frequency domain, and to obtain the wake function by a Fourier transformation. To obtain good accuracy with this method at short time delay (distance) between exciting and trailing charges, one needs to know the impedance up to very high frequencies. The impedance function consists essentially of resonances whose density increases rapidly with frequency. Hence their numerical calculation faces limitations caused by finite computing time and memory size similar to those in the time domain [Bane80b, Wang89].

2.1 Analytic and Numeric Calculation

Limitations in numerical calculations justify the development of analytical or semi-analytical methods described in the present book. The latter methods not only provide useful formulae for roughly estimating the impedance, but also give insight into the physics of wake field generation. In a few simple cases, analytical methods can provide explicit solutions. In many others, it is expedient to express the field as an expansion in infinite series in which each term can be handled separately. The result is a set of linear algebraic equations for the expansion coefficients instead of coupled partial differential equations of the field components. Methods and results obtained in this manner are described in following chapters. Realistic geometries usually are still more complicated than those for which such methods can be applied. Nevertheless, the analytic approach can be useful for a number of reasons:

- Analytic solutions can serve as comparison or check of numerical solutions.
- Analytic solutions are often complimentary to numerical ones and applicable in different regions of beam and structure parameters.
- Analytical results can give new insights into the wake field generation. The examination of the steps of cylindrical beam pipes [Hereward75, Kheifets86a, Palumbo94] is an example where such insights were gained.

More complicated structures can often be considered as combinations of simpler ones, and superposition of individual solutions usually is a good approximation to the coupling impedance of such structures, unless they are located so close together that interference between the EM fields of individual structures must be taken into account. This is the case for two opposite single steps, which become a cavity when they are close together. The transition from the first into the second case will be discussed in more detail in the section on single and double steps in Chapter 8.

In Fig. 2.1, we show as an example of EM field calculations the electric field lines, obtained with the numerical computer code *ABCI* [Chin88a], for a short Gaussian bunch passing on the axis of a circular cylindrical *pillbox cavity* with concentric side tubes, in 10 subsequent time intervals of equal lengths.

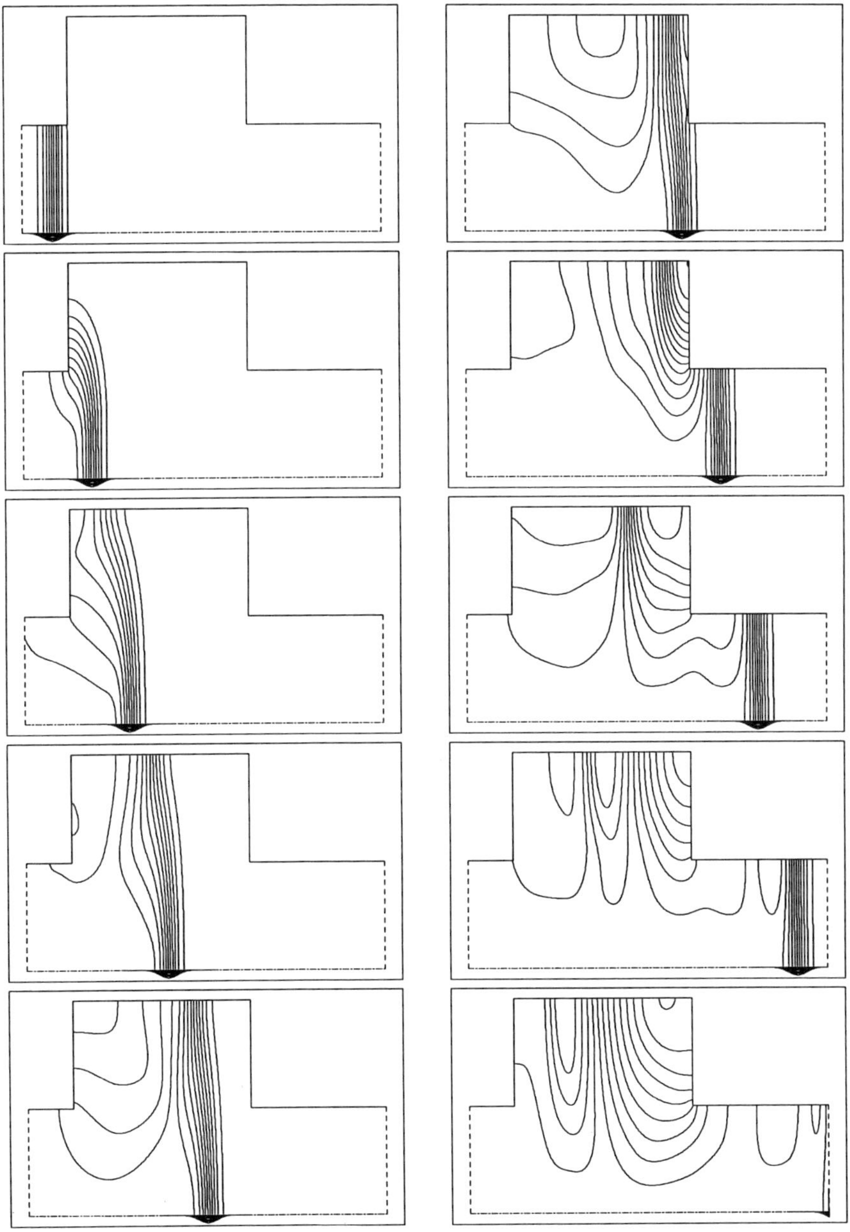

Figure 2.1: Electric field lines of Gaussian bunch passing a pillbox cavity with side tubes.

2.2 Frequency and Time Domain

The Maxwell equations are linear in the electric and magnetic field intensities, both in the time and in the frequency domain. Analytic solutions are usually simpler in the frequency domain, since a Fourier transformation, useful for real values of frequencies, or a Laplace transformation for complex frequencies, eliminate the explicit time dependence of the fields, and thereby decrease the number of derivatives in the equations. The price to pay for this simplification is the necessity to sum or integrate the solutions of the transformed equations over frequency when inverting to the time domain.

Nevertheless, such transformations are valuable for several reasons:

• First, the properties of the system in the frequency domain are of interest in themselves, since the excitation of fields is often produced by a monochromatic or quasi-monochromatic EM wave. In such a case, the system response is described more naturally in the frequency domain.

• For circular accelerators, there exist intrinsic frequencies – the revolution frequency of the reference particle, and the transverse (*betatron*) as well as the longitudinal (*synchrotron*) oscillation frequencies.

The system response to the excitation on integer harmonics of all these frequencies has direct bearing on the particle dynamics.

• Probably the most important consideration is the fact that the characteristic properties of a medium, i.e. its permittivity ε, permeability μ, and conductivity σ_c, are in general functions of frequency. When this dependence is not negligible, the frequency domain is the only possible way to find a solution.

2.3 The Fourier Transform

The Fourier transform of a function $f(t)$ and its inverse will be defined as:

$$\begin{aligned} \tilde{f}(\omega) &= \int_{-\infty}^{\infty} dt\, f(t)\, \exp(-j\omega t)\,, \\ f(t) &= \int_{-\infty}^{\infty} \frac{d\omega}{2\pi}\, \tilde{f}(\omega)\, \exp(j\omega t)\,. \end{aligned} \tag{2.1}$$

The position of the factor 2π is easy to remember when it is written inside the integral as $d\omega/2\pi = df$. Contrary to general practice, we write the differential directly after the integral. This is in particular useful with multiple integrals where one can immediately see the corresponding range of integration. The dimension of the Fourier transform of $f(t)$, which is written $\tilde{f}(\omega)$, has an additional factor s (seconds). For example, the Fourier transform of current $I(t)$, usually measured in Amperes, is the charge q, measured in Coulombs.

We use a decomposition of time dependent functions $f(t)$ into Fourier components proportional to $\exp(j\omega t)$. This choice of the time factor, and the notation $j = \sqrt{-1}$ for the imaginary unit, agree with the conventions used in electrical engineering, and are therefore more convenient for electrical quantities such as impedances. Instead of the frequency ω, we sometimes use the wave number $k = \omega/c$. Some authors use $k = \omega/\beta c$, where $\beta = v/c$, for the ultra-relativistic motion treated here the two conventions coincide.

For functions which are periodic in time, the Fourier transform degenerates into a Fourier series. Only the real parts should be used for quantities which by their physical meaning are real, such as EM fields. Quadratic Expressions, such as energy flow, are always defined as real.

2.4 Charge and Current Densities

The Maxwell equations constitute a system of inhomogeneous, linear, partial differential equations. The solution of such a system can be expressed by the sum of a particular solution for the inhomogeneous equations, and of a general solution of the homogeneous equations. The latter contains free constants which may be chosen to fulfil the initial or boundary conditions.

Conservation of charge is expressed by the *continuity equation*:

$$\nabla \cdot \mathbf{J} + \frac{\partial \rho}{\partial t} = 0. \tag{2.2}$$

Integrating the l.h.s over any volume and using the Green divergence theorem shows that the change of the total charge within the volume is equal to the net flow of charge through its surface.

For a point charge e moving with velocity $v = \beta c$ along the z axis, the charge density is given by

$$\rho(\mathbf{r}, t) = \frac{e}{2\pi r}\delta(r)\delta(z - vt) \ , \tag{2.3}$$

while the current density has only a z component $J_z(\mathbf{r}, t) = v\rho$. When the charge distribution is non-uniform in azimuth it is useful to represent it as a sum of modulated rings of charges. In each term then $\delta(r)$ has to be replaced by $\delta(r - a)\cos(m\theta)$. Such terms excite pure azimuthal modes.

The Fourier transforms of the charge and current densities of a point charge, given by Eq.(2.3) are:

$$\tilde{\rho}(\omega) = \frac{e}{2\pi}\delta(r)\mathrm{e}^{-jkz/\beta}, \quad \tilde{J}_z(\omega) = \frac{\beta ce}{2\pi}\delta(r)\mathrm{e}^{-jkz/\beta} . \tag{2.4}$$

The charge density of a bunch containing N_b particles can be obtained by summing the expressions in Eq.(2.3). The typical number of particles in a bunch is of the order of 10^{10} for electron machines, and up to 10^{12} for proton accelerators. Since $N_b \gg 1$, the sum over delta functions can be approximated by a smooth distribution called the *normalised line density* $\lambda(s)$, ($\int \lambda(s)ds = 1$), multiplied by the bunch charge $q_b = N_b e$. For proton machines, with negligible synchrotron radiation, the distribution usually remains the one injected. It is often approximately parabolic, with the bunch length specified either at the base, or better at half-height, where it is easier to measure. In order to simplify the calculations, sometimes a uniform (or "rectangular") line density is assumed, although it contains unrealistic sharp edges. In such a distribution the high-frequency content is exaggerated, which may lead to unphysical results.

In circular electron accelerators, the usually strong synchrotron radiation produces a distribution which is essentially Gaussian in energy. This is caused by two counteracting effects:

radiation damping:

beam particles with higher energies radiate stronger and therefore loose more energy than those with lower energy. The radiation also causes a decrease of energy of the whole beam, and the average energy loss has to be replaced by an applied rf voltage: thus only the energy spread decreases;

quantum excitations:

oscillations due to random fluctuations in the emission of photons limit the shrinking of the energy spread. The combination of these two effects leads to a Gaussian equilibrium distribution.

Particles in a beam bunched by an rf voltage perform *synchrotron oscillations* in energy and time - or rf phase - about a stable *synchronous phase* ϕ_s. The *transition energy* $\mathcal{E}_t$ is defined by the condition that the Lorentz factor $\gamma_t = \mathcal{E}_t/mc^2 = 1/\sqrt{\alpha}$, where $\alpha = \mathcal{E}\partial C/C\partial\mathcal{E}(> 0)$ is the *momentum compaction factor*. Above transition energy, the revolution time of the particles then increases with higher energy due to the lengthening of the trajectory, while its decrease due to the higher velocity near $v \approx c$ is negligible. Hence higher energy particles fall back in longitudinal position.

For phase stability, the synchronous phase angle below transition should be adjusted such that particles which arrive too early at the rf cavity see a smaller voltage than the synchronous particle, and thus receive less acceleration (for $V_{rf} \propto \sin\phi, \phi_s$ should lie between 0 and 90 degrees). This makes them fall back, while those which arrive later get more acceleration and catch up. Above transition, phase stability requires exactly the opposite (for a sinusoidal rf voltage ϕ_s should lie between 90 and 180 degrees). All electron rings and some high energy proton machines operate above transition.

Longitudinal and energy deviations of the particles are periodically exchanged by the *synchrotron oscillations*, and therefore the bunch length is related to the energy spread. The applied rf voltage is usually sinusoidal, but in many cases it can be considered linear over the length of the bunch when it is short compared to the rf wavelength. Then the charge density also becomes Gaussian and can be written:

$$\rho(\mathbf{r}, t) = \frac{N_b e}{\sqrt{2\pi}\sigma_s} \frac{\delta(r)}{2\pi r} \exp\left[-\frac{(z - vt)^2}{2\sigma_s^2}\right] . \tag{2.5}$$

The rms value of the bunch length is σ_s, while the *effective bunch length* l_b is equal $C_l \sigma_s$. The constant factor C_l cannot be defined in general, but has to be chosen for the particular problem considered. In some cases $C_l = 2\sqrt{2}$, in others $C_l \sim \pi$, sometimes even larger.

When the rms bunch length approaches zero, the distribution tends to a δ-function and we find again Eq.(2.3), but with the charge $N_b e$. Such a distribution is sometimes referred to as a *super-particle*.

2.5 The Maxwell Equations

The EM fields generated by charged particle beams, as well as those due to external sources, are generally described by the Maxwell equations which are given in many textbooks [Jackson75]. In MKSA units, they can be written[1]

$$\begin{aligned} \nabla \times \mathbf{H} &= \frac{\partial \mathbf{D}}{\partial t} + \mathbf{J} , \qquad \nabla \cdot \mathbf{D} = \rho , \\ \nabla \times \mathbf{E} &= -\frac{\partial \mathbf{B}}{\partial t} , \qquad \nabla \cdot \mathbf{B} = 0 , \end{aligned} \tag{2.6}$$

[1] In the US literature, the operator ∇ is usually written as *grad, div or curl* depending on its action on a scalar, or on a vector as a dot or cross product. In the German and Russian literature, *curl* is called *rot*.

to which one has to add the relations for the *electric displacement* $\mathbf{D} = \varepsilon\mathbf{E}$ and for the *magnetic induction* or *flux density* $\mathbf{B} = \mu\mathbf{H}$, where ε is the *permittivity* and μ the *permeability*. As the first pair of these equations shows, EM fields are excited by a charge density ρ and a current density $\mathbf{J}$. The latter may consist of a *convection current* density $\mathbf{J}_{conv} = \rho\mathbf{v}$ due to a moving charge with velocity $\mathbf{v}$ and - in a material with *conductivity* σ_c - of a *conduction current* density $\mathbf{J}_{cond} = \sigma_c\mathbf{E}$. The term $\partial\mathbf{D}/\partial t$ is also called *displacement current.*

In homogeneous, isotropic, continuous media with constant ε and μ, the Maxwell equations can be written in terms of the field strengths $\mathbf{H}$ and $\mathbf{E}$ only:

$$\begin{aligned} \nabla \times \mathbf{H} &= \varepsilon\frac{\partial \mathbf{E}}{\partial t} + \rho\mathbf{v} + \sigma_c\mathbf{E} \,, & \nabla \cdot \mathbf{E} = \frac{\rho}{\varepsilon} \,, \\ \nabla \times \mathbf{E} &= -\mu\frac{\partial \mathbf{H}}{\partial t} \,, & \nabla \cdot \mathbf{H} = 0 \,. \end{aligned} \tag{2.7}$$

Multiplying vectorially the operator ∇ with the second Maxwell equation, and substituting it into the first one, one can eliminate $\mathbf{H}$. With the *vector Laplacian* (differential) operator defined by $\Delta\,\mathbf{E} = \nabla\,(\nabla\cdot\mathbf{E}) - \nabla\times\nabla\times\mathbf{E}$, one thus obtains the *wave equations* for the electric, and in a similar manner for the magnetic field strengths:

$$\begin{aligned} \left[\Delta - \mu\varepsilon\frac{\partial^2}{\partial t^2}\right]\mathbf{E} &= \mu\frac{\partial \mathbf{J}}{\partial t} + \frac{1}{\varepsilon}\nabla\rho \,, \\ \left[\Delta - \mu\varepsilon\frac{\partial^2}{\partial t^2}\right]\mathbf{H} &= -\nabla\times\mathbf{J} \,. \end{aligned} \tag{2.8}$$

In the frequency domain, the derivatives with respect to time are replaced by the factor $j\omega$ and the wave equations are then called *Helmholtz equations.*

2.5.1 EM Potentials

Several descriptions of the EM fields in terms of potentials are possible, and we show here the *vector and scalar potentials*, *Hertz* and *Debye potentials.*

Vector and Scalar Potentials

Alternatively, the EM fields can be determined from a *vector potential* $\mathbf{A}$ and a *scalar potential* Φ:

$$\mathbf{E} = -\nabla\Phi - \frac{\partial \mathbf{A}}{\partial t} \,, \qquad \mathbf{B} = \nabla\times\mathbf{A} \,. \tag{2.9}$$

The EM potentials must satisfy the following system of the second order differential equations called *wave equations* for the vector and scalar potentials:

$$\begin{aligned}\left[\Delta - \mu\varepsilon\frac{\partial^2}{\partial t^2}\right]\mathbf{A} &= -\mu\mathbf{J} + \nabla\left[\nabla\cdot\mathbf{A} + \mu\varepsilon\frac{\partial\Phi}{\partial t}\right] , \\ \left[\Delta - \mu\varepsilon\frac{\partial^2}{\partial t^2}\right]\Phi &= -\frac{\rho}{\varepsilon} - \frac{\partial}{\partial t}\left[\nabla\cdot\mathbf{A} + \mu\varepsilon\frac{\partial\Phi}{\partial t}\right] . \end{aligned} \tag{2.10}$$

The divergence $\nabla\cdot\mathbf{A}$ can be chosen arbitrarily with a *gauge condition.* The most common choice is the *Lorentz gauge* for which

$$\nabla\cdot\mathbf{A} + \mu\varepsilon\frac{\partial\Phi}{\partial t} = 0 . \tag{2.11}$$

In this gauge, the above equations simplify considerably and the field components behave correctly under relativistic transformations, i.e. they are Lorentz invariant.

However, sometimes it is more convenient to chose the gauge condition $\nabla\cdot\mathbf{A} = 0$: this is called the *Coulomb gauge.* In this gauge, the vector potential is purely solenoidal, i.e. source-free, while the scalar potential is independent of time, and is given simply by the *Poisson equation* $\Delta\Phi = -\rho/\varepsilon$.

Hertz Potentials

In a region where there are no free charges or currents, it is often more convenient to use the *Hertz potentials*, which are usually written $\mathbf{\Pi}_E$ and $\mathbf{\Pi}_M$. The EM fields can be found from the relations [Stratton41]:

$$\begin{aligned}\mathbf{E} &= \nabla\times\nabla\times\mathbf{\Pi}_E - \mu\nabla\times\frac{\partial\mathbf{\Pi}_M}{\partial t} , \\ \mathbf{H} &= \nabla\times\left[\varepsilon\frac{\partial\mathbf{\Pi}_E}{\partial t} + \sigma\mathbf{\Pi}_E\right] + \nabla\times\nabla\times\mathbf{\Pi}_M . \end{aligned} \tag{2.12}$$

Debye Potentials

The *Debye potentials* which have been used to calculate the impedances due to small perturbations of simple structures (see Chapter 7).

When the EM field is excited in a longitudinally uniform pipe by a relativistic beam, it is convenient to separate its longitudinal and transverse parts. For an electric field $\mathbf{E}$, for example,

$$\mathbf{E}(\mathbf{r}) = \mathbf{E}_\perp(\mathbf{r}) + \hat{\mathbf{z}}E_z(\mathbf{r}) \, , \tag{2.13}$$

where $\hat{\mathbf{z}}$ is the unit vector in the longitudinal direction and the subscripts $\perp$ and z indicate the transverse and longitudinal components. A similar decomposition can be applied to the magnetic field $\mathbf{H}$, as well as $\mathbf{r} = \mathbf{r}_\perp + \hat{\mathbf{z}}s$. When all particles in a beam are travelling in the positive z-direction with the same constant velocity v, their longitudinal distribution does not depend on time t and the coordinate z separately, but only on their combination $s \equiv z - vt$. The excitation of the EM field by such a beam can be found by integrating the fields excited by a sheet charge of transverse distribution $\rho_\perp(\mathbf{r}_\perp)$ weighted by the longitudinal line density $\lambda(s)$. Hence, the problem of finding the EM field is restricted to solving the Maxwell equations with the source terms:

$$\rho(\mathbf{r}) = \rho_\perp(\mathbf{r}_\perp)\lambda(s) \, , \quad \mathbf{J}(\mathbf{r}) = \hat{\mathbf{z}}v\rho(\mathbf{r}) \, . \tag{2.14}$$

In this case, the dependence of the fields on t and z enters through the quantity s only, and hence it is more convenient to perform a Fourier transformation directly in the variable s. For the electric field one thus obtains:

$$\widetilde{\mathbf{E}}(\mathbf{r}_\perp, k) \equiv \int_{-\infty}^{\infty} ds \mathbf{E}(\mathbf{r}_\perp, s) \exp(-jks) \tag{2.15}$$

and a similar expression for the magnetic field. Here $k = \omega/v$ is the wave number.

The Fourier transformations lead to the following rules for differentiation:

$$\partial_t \widetilde{\mathbf{E}} = j\omega\widetilde{\mathbf{E}} \, , \quad \partial_z \widetilde{\mathbf{E}} = -jk\widetilde{\mathbf{E}} \, . \tag{2.16}$$

The Maxwell equations for the Fourier transforms can be written:

$$\begin{aligned}
\nabla_\perp \times \widetilde{\mathbf{E}}_\perp &= -j\omega\mu\widetilde{H}_z\hat{\mathbf{z}} \, , \\
\nabla_\perp \times \widetilde{\mathbf{H}}_\perp &= (j\omega\varepsilon\widetilde{E}_z + v\tilde{\rho})\, \hat{\mathbf{z}} \, , \\
\hat{\mathbf{z}} \times (-jk\widetilde{\mathbf{E}}_\perp - \nabla_\perp\widetilde{E}_z) &= -j\omega\mu\widetilde{\mathbf{H}}_\perp \, , \\
\hat{\mathbf{z}} \times (-jk\widetilde{\mathbf{H}}_\perp - \nabla_\perp\widetilde{H}_z) &= j\omega\varepsilon\widetilde{\mathbf{E}}_\perp \, , \\
\nabla_\perp \cdot \widetilde{\mathbf{E}}_\perp &= jk\widetilde{E}_z + \frac{\tilde{\rho}}{\varepsilon} \, , \\
\nabla_\perp \cdot \widetilde{\mathbf{H}}_\perp &= jk\widetilde{H}_z \, .
\end{aligned} \tag{2.17}$$

The last two equations allow the elimination of the longitudinal components of the fields from the first four equations:

$$\begin{aligned}
\nabla_\perp \times \widetilde{\mathbf{E}}_\perp &= -v\mu\hat{\mathbf{z}}\nabla_\perp \cdot \widetilde{\mathbf{H}}_\perp \,, \\
\nabla_\perp \times \widetilde{\mathbf{H}}_\perp &= v\varepsilon\hat{\mathbf{z}}\nabla_\perp \cdot \widetilde{\mathbf{E}}_\perp \,, \\
\hat{\mathbf{z}} \times \Big[- k\widetilde{\mathbf{E}}_\perp + \frac{1}{k}\nabla_\perp \Big(\nabla_\perp \cdot \widetilde{\mathbf{E}}_\perp - \frac{\tilde{\rho}}{\varepsilon}\Big)\Big] &= -\omega\mu\widetilde{\mathbf{H}}_\perp \,, \\
\hat{\mathbf{z}} \times \Big[- k\widetilde{\mathbf{H}}_\perp + \frac{1}{k}\nabla_\perp(\nabla_\perp \cdot \widetilde{\mathbf{H}}_\perp)\Big] &= \omega\varepsilon\widetilde{\mathbf{E}}_\perp \,, \qquad (2.18)
\end{aligned}$$

The transverse components of the electric field can be split into two parts [Morse53]:

$$\widetilde{\mathbf{E}}_\perp = \widetilde{\mathbf{E}}^{irr} + \widetilde{\mathbf{E}}^{sol} \,, \qquad (2.19)$$

where $\widetilde{\mathbf{E}}^{irr}$ is the *irrotational* field, for which $\nabla_\perp \times \widetilde{\mathbf{E}}^{irr} = 0$, and $\widetilde{\mathbf{E}}^{sol}$ is the *solenoidal* field, for which $\nabla_\perp \cdot \widetilde{\mathbf{E}}^{sol} = 0$.

Using these properties, Eqs. (2.18) become:

$$\begin{aligned}
\nabla_\perp \times \widetilde{\mathbf{E}}^{sol} &= -v\mu\hat{\mathbf{z}}\nabla_\perp \cdot \widetilde{\mathbf{H}}^{irr} \,, \\
\nabla_\perp \times \widetilde{\mathbf{H}}^{sol} &= v\varepsilon\hat{\mathbf{z}}\nabla_\perp \cdot \widetilde{\mathbf{E}}^{irr} \,, \\
\hat{\mathbf{z}} \times \Big[- k^2\widetilde{\mathbf{E}}^{irr} - k^2\widetilde{\mathbf{E}}^{sol} + \nabla_\perp \Big(\nabla_\perp \cdot \widetilde{\mathbf{E}}^{irr} - \frac{\tilde{\rho}}{\varepsilon}\Big)\Big] &= -k\omega\mu\left(\widetilde{\mathbf{H}}^{irr} + \widetilde{\mathbf{H}}^{sol}\right) \,, \\
\hat{\mathbf{z}} \times \Big[- k^2\widetilde{\mathbf{H}}^{irr} - k^2\widetilde{\mathbf{H}}^{sol} + \nabla_\perp(\nabla_\perp \cdot \widetilde{\mathbf{H}}^{irr})\Big] &= k\omega\varepsilon\left(\widetilde{\mathbf{E}}^{irr} + \widetilde{\mathbf{E}}^{sol}\right) \,. \qquad (2.20)
\end{aligned}$$

Applying the vector identity $\mathbf{a} \times \mathbf{b} \times \mathbf{c} = \mathbf{b}(\mathbf{a} \cdot \mathbf{c}) - \mathbf{c}(\mathbf{a} \cdot \mathbf{b})$ one obtains:

$$\begin{aligned}
\hat{\mathbf{z}} \times \widetilde{\mathbf{E}}^{sol} &= \mu v\widetilde{\mathbf{H}}^{irr} \,, \\
\hat{\mathbf{z}} \times \widetilde{\mathbf{H}}^{sol} &= -\varepsilon v\widetilde{\mathbf{E}}^{irr} \,, \\
\hat{\mathbf{z}} \times \Big[k^2\widetilde{\mathbf{E}}^{irr} - \nabla_\perp(\nabla_\perp \cdot \widetilde{\mathbf{E}}^{irr} - \tilde{\rho}/\varepsilon)\Big] &= v\mu\widetilde{\mathbf{H}}^{sol} \,, \\
\hat{\mathbf{z}} \times \Big[k^2\widetilde{\mathbf{H}}^{irr} - \nabla_\perp(\nabla_\perp \cdot \widetilde{\mathbf{H}}^{irr})\Big] &= -v\varepsilon\widetilde{\mathbf{E}}^{sol} \,. \qquad (2.21)
\end{aligned}$$

The first two of these equations can also be written in the form:

$$\widetilde{\mathbf{E}}^{sol} = -\mu v\hat{\mathbf{z}} \times \widetilde{\mathbf{H}}^{irr} \,, \qquad \widetilde{\mathbf{H}}^{sol} = \varepsilon v\hat{\mathbf{z}} \times \widetilde{\mathbf{E}}^{irr} \,. \qquad (2.22)$$

With the help of these expressions, we can eliminate $\widetilde{\mathbf{E}}^{sol}$ and $\widetilde{\mathbf{H}}^{sol}$ from the second two equations to obtain with the *transverse Laplacian* operator (acting on an irrotational vector or on a scalar) $\Delta_\perp \equiv \nabla_\perp \nabla_\perp$:

$$\begin{aligned}
\Delta_\perp \widetilde{\mathbf{E}}^{irr} - k^2(1 - v^2 \varepsilon\mu)\widetilde{\mathbf{E}}^{irr} &= \nabla_\perp \frac{\tilde{\rho}}{\varepsilon} , \\
\Delta_\perp \widetilde{\mathbf{H}}^{irr} - k^2(1 - v^2 \varepsilon\mu)\widetilde{\mathbf{H}}^{irr} &= 0 .
\end{aligned} \tag{2.23}$$

The Fourier transforms of the fields can be written in terms of the *Debye potentials* $\tilde{\Phi}$ and $\tilde{\Psi}$ for which:

$$\widetilde{\mathbf{E}}^{irr} = -\nabla_\perp \tilde{\Phi} , \qquad \widetilde{\mathbf{H}}^{irr} = -\nabla_\perp \tilde{\Psi} . \tag{2.24}$$

Adding two arbitrary constants, we obtain the wave equations:

$$\begin{aligned}
\Delta_\perp \tilde{\Phi} - k^2(1 - v^2 \varepsilon\mu)\tilde{\Phi} &= -\frac{\tilde{\rho}}{\varepsilon} + C_1 , \\
\Delta_\perp \tilde{\Psi} - k^2(1 - v^2 \varepsilon\mu)\tilde{\Psi} &= C_2 .
\end{aligned} \tag{2.25}$$

The constants C_1 and C_2 may be set to zero if $1 - v^2\varepsilon\mu \neq 0$, but this cannot be done for $v \to c$ in vacuum.

Once the Debye potentials are found, all field components can be calculated. From Eqs. (2.24) one obtains the irrotational parts, from the last two Eqs. (2.17) the longitudinal components:

$$E_z = jk(1 - v^2\varepsilon\mu)\tilde{\Phi} + j\frac{C_1}{k} , \quad H_z = jk(1 - v^2\varepsilon\mu)\tilde{\Psi} + j\frac{C_2}{k} , \tag{2.26}$$

and the solenoidal parts from Eqs. (2.22):

$$\begin{aligned}
\widetilde{\mathbf{E}}^{sol} &= v\mu\hat{\mathbf{z}} \times \nabla_\perp \tilde{\Psi} , \\
\widetilde{\mathbf{H}}^{sol} &= -v\varepsilon\hat{\mathbf{z}} \times \nabla_\perp \tilde{\Phi} .
\end{aligned} \tag{2.27}$$

2.5.2 Separable Coordinate Systems

The wave equation is a partial differential equation, in which the independent variables can be separated only in a very limited number of coordinate systems. In addition to the Cartesian system, the most important one is the (circular) cylindrical one, it corresponds most closely to the geometry of many accelerator components which often have circular cross section, such as beam pipes, rf cavities or kicker tanks etc.

In circular cylindrical coordinates r, θ, z, the differential operations can be written:

$$\begin{aligned}
\nabla\, \Phi &= \frac{\partial \Phi}{\partial r}\hat{\mathbf{r}} + \frac{1}{r}\frac{\partial \Phi}{\partial \theta}\hat{\theta} + \frac{\partial \Phi}{\partial z}\hat{\mathbf{z}} \\
\nabla \cdot \mathbf{A} &= \frac{1}{r}\frac{\partial r A_r}{\partial r} + \frac{1}{r}\frac{\partial A_\theta}{\partial \theta} + \frac{\partial A_z}{\partial z}
\end{aligned} \tag{2.28}$$

$$\nabla \times \mathbf{A} = \left(\frac{1}{r}\frac{\partial A_z}{\partial \theta} - \frac{\partial A_\theta}{\partial z}\right)\hat{\mathbf{r}} + \left(\frac{\partial A_r}{\partial z} - \frac{\partial A_z}{\partial r}\right)\hat{\theta} + \left(\frac{1}{r}\frac{\partial r A_\theta}{\partial r} - \frac{1}{r}\frac{\partial A_r}{\partial \theta}\right)\hat{\mathbf{z}}\ .$$

It is important to distinguish between the *scalar wave equation*, e.g. for the scalar potential Φ, with the *scalar Laplacian* operator:

$$\Delta\, \Phi = \frac{1}{r}\frac{\partial}{\partial r}\left(r\frac{\partial \Phi}{\partial r}\right) + \frac{1}{r^2}\frac{\partial^2 \Phi}{\partial \theta^2} + \frac{\partial^2 \Phi}{\partial z^2} \tag{2.29}$$

and the *vector wave equation*, e.g. for the magnetic field Eq. (2.8) or for the vector potential $\mathbf{A}$, with the *vector Laplacian* operator given by

$$\Delta\, \mathbf{A} = \nabla(\nabla \cdot \mathbf{A}) - \nabla \times \nabla \times \mathbf{A}\ . \tag{2.30}$$

In (circular) cylindrical coordinates, the radial and azimuthal components of the vector wave equation are coupled, i.e. they contain not only the corresponding components of the vector, but also the other transverse component, since the unit vectors $\hat{\mathbf{r}}$ and $\hat{\theta}$ depend on the angular coordinate: $\partial\hat{\mathbf{r}}/\partial\theta = \hat{\theta}$ and $\partial\hat{\theta}/\partial\theta = -\hat{\mathbf{r}}$. The longitudinal ($z$) component satisfies the scalar wave equation. Hence it is advantageous to express the radial and azimuthal vector components as derivatives of the longitudinal component, which can be done using the Maxwell equations.

The dependence on the coordinate θ of the scalar potential Φ and of the longitudinal component of the vector potential A_z can be expanded into a series of *azimuthal modes* either in (real) circular functions $\sin m\theta$ and $\cos m\theta$, or in complex exponentials $\exp(jm\theta)$. The *azimuthal mode number* m are integer since the potentials must be periodic with period 2π in the azimuthal angle.

Other separable coordinate systems, such as spherical, oblate or prolate spheroidal, or parabolic cylindrical ones, are rarely used in accelerator theory.

Elliptical cylindrical coordinates could be of some practical interest, and have been used in a few papers to approximate e.g. the resistive wall impedance of oval vacuum chambers [Ruggiero67, Gluckstern93b].

2.6 Boundary Conditions

The arbitrariness in the solutions of the Maxwell equations is removed by imposing boundary conditions on the EM fields. These conditions depend on the properties of the materials used. We consider two cases commonly encountered.

2.6.1 Infinite Conductivity

For a perfectly conducting surface S, the boundary conditions can be formulated in the following way. With a unit vector $\mathbf{n}$, orthogonal to S and directed to the inside of the enclosed volume, the electric field strength $\mathbf{E}$ should satisfy the condition:

$$\mathbf{n} \times \mathbf{E} = 0 \quad \text{on } S, \tag{2.31}$$

i.e., the tangential component of the electric field vanishes. Physically this means that the charge and current, which excite the EM field, induce surface charge and current which just compensate the normal component of the electric field inside the metal. With negligible resistance, the tangential electric field must vanish since otherwise the current at the surface would become infinite.

Similarly, the magnetic induction $\mathbf{B}$ must satisfy the condition:

$$\mathbf{n} \cdot \mathbf{B} = 0 \quad \text{on } S, \tag{2.32}$$

i.e. the normal component of the magnetic induction should vanish. This condition is due to the fact that any time dependent normal component of the magnetic induction would cause the appearance of a tangential component of the electric field, which is forbidden by Eq.(2.31).

2.6.2 Finite Conductivity

For a metal with a large but finite conductivity $\sigma_c = 1/\varrho_c$, the effective length over which the field penetrates inside the metal is of the order of the skin depth

$$\delta_s = \sqrt{\frac{2\varrho_c}{\mu\omega}} \,. \tag{2.33}$$

It is usually much smaller than the radius of curvature of the relevant metal surface, and hence the field near the surface can be regarded as that of a plane wave. The normal component of the electric field inside the metal is zero. However, its tangential, two-component vector E_t does not vanish and is, to first order, related to the magnetic field by the relation

$$\mathbf{E}_t = \sqrt{\frac{\mu}{\varepsilon_c}} \mathbf{H}_t \times \mathbf{n} \text{ on } S. \tag{2.34}$$

which is called the *Leontovich condition* [Leontovich48] in the Russian literature. The *surface impedance* $\zeta = \sqrt{\mu/\varepsilon_c}$ is defined with the *complex permittivity* $\varepsilon_c = \varepsilon + \sigma_c/j\omega$. For a good conductor with small the resistivity ϱ_c such as a metal, for which the conductivity $\sigma_c \gg \varepsilon\omega$, it can be written:

$$\zeta \simeq \sqrt{\frac{j\omega\mu}{\sigma_c}} = (1+j)\,\frac{\varrho_c}{\delta_s} \,. \tag{2.35}$$

The boundary conditions described above are obtained under the assumption that the shape of the wall is continuous and differentiable with respect to the coordinate z. While these assumptions are usually valid for realistic surfaces, at least the second one may fail for certain idealised models of the geometry. For analytical models, abrupt steps and corners with sharp angles are standard approximations. The formal solutions of the Maxwell equations for the EM fields may be discontinuous at such exceptional points. However, the discontinuity must be mild enough that the integral over the total field energy remains finite [Meixner49], [Von Bladel85]. This criterion, sometimes called the *edge condition*, should be kept in mind when obtaining numerical results. For example, it may determine the minimum number of terms which have to be retained when truncating an infinite series.

For open, unterminated structures, such as beam pipes, the EM fields may propagate towards infinity. In addition to the usual boundary conditions, such fields must satisfy the *radiation condition* that the waves are only going outwards from a structure, and are not coming back from infinity [Jackson75].

Boundary conditions for the Fourier components of EM fields are in general the same as those for the EM fields in the time domain.

2.7 Green Function and Reciprocity Theorem

The solutions of the *wave equations* for EM fields or potentials in the time domain, or the corresponding *Helmholtz equations* in the frequency domain, can be expressed in terms of a *Green function* for any particular geometry.

2.7.1 The Green Function

A Green function is the solution of a linear differential equation with delta-functions on the r.h.s., whose arguments are the differences between the *field coordinate* $\mathbf{r}$ and the *source coordinate* $\mathbf{r}'$. For the wave equation, in general $G(\mathbf{r}, \mathbf{r}', t, t')$ also depends on times t and t', while its Fourier transform $\widetilde{G}(\mathbf{r}, \mathbf{r}', \omega)$ depends on frequency ω or on the wave number $k_c = \omega(\mu_c\varepsilon_c)^{1/2}$, which is complex in a conducting or lossy medium. These functions have to satisfy the following two equations:

$$\begin{aligned} \Delta G - \mu_c\varepsilon_c \frac{\partial^2}{\partial t^2} G &= \delta(\mathbf{r}-\mathbf{r}')\delta(t-t') \ , \\ \Delta\widetilde{G} + k_c^2\widetilde{G} &= \delta(\mathbf{r}-\mathbf{r}') \ . \end{aligned} \tag{2.36}$$

The Green function can be written as a product of a spatial and a time dependent part $G(\mathbf{r}, \mathbf{r}', t, t') = G(\mathbf{r}, \mathbf{r}')\delta(t-t')$, where the spatial part must satisfy the boundary conditions on the surface S of the volume V of the geometry of the problem under consideration. If any part of the surface extends to infinity, it should furthermore satisfy the *radiation condition*, i.e. the requirement that there should only be outgoing waves, vanishing at infinity.

Since the integration over $\delta(t-t')$ can be performed explicitly, solutions of the wave equations Eqs. (2.10) can be written as integrals over the volume V. The integrands are products of the spatial part of the Green function $G(\mathbf{r}, \mathbf{r}')$ and the charge density $\rho(\mathbf{r}, t')$ or the current density $\mathbf{J}(\mathbf{r}, t')$, which are evaluated at the *retarded time* $t' = t - |\mathbf{r}-\mathbf{r}'|/c$, in order to take into account the finite velocity c of field propagation:

$$\begin{aligned} \Phi(\mathbf{r}, t) &= \textstyle\int_V d\mathbf{r}' \frac{\rho(\mathbf{r}', t')}{\varepsilon_c} G(\mathbf{r}, \mathbf{r}') \ , \\ \mathbf{A}(\mathbf{r}, t) &= \textstyle\int_V d\mathbf{r}' \mu_c \mathbf{J}(\mathbf{r}', t') G(\mathbf{r}, \mathbf{r}') \ . \end{aligned} \tag{2.37}$$

Solutions for the Fourier transforms $\widetilde{\Phi}(\mathbf{r})$ and $\widetilde{\mathbf{A}}(\mathbf{r})$ are expressed by similar integral expressions in which G is replaced by $\widetilde{G}$. However, in general it is quite difficult to find exact expressions for the Green function.

In *free space*, the spatial Green function can be written explicitly as

$$G(\mathbf{r}, \mathbf{r}') = \frac{1}{|\mathbf{r} - \mathbf{r}'|} \; . \tag{2.38}$$

The Fourier transform of the free space Green function is:

$$\widetilde{G}(\mathbf{r}, \mathbf{r}') = \frac{\exp(-jk|\mathbf{r} - \mathbf{r}'|)}{4\pi|\mathbf{r} - \mathbf{r}'|} \; . \tag{2.39}$$

Note that the dimensions of Green functions are m^{-1}, in correspondence with Eq. (2.36).

The Green function can also be represented in the form:

$$\widetilde{G}(\mathbf{r}, \mathbf{r}') = -\frac{j}{8\pi} \int_{-\infty}^{\infty} dp \; \widetilde{G}_p(r, r') \; \exp[-jp(z - z')] \; , \tag{2.40}$$

where

$$\widetilde{G}_p(r, r') = \begin{cases} \sum_{m=0}^{\infty} \varepsilon_{m0} J_m(\nu r) \; H_m^{(1)}(\nu r') \cos m\theta & \text{for} \quad r' > r \\ \sum_{m=0}^{\infty} \varepsilon_{m0} J_m(\nu r') \; H_m^{(1)}(\nu r) \cos m\theta & \text{for} \quad r' < r \; , \end{cases} \tag{2.41}$$

for $k > 0$, while for $k < 0$ it is defined by $\widetilde{G}_p(k) = \widetilde{G}_p^*(-k)$. Here the *Neumann symbol* $\varepsilon_{m0} = 1$ for $m = 0$, and 2 for all other m, and J_m and $H_m^{(1)}$ are Bessel and Hankel functions of order m. The function $\nu(k, p)$ has a branch cut along the negative axis in the plane of its argument, which we define in the following way:

$$\nu = \sqrt{k^2 - p^2 - 2jk\epsilon} \; , \qquad \epsilon \to 0 \; . \tag{2.42}$$

The Hankel functions $H_m^{(1)}$ of purely imaginary argument are related to the modified Bessel functions by $H_m^{(1)}(jx) = (2/\pi)(-j)^{m+1} K_m(x)$ [Gradshteyn80].

In general, solutions of the wave equations in frequency domain, fulfilling boundary conditions at the surface of a finite volume V, exist only for discrete *eigenfrequencies*. These *eigenfunctions* $\widetilde{\mathbf{H}}_n(\mathbf{r})$, with (composite) index n constitute a full set of orthogonal functions. They can be normalised by division with the coefficients $N_n^{1/2}$

$$N_n = \int_V d\mathbf{r} \; \widetilde{\mathbf{H}}_n(\mathbf{r}) \cdot \widetilde{\mathbf{H}}_n^*(\mathbf{r}) \tag{2.43}$$

to obtain the normalised functions $\widetilde{\mathbf{h}}_n = \widetilde{\mathbf{H}}_n / N_n^{1/2}$. Then they fulfil the *orthogonality relations*

$$\int_V d\mathbf{r}\ \widetilde{\mathbf{h}}_n(\mathbf{r}) \cdot \widetilde{\mathbf{h}}_m^*(\mathbf{r}) = \delta_{nm} \ , \tag{2.44}$$

(where δ_{nm} is the Kronecker symbol $\delta_{nm} = 1$ if $n = m$, $\delta_{nm} = 0$ if $n \neq m$), as well as the *completeness condition*

$$\sum_n \widetilde{\mathbf{h}}_n(\mathbf{r}) \cdot \widetilde{\mathbf{h}}_n^*(\mathbf{r}) = \delta(\mathbf{r} - \mathbf{r}') \ . \tag{2.45}$$

Using these properties of the normalised eigenfunctions, one can find the following representation of the Green function:

$$\widetilde{G}(\mathbf{r} - \mathbf{r}') = \sum_n \frac{\widetilde{\mathbf{h}}_n(\mathbf{r}) \cdot \widetilde{\mathbf{h}}_n^*(\mathbf{r}')}{k^2 - k_n^2} \ . \tag{2.46}$$

2.7.2 Lorentz Reciprocity Theorem

Reciprocity is a general property of self-adjoint, integro-differential linear operators. For our purposes though it is sufficient to establish a particular form of reciprocity for the Maxwell equations, often called *Lorentz Reciprocity Theorem*

We consider a volume V, bounded by the surface S, in which exist two independent sources of EM fields $\tilde{\rho}_b, \tilde{\mathbf{J}}_b$ (subscript b for *beam*), and $\tilde{\rho}_e, \tilde{\mathbf{J}}_e$ (subscript e for *external*).

The Fourier transforms of the EM fields associated with the beam current $\widetilde{\mathbf{E}}_b$, $\widetilde{\mathbf{H}}_b$, and those produced by the external sources $\widetilde{\mathbf{E}}_e$, $\widetilde{\mathbf{H}}_e$, must fulfil the Maxwell equations. We first take the difference $\widetilde{\mathbf{E}}_b \times \widetilde{\mathbf{H}}_e - \widetilde{\mathbf{E}}_e \times \widetilde{\mathbf{H}}_b$, calculate its divergence, use the vector identity $\nabla \cdot (\mathbf{A} \times \mathbf{B}) = \mathbf{B} \cdot (\nabla \times \mathbf{A}) - \mathbf{A} \cdot (\nabla \times \mathbf{B})$ [Jackson75], and finally use Maxwell equations to obtain:

$$\nabla \cdot (\widetilde{\mathbf{E}}_b \times \widetilde{\mathbf{H}}_e - \widetilde{\mathbf{E}}_e \times \widetilde{\mathbf{H}}_b) = (\widetilde{\mathbf{E}}_e \cdot \tilde{\mathbf{J}}_b - \widetilde{\mathbf{E}}_b \cdot \tilde{\mathbf{J}}_e) \ . \tag{2.47}$$

We integrate both sides, use the divergence theorem [Jackson75] to convert the volume integral on the l.h.s. into a surface integral, where $d\mathbf{S} = \hat{\mathbf{n}} dS$ is a vector surface element in the direction of the normal to it, and thus obtain the *Lorentz reciprocity theorem* [Collin50]:

$$\int_S d\mathbf{S} \cdot \left(\widetilde{\mathbf{E}}_e \times \widetilde{\mathbf{H}}_b - \widetilde{\mathbf{E}}_b \times \widetilde{\mathbf{H}}_e \right) = \int_V dv \left(\widetilde{\mathbf{E}}_b \cdot \tilde{\mathbf{J}}_e - \widetilde{\mathbf{E}}_e \cdot \tilde{\mathbf{J}}_b \right) . \tag{2.48}$$

The term $\tilde{\mathbf{J}}_b$ includes the beam current, as well as all currents induced by it at conducting surfaces in the volume. For rigour, the term $\tilde{\mathbf{J}}_e$ should include not only the current resulting from the application of an external voltage V_e, but also any *change of the beam current* produced by it. However, one usually assumes that $\tilde{\mathbf{J}}_e$ is produced only by an external voltage, neglecting the influence of the beam on the external fields, as well as the perturbation which the beam might experience by those fields.

In some cases, the l.h.s. of the last equation becomes vanishingly small. For the r.h.s. in the first approximation, the integration of the current densities over transverse coordinates yields the corresponding currents $\tilde{\mathbf{J}}_e, \tilde{\mathbf{J}}_b$, while the integration of the electric fields over the longitudinal coordinate gives the potentials $\tilde{V}_e, \tilde{V}_b$. Then, in this approximation, the reciprocity relation takes a much simpler form, allowing many practical applications:

$$\tilde{V}_b \, \tilde{\mathbf{I}}_e = \tilde{V}_e \, \tilde{\mathbf{I}}_b \; . \tag{2.49}$$

Dividing by both currents one finds that, the reciprocity theorem in this form is equivalent to matching of the impedances of the source to the external fields. A particular form of the Lorentz reciprocity theorem for the case of two beams moving in opposite directions through a structure is derived in Chapter 4.

2.8 Restrictions and Limitations

Although the limits of integration in Eq.(2.1) are formally infinite, the range of frequencies essential for a particular problem is limited by the length of the bunch and the geometry of the vacuum chamber, as well as by the physical conditions on its surface. The "macroscopic" description of EM fields breaks down for wave lengths which are comparable to distances between atoms a_0. The frequencies are also limited by the strong dispersion of the permittivity of metals, which again happens when the wave length approaches a_0. In the following we will assume that the inequality $\omega \ll c/a_0$ is always fulfilled. However, this may not be sufficient for the macroscopic description to be valid.

The frequencies relevant for the present approach extend well above the *cutoff frequency* of a beam pipe of radius a, but are limited to $ka \ll \gamma$. This range is sufficient for studying the stability of even the shortest bunches used in the design of high-energy particle accelerators. Below the cutoff frequency of the beam pipe, the impedance is given by a small number of narrow, resonant peaks. At higher frequencies, the peaks become wider and overlap, and the impedance becomes smooth. For extremely high frequencies, when $ka > \gamma$, the impedance falls off exponentially.

In general, the change of particle energy is accompanied by a change of its velocity v and, correspondingly, its current. In this case, self-consistent solutions for the EM fields and the particle motion should be found. Problems of this kind are important mainly for low-energy beams with non-relativistic particle velocities, e.g. in particle sources and guns and for the injection of proton or ion beams. There exist several computer programs, called *particle-in-cell* codes, which solve this type of problem. However, they are outside the scope of the present book and will not be considered here.

At ultra-relativistic energies, the velocity factor β differs from unity by $1/2\gamma^2$. Such a small difference can usually be neglected. Throughout our discussion we will assume that the particle velocity is constant and does not change during the passage through a structure. Therefore, also the beam current remains constant. We will also neglect the small oscillations that particles perform while moving in an accelerator. In other words, we assume that the velocity vector $\mathbf{v}$ is constant for particles travelling through the structure under consideration, and is directed parallel to the axis of the structure, which is usually taken to be straight, and is called the s- or z- axis of the coordinate system chosen. The particular case of a beam circulating in a toroidal chamber will be considered in Chapter 10.

In most cases, the resistivity of metal walls can be neglected. It will be assumed to be different from zero only in Chapter 6 where the *resistive wall impedance* is discussed. In the ultra-relativistic limit, the corrections due to non-zero resistivity are usually very small. The assumption of perfect conductivity permits much simpler boundary conditions on the EM fields, which makes the derivations shorter and clearer.

For the range of problems which are considered in this book, quantum effects are quite negligible. They play only an indirect role in determining the equilibrium dimensions of electron bunches, and thus the range of frequencies which has to be considered for a particular problem. However, the bunch length will be taken as a given parameter in the present approach.

Chapter 3

Wake Functions and Wake Potentials

The motion of particles can be most easily visualised in the *time domain*. This description is also well adapted to *simulation* of the coherent stability of particle motion, which receives increasingly more attention in the design of accelerators with the advent of more powerful and faster computers. Several computer codes for direct numerical evaluation of wake fields in the time domain are now available, and can be used just as easily as those which compute resonant modes in the frequency domain.

Wake functions can be defined in principle for particle velocities $v < c$, e.g. for closed cavities, or for more general geometries if the initial conditions are properly stated. However, at least for high-energy particle accelerators, it is usually quite sufficient to compute them in the limit $v \to c$. In this way they become pure properties of the structures surrounding the beam, and are independent of all beam parameters. This assumption also justifies neglecting the reaction of the wake fields on the beam trajectory.

In this chapter we assume that the walls are perfectly conducting - a quite good approximation for most metals used for vacuum chambers in accelerators. The structures investigated here are straight, i.e. we ignore the curvature of the beam trajectory in circular accelerators. This is at least partially justified by the fact that most structures presenting impedances, such as rf cavities or kicker tanks, as well as bellows and collimators, are usually located in straight sections, while curved chambers are generally quite smooth. An exception to this are the impedances due to synchrotron radiation from particles on curved trajectories, which will be discussed in Chapter 10.

3.1 Wake Functions

In order to determine the motion of charged particles in high-energy accelerators, it is usually sufficient to study only their integrated interaction with the EM fields. The *wake functions* are defined as integrals over the normalised forces due to the EM fields excited in a structure by a *point charge* q and evaluated at a distance s behind it. The charge is assumed to move unperturbed on a straight trajectory, with constant velocity $v \approx c$. In general, we shall distinguish between *longitudinal and transverse wake functions* according to the direction of the forces integrated. Since the wake functions are excited by a point charge (or delta-function distribution), they are also the *Green functions* for a particular structure. Therefore we will call them $G(s)$, to distinguish them from the *wake potentials* $W(s)$ due to a finite particle distribution, which will be discussed in Sections 3.4 and 3.5.

3.1.1 Properties of Wake Functions

The assumption $v = c$ is usually a very good approximation for high-energy particles in accelerators and storage rings, in particular for light particles like electrons, but also for protons at high enough energy. A direct consequence of this is the vanishing of the wake function everywhere in front of the exciting charge. This is often referred to as *causality* in the accelerator literature [Roos69] - but it simply expresses the fact that the EM fields, which propagate at most with velocity of light, cannot get ahead of particles when they move - at most - with the same speed. In reality, since particles must always be slower than light velocity, the fields can get arbitrarily far ahead in sufficiently long time - without violating any causality principle. However, the advanced fields are usually weakened by wall dissipation and/or by reflections from obstacles in the beam pipe, and therefore can be neglected in most cases. An exception to this are radiation fields from particles on curved trajectories, which can precede the particles. Radiation wakes and impedances will be discussed in Chapter 10.

Here we assume that the wake function must vanish in front of a point charge, i.e. $G_{\parallel}(0^-) = 0$ for $v = c$ where the argument 0^- indicates that the zero value is approached from the region in front of the charge. Since $G_{\parallel}(s)$ thus changes discontinuously from zero in front to the finite value $G_{\parallel}(0^+)$ just behind the particle, it takes half that value at exactly its own position, i.e. $G_{\parallel}(0) = (1/2)G_{\parallel}(0^+)$. In physical terms, this means that the centre of a bunch sees only the effect of half the charges, which are in front of it, even in the limit of vanishing bunch length. This property is sometimes referred to as the *fundamental theorem of beam loading* [Wilson87]. A particle traversing a passive structure cannot gain energy from its own field, thus the wake function $G_{||}(0)$ right at the particle position, $s = 0$, must cause a loss of energy. We define $G_{||}(0)$ as positive in this book[1].

Another important property is the observation that the first maximum of the wake function is the largest extremum, i.e. $|G_{||}(\tau)| \leq G_{||}(+0)$. Therefore it is not possible to accelerate a second particle on the same trajectory to more than twice the energy of the first one before it has lost all its energy.

3.1.2 Numerical Evaluation

Analytic evaluation of the wake functions is possible only in very few, simplified geometries, such as closed cylindrical cavities. In most practical cases one has to revert to numerical methods using computers. One can distinguish between computer codes working directly in the time domain, and those which first compute resonant modes in the frequency domain, from which the wake fields can be subsequently obtained.

Computer codes which solve the Maxwell equations in the time domain can only find fields excited by distributions of finite length, i.e. *bunches* of particles. In all these codes the bunch length must be a multiple (typically 5 times or more) of the longitudinal mesh size, and hence the wake functions of point charges cannot be computed exactly. One can obtain only a rough approximation to the wake function by determining the wake potential of a very short bunch. This requires correspondingly small mesh sizes, but - in particular for large structures - the number of mesh points is limited by computer time and memory. The behaviour of the wake function just behind the point charge can only be obtained with somewhat arbitrary procedures like eliminating the wake in front of the bunch centre, or by adding it to the wake behind. In the frequency domain the wake functions can be approximated by computing a sufficiently large number of resonances of a structure. However, this technique is also limited, since only a finite number of resonances can be computed in any given time. An improvement of the results is possible by using the *optical resonator model* which extends the discrete resonances by an

[1]Sometimes the opposite convention is used, then a decelerating wake is counted negative, e.g. in the numerical codes *MAFIA* or *ABCI*, with which some figures in this book have been prepared.

appropriately chosen continuum [Bane80b, Brandt82]. Based on an asymptotic high-frequency dependence ($\propto \omega^{-3/2}$) of the loss factor, it gives quite satisfactory results for periodic arrays of cavities even with a limited number of discrete resonances.

Wake potentials are convolutions of wake functions with the particle distribution, hence it is tempting to try to obtain the wake functions simply by a *de*-convolution. Unfortunately, this cannot be performed without severe loss of accuracy occurring in the regions outside the bunch proper, where one has to divide by extremely small values. Therefore the unavoidable numerical noise is amplified by extremely large factors. Another possibility for obtaining the wake function would be to apply a Fourier transformation to the wake potential, divide the resulting *effective impedance* by the line-density transform, and then convert back to the time domain. The weakness of this method is that in the frequency domain, the loss of accuracy corresponds to the fact that the impedance seen by longer bunches does not contain the high-frequency information required for the reconstruction of the impedance of shorter ones.

3.2 Longitudinal Wake Function

The wake functions are generally obtained by integration of the EM forces acting on a test charge (at a distance s behind the exciting one) at the transverse offset $\mathbf{r}_e$ over its trajectory assumed to be parallel to the axis of the structure. Since the magnetic force is perpendicular to the direction of the particle motion it does not contribute, and the *longitudinal wake function* is obtained by integrating over the electric field component E_z normalised by the charge q:

$$G_{\|}(\mathbf{r_b}, \mathbf{r_e}, s) = -\frac{1}{q} \int_{-\infty}^{\infty} dz E_z(\mathbf{r_b}, \mathbf{r_e}, z, t = \frac{z+s}{c}). \tag{3.1}$$

The transverse offset of the exciting charge is given by $\mathbf{r}_b$. The minus sign makes the wake function positive for a decelerating electric field component. The dimensions of the wake function are $V/C = \Omega/s$, for accelerator structures the units $V/pC = 10^{12} V/C$ are commonly used.

The longitudinal wake function is defined as the change of energy ΔE of a test charge of magnitude e trailing at a distance s behind a point charge q, i.e. $\Delta\mathcal{E} = -eqG_{\|}(s)$.

3.2.1 Rotationally Symmetric Structures

For structures with axial symmetry, it is most appropriate to use circular cylindrical coordinates r and θ, and to expand the EM fields into multi-polar components. In general E_z, E_r, and B_θ are proportional to $\cos m\theta$, and B_z, B_r, and E_θ to $\sin m\theta$. The index m is called the *azimuthal mode number*: $m = 0$ is the azimuthally uniform monopole mode, $m = 1$ are the dipole modes proportional to $\cos\theta$ or $\sin\theta$, $m = 2$ are the quadrupole modes proportional to $\cos 2\theta$ or $\sin 2\theta$, etc. In most cases, however, the higher modes with $m \geq 2$ are weak and thus rarely of practical interest.

The longitudinal wake function can also be expressed as an expansion in azimuthal modes:

$$G_{\parallel}(r_b, r_e, \theta_b, \theta_e, s) = \sum_{m=0}^{\infty} \varepsilon_{m0} r_b^m r_e^m \cos m(\theta_e - \theta_b) G_{\parallel}{}^{(m)}(s) , \tag{3.2}$$

where ε_{m0} is the Neumann symbol, $\varepsilon_{m0} = 1$ for $m = 0$, $\varepsilon_{m0} = 2$ for $m \neq 0$. Here r_e and θ_e give the radial and azimuthal offsets of the exciting charge, and r_b and θ_b of the test charge. In the usual definition of the wake function, the test charge follows the same trajectory as the exciting one: $r_b = r_e$ and $\theta_e = \theta_b$. The functions $G_{\parallel}{}^{(m)}(s)$ express the dependence on the longitudinal distance s between the two charges. Since the dimension of the wake function is V/C, the dimensions of the m-th azimuthal component $G_{\parallel}{}^{(m)}(s)$ are in general V/Cm^{2m}. For beams near the axis, the strongest wake usually is the uniform, $m = 0$ mode, while all higher order terms are quite small due to the factor $(r/b)^{2m}$, where b refers to the radius of the structure. When one refers to the longitudinal wake function, one usually means simply $G_{\parallel}{}^{(0)}(s)$ and leaves off the superscript 0.

3.2.2 Generalised Integration Path

As will be shown in the Appendix to this chapter, the integral over the EM force becomes independent of the transverse offset of the charge for structures with axial symmetry and for $v \to c$. Therefore in Eq. (3.1) the integration path may be chosen with a displacement equal to the radius of the beam pipe. Since the wall material is assumed to be perfectly conducting, the longitudinal electric field component vanishes on its surface, and the integration interval can thus be limited[Weiland83a] to the finite length of the *cavity gap* g at the radius of the beam pipe (see Fig. 2.1).

This technique is very useful for many structures such as cavities which are everywhere larger than the beam pipe. This gives both increased accuracy and shorter computation times than *direct integration* along the axis, which requires adding very long beam pipes before the wake fields become negligible. However, the integration at the beam pipe radius fails when materials such as collimators or scrapers extend into the beam pipe.

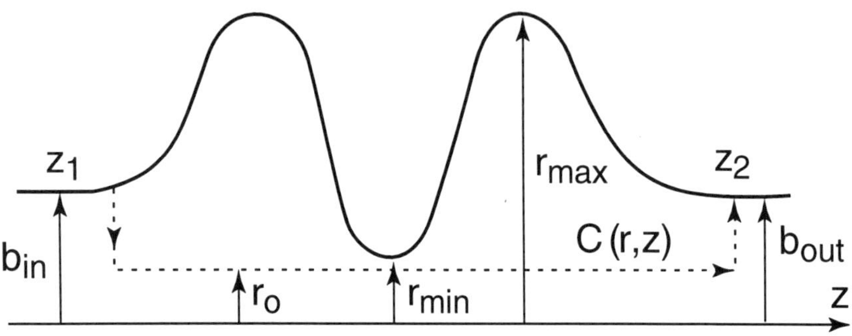

Figure 3.1: Integration path around an object protruding into the beam pipe.

In the Appendix we will further describe how a reduction to a finite integration interval is achieved even for geometries extending into the beam pipe: by integrating not simply the longitudinal electric field, but a properly chosen combination of EM field components over a deformed integration path [Napoly91]. One can apply this technique also to structures which have beam pipes with unequal radii at either side, which is illustrated in Fig. 3.1. This technique has been incorporated into the computer code *ABCI* for wake potential calculation [Napoly93].

3.2.3 Wake Function in a Pillbox Cavity

In a closed, circular-cylindrical cavity - commonly called *pillbox cavity* - the wake fields can be found analytically [Papiernik74]. When the cavity radius is so large that reflections from the cylindrical walls do not arrive within the period of interest, the cavity can be treated as two parallel plates [Chao75] and the solution can be expressed in closed form. However, solutions for cavities of finite radius can also be found in terms of functions defined by an infinite series [Zotter79a].

Rather than computing the wake function excited by a delta function pulse, it is preferable to determine the *step-function wake potential* $W_s(s)$, since the convergence of the resulting infinite series is better.

The wake function can then be obtained as the derivative of the step function wake potential. For particle beams with $v = c$, the step-function response is given by a double series over all radial and azimuthal resonant modes of the cavity of length g and radius d [Papiernik74]:

$$W_s(s) = \frac{2}{\pi} Z_0 \sum_{p=-\infty}^{\infty} \sum_{n=1}^{\infty} \frac{1}{j_{0n}^2 J_1^2(j_{0n})} \times$$

$$\begin{cases} \dfrac{2\sin(ks)}{kg} - (-)^p \dfrac{\sin(ks+kg)-\sin(kg)}{kg} - \dfrac{\sin(\pi ps/g)}{\pi p} & \text{for } 0 < s < g \\ \dfrac{2\sin(ks)}{kg} - (-)^p \dfrac{\sin(ks+kg)}{kg} - \dfrac{\sin(ks-kg)}{\pi p} & \text{for } s > g\ , \end{cases} \tag{3.3}$$

where the propagation constant $k = k_{np}$ is given by

$$k_{np} = \sqrt{\left(\frac{j_{0n}}{d}\right)^2 + \left(\frac{\pi p}{g}\right)^2}\ . \tag{3.4}$$

The series over p can be summed using the expression [Papiernik74]

$$\sum_{p=-\infty}^{\infty} \frac{\sin(y\sqrt{\beta^2+\pi^2 p^2})}{\sqrt{\beta^2+\pi^2 p^2}} \cos(\pi p z) = \begin{cases} J_0(\beta\sqrt{y^2-z^2}) & 0 \le z < y < 1 \\ 0 & 0 < y < z \le 1 \end{cases} \tag{3.5}$$

for any value of β. Now the series over n can be recognised as the Fourier-Bessel expansion of the logarithmic function:

$$\ln x = -2 \sum_{n=1}^{\infty} \frac{J_0(j_{0n}x)}{j_{0n}^2\, J_1^2(j_{0n})}\ , \tag{3.6}$$

which is valid [2] for $0 < x < 2$. This restriction of the range is important as the expression for the wake function in closed form is valid only for times before the arrival on the axis of reflections from the cylindrical wall. Hence, the wake function in closed form refers to the passage of a particle through two infinite parallel plates. These results are shown in Figs. 3.2.

[2] The validity of this expansion is given as $0 < x \le 1$ in the literature [Watson66], but it is actually twice as large.

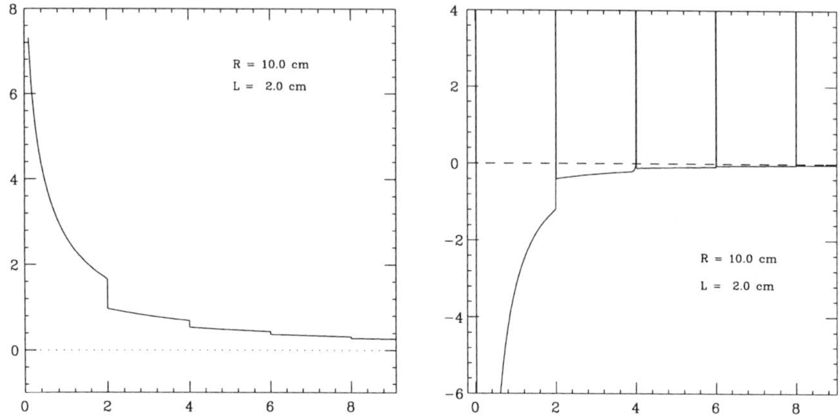

Figure 3.2: Step function response and wake function in pillbox cavity versus ratio s/g ($L = g, R = d$).

The solution for $0 < s < \min(2g, \sqrt{g^2 + 2d^2} - g)$ becomes

$$W_s(s) = \frac{2Z_0}{\pi} \ln\left(1 + \frac{2g}{s}\right) . \tag{3.7}$$

For $d \to \infty$, it is possible to extend the range of validity of the summation over p to get:

$$W_s(s) = \frac{2Z_0}{\pi} \ln\left(1 + \frac{1}{s/2g + [s/2g]}\right) , \tag{3.8}$$

where the square brackets indicate that only the integer part of the enclosed expression should be taken.

When $x > 2$ the reflections from the cylindrical wall are included. For a finite cavity radius, the step-function wake can be written in terms of the function

$$f_0(x) = -2 \sum_{n=1}^{\infty} \frac{J_0(j_{0n}x)}{j_{0n}^2 J_1^2(j_{0n})} , \tag{3.9}$$

which equals $\ln x$ only for $0 < x < 2$, but can be evaluated readily enough for any value of x by computer.

3.2.4 Resonator Wake Function

We use the term *resonator* to describe an object with a single resonance, for which the impedance is given by (cf. Eq.. (1.21):

$$Z_{\|}(\omega) = \frac{R_s}{1 + jQ_r(\omega/\omega_r - \omega_r/\omega)} , \tag{3.10}$$

where R_s is the *shunt impedance*, Q_r the quality factor, and ω_r the resonant frequency. One of the useful properties of the resonator impedance is the fact that it can be decomposed into linear fractions. This permits explicit evaluation of many integral expressions, in particular, the wake function can be found from its Fourier transform:

$$G_{\|}(\tau) = \frac{R_s}{4\pi j Q'_r} \int_{-\infty}^{\infty} d\omega \exp(j\omega\tau) \left(\frac{\omega_1}{\omega - \omega_1} - \frac{\omega_2}{\omega - \omega_2} \right) , \tag{3.11}$$

where $\tau = s/c$, the *reduced quality factor* is $Q'_r = \sqrt{Q_r^2 - 1/4}$, and poles of the frequency in the complex plane are $\omega_{1,2} = (\omega_r/Q_r)(j/2 \pm Q'_r)$. This integral can be evaluated with the residue theorem:

$$\oint d\omega \frac{f(\omega)}{\omega - z_1} = 2\pi j f(z_1) \tag{3.12}$$

for z_1 inside the closed region of integration, and zero otherwise. To evaluate the contour integral, the path of integration along the real axis can be closed with a large semi-circle, as shown in Fig. 3.3. The semi-circle must be chosen in such a way that its contribution to the integral vanishes when its radius tends to infinity.

In front of the point charge, i.e. for negative values $\tau < 0$, one must therefore close the path of integration in the lower half of the complex plane (Im $\omega < 0$), where the real part of the exponent Re $(j\omega\tau) = -\tau \mathrm{Im}\ \omega$ is negative. There the integrand vanishes sufficiently fast that its contribution to the integral tends to zero. Since both poles lie in the upper half plane, Im $(\omega_{1,2}) > 0$, they do not contribute to the integral in this case, and the wake function in front of the point charge is zero as required.

On the other hand, behind the point charge ($\tau > 0$), the integration path has to be closed in the upper half plane, and both poles contribute with their residues:

$$G_{\|}(\tau) = \frac{R}{2Q'_r} \left[\omega_1 \exp(j\omega_1\tau) - \omega_2 \exp(j\omega_2\tau) \right] . \tag{3.13}$$

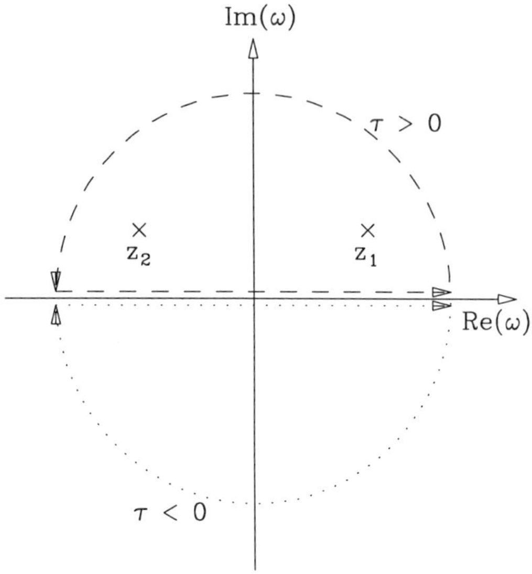

Figure 3.3: Integration path for wake in the complex frequency plane.

Usually $Q_r \geq 1/2$, thus Q_r' is real and the second pole is the negative complex conjugate of the first one $\omega_2 = -\omega_1^*$. The terms in the square brackets become the sum of two complex conjugates, and the result can be written more simply

$$G_{\|}(\tau) = \frac{R_s}{Q_r'} \mathrm{Re}\Big\{\omega_1 \exp(j\omega_1\tau)\Big\} \ . \tag{3.14}$$

In purely real terms, this expression can be written:

$$G_{\|}(\tau) = \frac{\omega_r R_s}{Q_r} \exp\left(-\frac{\omega_r\tau}{2Q_r}\right)\left(\cos\omega_r'\tau - \frac{1}{2Q_r'}\sin\omega_r'\tau\right) \ . \tag{3.15}$$

It corresponds to a damped oscillation at the modified resonant frequency $\omega_r' = \omega_r Q_r'/Q_r$ shown for broad-band resonators of various quality factors in Fig.3.4a. For narrow-band resonators, $Q_r \gg 1$, one finds $Q_r' \to Q_r$ and the *sine* term can be neglected. This gives an approximate result:

$$G_{\|}(\tau) \approx \frac{\omega_r R_s}{Q_r} \exp\left(-\frac{\omega_r\tau}{2Q_r}\right) \cos\left(\omega_r\tau\right) \ . \tag{3.16}$$

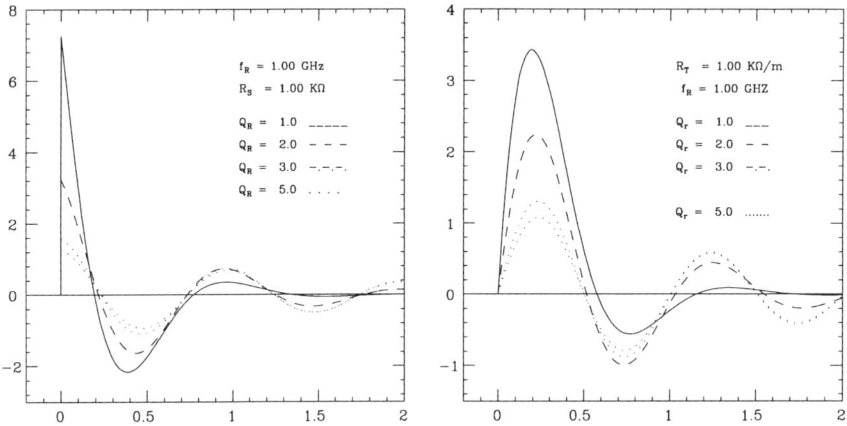

Figure 3.4: Longitudinal and transverse wake functions of resonators with various quality factors versus distance τ [ps].

For $Q_r < 1/2$, one has to use the general expression Eq. (3.13. Then both poles are *purely imaginary*: $\omega_{1,2} = (j\omega_r/Q_r)(1/2 \pm Q_r'')$ with $Q_r'' = \sqrt{1/4 - Q_r^2}$, hence again Im $(\omega_{1,2}) > 0$. The expression in purely real functions is obtained from Eq. (3.15) by replacing Q_r' by Q_r'', and the circular functions by hyperbolic ones. The wake function thus remains real for any values of Q_r.

3.3 Transverse Wake Functions

There exist a number of different definitions for the *transverse wake function*. The most common one is quite analogous to that of the longitudinal wake function, i.e. the integral over the transverse EM forces along a straight path at the distance s behind an exciting point charge travelling with constant velocity $v \approx c$, and divided by the value of the charge:

$$\mathbf{G}_\perp = \frac{1}{q} \int_{-\infty}^{\infty} dz (\mathbf{E} + \mathbf{v} \times \mathbf{B})_\perp . \tag{3.17}$$

This definition can be used for structures without any symmetry, but in general the wake function depends strongly on the location of the integration path, and in particular vanishes at a plane or axis of symmetry. Unlike the longitudinal wake function, the transverse wake function starts with zero value at the location of the point bunch, hence it is not discontinuous at the origin.

In general, the transverse wake function is a vector with two components, often in the horizontal and vertical directions. In symmetric structures, the horizontal wake function is found for a purely horizontal offset, and the vertical wake for a purely vertical offset. However, cross terms may appear for structures which are not axially symmetric, i.e. a vertical component of the wake function may also be produced by a horizontal offset and vice versa.

The more common structures used in accelerators, such as cavities or vacuum chambers, usually have one or more symmetry planes or even an axis of rotation. Near an axis of symmetry, the wake function of the most important deflecting (*dipole*) force is proportional to the displacement of the trajectory. It is therefore quite common to (re-)define the transverse wake function by normalising the integral over the force with the dipole moment of the charge $q\xi$, rather than by the charge itself, i.e. to divide $G_\perp$ by the displacement ξ:

$$\mathbf{G}'_\perp = \frac{1}{q\xi}\int_{-\infty}^{\infty} dz\Big(\mathbf{E} + \mathbf{v}\times\mathbf{B}\Big)_\perp . \tag{3.18}$$

We distinguish $G'_\perp$ as the *dipole* transverse wake function with dimensions are V/Cm. This definition has the advantage, that $G'_\perp$ is independent of the location of the integration path near the axis of symmetry, where the trajectories of particle beams are usually situated.

In structures with 2 planes of symmetry, e.g. with elliptic or rectangular cross section, the vector components of the dipole wake function are in general not equal, and cross terms will have to be taken into account. In Cartesian coordinates one then finds $G_x = G'_{xx}x_b + G'_{xy}y_b$ and $G_y = G'_{yx}x_b + G'_{yy}y_b$ with the *tensor components* [Petracca95a]:

$$\begin{aligned}
G'_{xx}(s) &= \frac{1}{qx_b}\int_{-\infty}^{\infty} dz\Big[E_x(z,t) - v_z B_y(z,t)\Big]_{x=x_b}^{y=0} ,\\
G'_{xy}(s) &= \frac{1}{qy_b}\int_{-\infty}^{\infty} dz\Big[E_x(z,t) - v_z B_y(z,t)\Big]_{x=0}^{y=y_b} ,\\
G'_{yx}(s) &= \frac{1}{qx_b}\int_{-\infty}^{\infty} dz\Big[E_y(z,t) + v_z B_x(z,t)\Big]_{x=x_b}^{y=0} ,\\
G'_{yy}(s) &= \frac{1}{qy_b}\int_{-\infty}^{\infty} dz\Big[E_y(z,t) + v_z B_x(z,t)\Big]_{x=0}^{y=y_b} ,
\end{aligned} \tag{3.19}$$

where the EM fields are evaluated at the time $t = (z+s)/c$. For structures with less symmetry it is necessary to use the general definition of the wake function Eq. (3.21).

3.3.1 Rotationally Symmetric Structures

In a structure with axial symmetry, the expansion of the EM fields into azimuthal modes when excited by a point bunch with offset r_e, yields for the transverse wake function:

$$\mathbf{G}'_\perp(r,\theta,s) = \sum_{m=1} r_b^{m-1} r_e^m \Big[\hat{r}\cos(m\theta) - \hat{\theta}\sin(m\theta)\Big] G_\perp{}^{(m)}(s) \;, \tag{3.20}$$

where r_e and r_b are the radial displacements of the exciting and the test particles, $\hat{r}$ and $\hat{\theta}$ are unit vectors, and θ is the angle between the two displacements. A test particle with the same offset as the exciting one thus sees a purely radial wake field. At any other transverse position, the direction and the value of the wake field remain the same, and are given by the function $G_\perp{}^{(m)}$ which depends only on the longitudinal distance $s = c\tau$ between the exciting and the test particles.

The transverse wake field is related to the longitudinal field having the same azimuthal dependence - i.e. with the same mode number m- through the Maxwell equations. This is usually expressed by the Panofsky-Wenzel theorem [Panofsky56] which can be stated simply that the longitudinal derivative of the transverse wake equals the transverse derivative (*gradient*) of the longitudinal one. This theorem and the conditions for its validity are discussed in the Appendix to Chapter 4.

3.3.2 Transverse Resonator Wake Function

Similar to the longitudinal case, the *transverse wake function* can be found from the Fourier transform of the transverse impedance

$$Z_\perp(\omega) = \frac{\omega_r}{j\omega}\frac{R_\perp}{1 + jQ_r(\omega/\omega_r - \omega_r/\omega)} \;, \tag{3.21}$$

where $R_\perp$ is the transverse resistance (in Ω/m, cf. next Chapter). Then

$$G_\perp{}'(\tau) = -\frac{\omega_r R_\perp}{4\pi Q'_r}\int_{-\infty}^{\infty} d\omega \exp(j\omega\tau)\left(\frac{1}{\omega - \omega_1} - \frac{1}{\omega - \omega_2}\right) \;, \tag{3.22}$$

where the definitions for Q'_r and $\omega_{1,2}$ were given after Eq. (3.11). However, now it is necessary to divide by an additional factor j to take into account that the wake acts only on the slope, and not on the trajectory itself - corresponding to a phase shift of $\pi/2$ or a factor j in the frequency domain.

This integral can again be evaluated with the residue theorem stated in Eq. (3.12), after closing the path of integration along the real axis with a large semi-circle whose radius tends to infinity. Again the result is zero in front of the point charge ($\tau < 0$), where the path must be closed in the lower half of the complex plane for convergence.

Behind the point charge, ($\tau > 0$), the integration path should be closed in the upper half plane, and both residues will contribute to yield:

$$\begin{aligned} G_\perp{}'(\tau) &= \frac{\omega_r R_\perp}{2jQ'_r}\Big[\exp(j\omega_1\tau) - \exp(j\omega_2\tau)\Big] \\ &= \frac{\omega_r R_\perp}{Q'_r}\mathrm{Im}\Big\{\exp(j\omega_1\tau)\Big\}\,. \end{aligned} \tag{3.23}$$

The second form is valid for $Q_r \geq 1/2$, for which case Q'_r is real and $\omega_2 = -\omega_1^*$. Then the transverse wake function of a resonator can be written more simply

$$G'_\perp(\tau) = \frac{\omega_r R_\perp}{Q'_r}\exp\left(-\frac{\omega_r\tau}{2Q_r}\right)\,\sin\omega'_r\tau\,, \tag{3.24}$$

where $\omega'_r = (Q'_r/Q_r)\omega_r$ is the *modified resonant frequency*. The expression for the transverse case is somewhat simpler than that for the longitudinal one, Eq. (3.15), where the exponential damping term is multiplied with the difference of two circular functions. The expression for $Q_r < 1/2$ is again obtained by replacing Q'_r with Q''_r, and the circular *sine* by the hyperbolic *sinh* function. The transverse wake function for broad-band resonators with various quality factors is shown in Fig.3.4b.

The transverse wake function starts at the origin, increases to a first maximum and, then oscillates with a damping decrement of $\omega_r/2Q_r$. The first zero crossing is at a time $\pi/\omega'_r = 1/f_r\sqrt{4 - 1/Q_r^2} \approx 1/2f_r$ for $Q_r > 1/2$. For example, for a resonant frequency of 1 GHz the zero crossing lies at about 1/2 ns (corresponding to a distance of 15 cm). The derivative of the transverse wake function of a resonator is proportional to the longitudinal one:

$$\frac{dG'_\perp(\tau)}{d\tau} = \frac{\omega_r R_\perp}{Q'_r}\mathrm{Re}\Big\{\omega_1\exp(j\omega_1\tau)\Big\} = \omega_r\frac{R_s}{R_\perp}G_\parallel(\tau)\,. \tag{3.25}$$

In particular this shows that the maximum of the transverse wake function coincides with the zero crossing of the longitudinal one.

3.4 Longitudinal Wake Potentials

Similar to the wake functions, the *wake potentials* are defined as integrals over the EM forces exerted by wake fields at the position of a test charge following on the same trajectory. However, the exciting charge is now a bunch of particles, of finite length, and the distance to the test charge $s = v\tau$ is measured from the bunch centre. Thus the wake functions are the wake potentials of a delta-function distribution, and can be considered *Green functions* for the wake potentials of a finite distribution in the structure considered.

For a bunch of arbitrary shape, the wake potential can in principle be found from the convolution of the wake function $G_{\|}(\tau)$ with the normalised line density $\lambda(\tau)$:

$$W_{\|}(\tau) = \int_{-\infty}^{\tau} dt\; G_{\|}(\tau - t)\; \lambda(t) = \int_{0}^{\infty} dt\; G_{\|}(t)\; \lambda(\tau - t)\;. \tag{3.26}$$

Since the wake function of an ultra-relativistic particle vanishes in front of it ($t < 0$), the integration can be limited to positive time, from 0 to ∞. In practice one replaces the convolution integral with a sum over the wakes of a finite number of *super-particles* of negligible extension. Typically a few thousand super-particles are used to get a good description of the distribution function. The wake potential for a particular geometry is then obtained by adding the individual wakes multiplied by the local densities. This can be further simplified by grouping or *binning* the particles in adjacent intervals, and replacing each bin by a single super-charge at its centre. However, the procedure is still lengthy, and the results of simulation programs using this technique [Brandt82] depend on the number of super-particles chosen as well as on the bin widths selected.

Furthermore, due to numerical difficulties mentioned before, it is difficult to determine wake functions of point charges - except for a few, over-simplified geometries which can be solved analytically, such as cylindrical *pillbox* cavities. For arbitrary geometries, the wake functions can hardly ever be obtained to sufficient accuracy, and it is preferable to directly compute the wake potentials of finite bunches. Since this computation is also rather time consuming, it is often not practical to repeat it many times, as would be necessary for simulation of particle motion over many turns, when the distribution changes continuously.

3.4.1 Expansion in Orthogonal Polynomials

In principle, many of these problems can be avoided by expressing the distributions as (infinite) sums over properly chosen expansion functions. One can then store the pre-calculated wake potentials in tables to speed up simulation of particle motion.

An arbitrary distribution can always be decomposed into an infinite series of orthogonal polynomials, and the coefficients are obtained quite rapidly using the orthogonality of the expansion functions. The wake potentials of the higher order polynomials can be computed directly, or derived by numerical differentiation of the wake potential of the corresponding basis function [Wang89]. However, accuracy is lost by this procedure at each step, and the order of the polynomials must be limited to a low number. Similar limitations apply to the more lengthy, direct calculation of wakes of rapidly oscillating higher order polynomials. The limited accuracy of the wakes requires truncation of the expansion at a relatively low order, hence the approximation to the distribution is not always very good. In particular, the positive definite line density is often approximated with a function which may become locally negative.

For line densities which are not too different from a Gaussian, which is usually the case for electron bunches, an expansion into *Hermite polynomials* is most natural [Siemann83, Nys86] since their orthogonality integrals have a Gaussian *weight function.* For the same reason, for proton bunches it is better to expand into Legendre polynomials with a parabolic weight function.

The wake potential of a Gaussian bunch of a given length σ is given by

$$W_\sigma(\tau) = \frac{1}{\sqrt{2\pi}\sigma} \int_0^\infty dt\; G_{\|}(t)\; \exp\left[-\frac{(\tau-t)^2}{2\sigma^2}\right] . \tag{3.27}$$

The wake potential of a Gaussian bunch with length σ' in the same structure can be obtained by the convolution [Bane89]:

$$W_{\sigma'}(\tau) = \frac{1}{\sqrt{2\pi}\sigma_d} \int_0^\infty dt W_\sigma(t) \exp\left[-\frac{(\tau-t)^2}{2\sigma_d^2}\right]. \tag{3.28}$$

where $\sigma_d^2 = \sigma'^2 - \sigma^2$. When $\sigma' < \sigma$, the exponent becomes positive and the integral diverges. Therefore one cannot obtain the wake potential of a shorter Gaussian bunch from that of a longer one.

A similar convolution can be used to obtain the wake potentials of Hermite polynomials of n-th order H_n, which can be expressed by the derivatives of a Gaussian. The wake potentials are given by the integrals [Weiland80b]:

$$W_{\sigma' n}(\tau) = \frac{1}{\sqrt{2\pi}\sigma_d} \int_0^\infty dt \; W_\sigma(t) \; H_n\left(\frac{\tau - t}{\sqrt{2}\sigma_d}\right) \exp\left[-\frac{(\tau - t)^2}{2\sigma_d^2}\right], \tag{3.29}$$

again valid for $\sigma' > \sigma$. This expansion into Hermite polynomials has been used successfully for the simulation of collective effects in single bunches [Siemann83, Nys87]. However, this method has a problem due to the poor convergence of the expansion of the distribution function into Hermite polynomials. In particular, the approximated particle density is no longer everywhere positive. Secondly, since for practical reasons the series has to be truncated at not too high orders, sometimes this method leads to an insufficient approximation. Finally, it is necessary to compute and store a rather large number of 2-dimensional wake potential tables, requiring many lengthy interpolations.

3.4.2 Expansion into Finite Basis Functions

The limitations of using orthogonal polynomials are removed by using simpler expansion functions, such as rectangular or triangular base [Sabbi95], whose centres are displaced from term to term. If one chooses expansion functions of constant width, only a single, one-dimensional table is required. In practice, one should adapt the width of the base to the expected range of bunch lengths, and may thus still need a number of such tables. While these functions are no longer orthogonal, the expansion of an arbitrary distribution into such bases can be obtained even more rapidly simply from the values of the distribution at constant intervals.

Expansion into rectangular bases yields a rather rough approximation of the distribution with discontinuous steps, and the computation of the wake potentials is badly converging due to the infinite slopes of the sides of a rectangular distribution. A base consisting of (equilateral) triangles yields a better, piecewise linear approximation to the distribution function. It has no infinite slopes or discontinuities, only a sudden change of slope at its pick, so the convergence of wake potential computation is improved.

A fine subdivision of the distribution function is possible with such bases, because of the high speed of computing the expansion coefficients and the simpler, one-dimensional tables. These methods have largely increased both the speed and the accuracy of computer simulation of single bunch stability in circular particle accelerators and storage rings.

3.4.3 Gaussian Bunch in a Resonator

The wake potential of a Gaussian bunch in a structure described by a resonator impedance can be found analytically from its convolution with the wake function given by Eq. (3.14):

$$W_{\parallel}(\tau) = \frac{1}{\sqrt{2\pi}}\frac{R_s}{\sigma Q_r'}\mathrm{Re}\left\{\omega_1 \int_0^\infty dt \exp\left[-\frac{(\tau-t)^2}{2\sigma^2} + j\omega_1 t\right]\right\} . \tag{3.30}$$

This integral can be evaluated using the w-*function*, related to the error function of complex argument [Abramowitz65]. With the definition

$$\int_0^\infty dt \exp\left(-a^2t^2 + jzt\right) = \frac{\sqrt{\pi}}{2a}\,\mathrm{w}\left(\frac{z}{2a}\right) \tag{3.31}$$

found in the same reference, one obtains

$$W_{\parallel}(\tau) = \frac{R_s}{2Q_r'}\exp\left(-\frac{\tau^2}{2\sigma^2}\right)\mathrm{Re}\left\{\omega_1 \mathrm{w}\left(\frac{z_1}{\sqrt{2}}\right)\right\} , \tag{3.32}$$

where

$$z_1 = \omega_1\sigma - \frac{j\tau}{\sigma} = \frac{Q_r'}{Q_r}\omega_r\sigma + j\left[\frac{\omega_r\sigma}{2Q_r} - \frac{\tau}{\sigma}\right] . \tag{3.33}$$

These expressions are readily evaluated by computer, however, the w-function becomes ill-behaved when the imaginary part of its argument becomes negative. This happens when $\tau > \omega_r\sigma^2/2Q_r$, and one should then use the relation $\mathrm{w}(z^*) = 2\exp(-z^{*2}) - \mathrm{w}^*(z)$ to avoid numerical problems. The longitudinal wake potentials of Gaussian bunches of various lengths are shown in Figs. 3.5a. For a pillbox cavity with side tubes, the $m = 0$ wake potential computed numerically[3] is shown in Figs. 3.6a.

Near the centre $|\tau| < \sigma$ of a short bunch $|\omega_1\sigma| = \omega_r\sigma \ll 1$, one can use the small argument expansion $\mathrm{w}(z) \approx 1 + 2jz/\sqrt{\pi}$ to get in first order:

$$W_{\parallel}(\tau) \approx \frac{\omega_r}{2}\frac{R_s}{Q_r}\left[1 - \sqrt{\frac{2}{\pi}}\left(\frac{\omega_r\sigma}{2Q_r} - \frac{\tau}{\sigma}\right)\right] . \tag{3.34}$$

[3] We repeat that the numerical codes used for computation of wake potentials have the opposite sign convention: a decelerating wake being counted as negative, from the one adopted in this book.

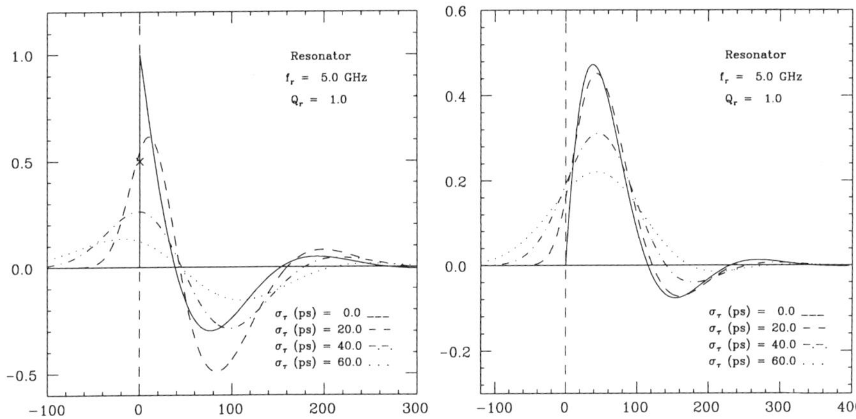

Figure 3.5: Longitudinal and transverse wake potentials of Gaussian bunches in broad-band resonators versus distance τ [ps].

Since both terms in brackets were assumed to be small, the wake potential at the centre of a short bunch is approximated by:

$$W_{\parallel}(0) = \frac{R_s}{2Q_r'} \operatorname{Re}\left\{\omega_1 \mathrm{w}\left(\frac{\omega_1 \sigma}{\sqrt{2}}\right)\right\} \approx \frac{\omega_r R_s}{2Q_r} \,. \tag{3.35}$$

At finite current levels, the induced EM fields will also change the equilibrium line density (*potential well effect*). Depending on the type of impedance, this may lead to a lengthening (inductive), shortening (capacitive), or asymmetric deformation (resistive) of the distribution. At higher current levels, *turbulent* instabilities widen also the energy spread of a bunch, and thus lead to even more bunch lengthening. Bunch shape deformations can also occur due to non-linearity of the rf voltage which is usually sinusoidal. These deformations become stronger for longer bunches. It is important to use expansions which can cope with all these various deformed distribution functions without excessive loss of accuracy.

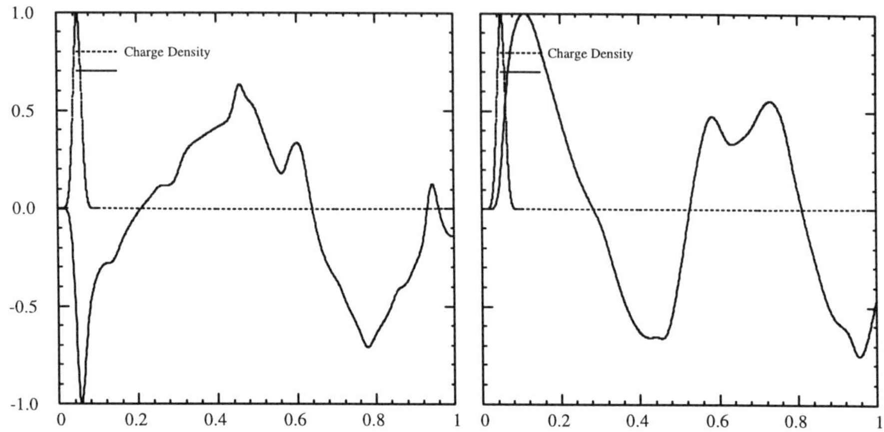

Figure 3.6: Wake potentials for Gaussian bunch ($\sigma = 1$ cm) in a pillbox cavity with beam tubes, versus distance: a) longitudinal $m = 0$ (vertical scale 0.32 V/pC), b) transverse $m = 1$ (vertical scale 2.14 V/pCm).

3.5 Transverse Wake Potentials

If the transverse wake function for a particular geometry is known, the *transverse wake potential* can again be found as its convolution with the particle distribution $\lambda(\tau)$:

$$W_\perp(\tau) = \int_0^\infty dt\ G'_\perp(t)\lambda(\tau - t) \ , \tag{3.36}$$

when one assumes a constant displacement of the bunch from the axis. However, when different parts of the bunch have varying displacements $\xi(\tau)$, the distribution function is replaced by the first moment $\xi(\tau)\lambda(\tau)$, and the result is divided by the average displacement $\bar{\xi}$. In all other respects the calculation of the transverse wake potential is analogous to that of the longitudinal wake potential:

$$W_\perp(\tau) = \frac{1}{\bar{\xi}} \int_0^\infty dt\ G'_\perp(t)\xi(\tau - t)\lambda(\tau - t) \ . \tag{3.37}$$

This expression, with the integral replaced by a finite sum, has been used in simulation programs when the bunches are modelled by a (large) number of super-particles with varying offsets. Note that the first moment $\xi\lambda$ is not positive definite, but may change sign several times for bunches oscillating transversely.

A numerical example of the transverse wake potential of a Gaussian bunch of finite length in a pillbox cavity with side tubes is illustrated in Fig. 3.6b.

3.5.1 Gaussian Bunch in a Resonator

The transverse wake potential is found from the convolution of the transverse wake function of a resonator with a Gaussian bunch distribution. For $Q_r \geq 1/2$ one obtains

$$W_\perp(\tau) = \frac{1}{\sqrt{2\pi}\sigma}\frac{\omega_r R_\perp}{Q'_r}\mathrm{Im}\int_0^\infty dt \exp\left[-\frac{(\tau-t)^2}{2\sigma^2} + j\omega_1 t\right] , \tag{3.38}$$

where the complex frequency ω_1 is defined following Eq. (3.11). This integral can be evaluated also in terms of the complex error function

$$W_\perp(\tau) = \frac{\omega_r R_\perp}{2Q'_r} \exp\left(-\frac{\tau^2}{2\sigma^2}\right) \mathrm{Im}\left\{\mathrm{w}\left(\frac{z_1}{\sqrt{2}}\right)\right\} , \tag{3.39}$$

with z_1 given above by Eq. (3.33). The transverse wake potentials of Gaussian bunches of various lengths are shown in Figs. 3.5b.

For a short bunch $\omega_r\sigma \ll 1$, one can again approximate the wake potential near the centre $\tau < \sigma$:

$$W_\perp(\tau) \approx \frac{\omega_r R_\perp}{\sqrt{2\pi}Q_r}\omega_r\sigma . \tag{3.40}$$

The expression is proportional to the rms bunch length σ, and is independent of the displacement τ from the bunch centre.

3.5.2 Focusing Effects

In rotationally symmetric structures, transverse wake potentials always deflect transversely displaced bunches further away from the axis, i.e. they are *defocusing*. However, in structures without this symmetry, such as rectangular beam pipes or cavities, the dipole wake potentials may also be *focusing*, i.e. they deflect subsequent particles back towards the axis, at least in one plane. The quadrupolar or higher order effects may lead to negative gradients of the transverse loss factors. The focusing properties of flat slots in cavities with circular cross section - or equivalently of circular holes in cavities with oval cross section - have been proposed to focus the beams in the *Compact Linear*

Collider (CLIC) [Schnell91]. The focusing produced by special structures are also the principle of *radio-frequency quadrupoles* or *RFQ*s [Puglisi87] which are presently often used for pre-acceleration of particles up to several MeV.

Appendix 3.A Generalised Integration Contour

Here we consider an axially symmetric structure, such as a cavity between two beam tubes extending to $\pm\infty$, which have perfectly conducting walls. As will be shown below, when the beam velocity is taken to be ultra-relativistic, i.e. $v \approx c$, the integrals for the wake potentials become independent of a parallel displacement of the path of integration. In particular, the path may be shifted to the beam tube surface where the tangential electric and the normal magnetic field components vanish. Then one needs to integrate only over the finite interval of the *cavity gap*, i.e. the region of the structure between the input and exit beam tubes. For the numerical computation of wake functions this is much faster - and at the same time more accurate - than *direct integration* along the axis of symmetry. However, this method fails when:

a) the structure extends to a radius smaller than the beam tube;

b) the beam tubes have unequal radii at the two sides of the structure.

It is possible to generalise the integration contour to remove the first restriction, and - at least for the $m = 0$ wakes - also the second one [Napoly91].

3.A.1 Independence of Integration Path

The generalised method of contour integration will be derived first for the $m = 0$ mode, excited by a point charge q travelling on the axis of symmetry of the structure. Only the field components E_r, E_z, and B_θ are then different from zero. The wake function has a single, longitudinal component, given by the integral over the normalised EM field at distance r from the axis:

$$G_{\|}(r,s) = -\frac{1}{q}\int_{-\infty}^{\infty} dz\ E_z\left(r, z, t = \frac{s+z}{v}\right) . \tag{3.41}$$

The total EM fields can be considered as superpositions of the unperturbed, direct fields due to the space charge of the bunch

$$B_\theta^{(0)}(r,s) = \frac{Z_0 q}{2\pi r}\delta(s) ; \qquad E_r^{(0)}(r,s) = c\ B_\theta^{(0)}(r,s) , \tag{3.42}$$

and the *radiated fields*, induced by the presence of a structure, i.e. $\mathbf{E}_{tot} = \mathbf{E}^{(0)} + \mathbf{E}$ and similarly for $\mathbf{B}$. The integration in Eq. (3.41) is performed only over the radiated fields induced by the presence of a non-uniform structure.

The radiated fields have to satisfy the homogeneous Maxwell equations, i.e. without source terms. The azimuthal component of the second Maxwell equation $\nabla \times \mathbf{E} = -\partial \mathbf{B}/\partial t$ yields:

$$\frac{\partial}{\partial r} E_z \left(r, z, \frac{s+z}{v} \right) = \frac{\partial E_r}{\partial z} + \frac{\partial B_\theta}{\partial t} \,. \tag{3.43}$$

Converting to total derivatives with $d/dz = \partial/\partial z + (1/v)\partial/\partial t$ yields

$$\frac{\partial}{\partial r} E_z = \frac{d}{dz}(E_r + vB_\theta) - \frac{1}{v}\frac{\partial E_r}{\partial t} - v\frac{\partial B_\theta}{\partial z} \,. \tag{3.44}$$

From the radial component of the Maxwell equation $\nabla \times \mathbf{B} = \varepsilon_0 \mu_0 \partial \mathbf{E}/\partial t$ one obtains $\partial E_r/\partial t = -c^2 \partial B_\theta/\partial z$, and thus:

$$\frac{\partial}{\partial r} E_z \left(r, z, \frac{s+z}{v} \right) = \frac{d}{dz}(E_r + vB_\theta) - \frac{1}{v\gamma^2}\frac{\partial E_r}{\partial t} \,, \tag{3.45}$$

For $v \approx c$ the term containing $1/\gamma^2$ may be neglected. Then the r.h.s. becomes a total derivative which can be integrated to yield:

$$\frac{\partial}{\partial r} G_{\parallel}(s, r) = -\frac{1}{q} \left[(E_r + cB_\theta)|_{+\infty} - (E_r + cB_\theta)|_{-\infty} \right] \,. \tag{3.46}$$

Since all radiated fields have to vanish at infinity, one finds $\partial G_{\parallel}(r, s)/\partial r = 0$, i.e. the wake function does not depend on a parallel displacement of the path of integration as long as it remains inside the beam tube without meeting a material boundary. If the radius $r(z)$ of the cavity is everywhere larger than that of the beam pipes, the path of integration can be chosen along the beam pipe surface where $E_z = 0$, and thus their contribution vanishes.

3.A.2 Integration across a Cavity-like Structure

When the beam entrance and exit pipes have different radii, it is necessary to generalise the integration path to follow the cavity surface at least partially. For this we define the *auxiliary function*

$$\bar{G_{\parallel}}(s, r_1) = -\frac{1}{q} \int_{z_1(r_1)}^{z_2(r_1)} dz \; E_z \left(r = r_1, z, t = \frac{s+z}{c} \right) \,, \tag{3.47}$$

where $z_{1,2}(r_1)$ are two points on the boundary of the structure at radius r_1 as shown in Fig. 3.7. When $r_1 < \min(b_{in}, b_{out})$, these boundary points move to $\pm\infty$, and the auxiliary function equals the wake function $\bar{G_{\parallel}}(s, r_1) = G_{\parallel}(s)$.

Taking the radial derivative of this auxiliary function yields:

$$\frac{d}{dr_1}\bar{G}_{\parallel}(r_1, s; z_1, z_2) = \frac{\partial}{\partial r_1}\bar{G}_{\parallel}(r_1, s; z_1, z_2) - \frac{1}{q}\left[E_z(r_1, z_1)\frac{dz_1}{dr_1} - E_z(r_1, z_2)\frac{dz_2}{dr_1}\right] . \tag{3.48}$$

Substituting Eq. (3.46) for the partial derivative one obtains a total differential. The integral of a total differential over any closed path vanishes. We will chose

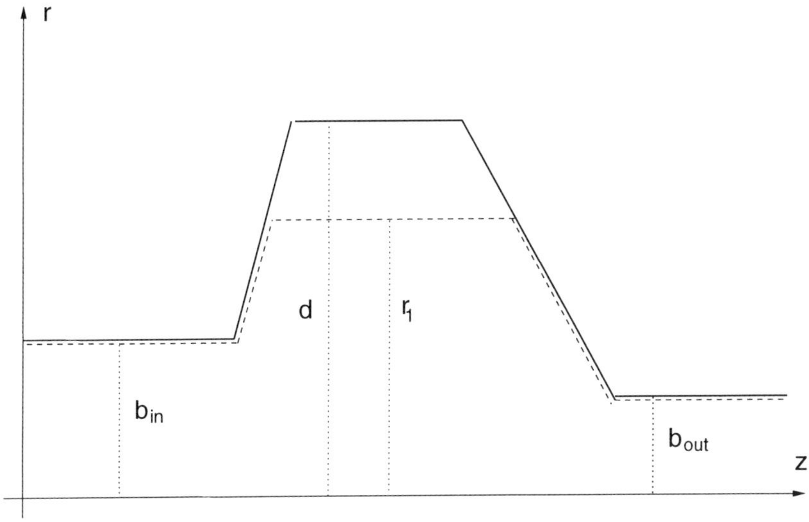

Figure 3.7: Integration path along and across cavity wall.

the path of integration along the z-axis from $-\infty$ to $+\infty$, which yields the wake function $G_{\parallel}(s)$, and back along the contour $\bar{C}$ consisting of the exit beam pipe b_{out}, the cavity surface $\bar{z}(r)$, cutting across the cavity at $r = r_1 \geq \sup(b_{in}, b_{out})$, and back along the entrance beam pipe b_{in} to $-\infty$. This yields:

$$\bar{G}_{\parallel}(r_1, s) = G_{\parallel}(s) - \frac{1}{q}\int_{\bar{C}}[(E_r + c\ B_\theta)dr + E_z dz] . \tag{3.49}$$

At a perfectly conducting (metallic) boundary, the total fields - i.e. including the space-charge terms given in Eq. (3.42) - have to fulfil the relation $E_r dr + E_z ds = 0$. Substituting Eqs. (3.42) and (3.47) one obtains:

$$G_{\parallel}(s) = \frac{1}{q}\int_{z_1}^{z_2} dz\ E_z(r_1, z) + \frac{c}{q}\int_{\bar{z}} dr\ B_\theta^{(tot)}(r, s) + \frac{1}{\pi\varepsilon_0}\ln\frac{b_{out}}{b_{in}}\delta(s) . \tag{3.50}$$

For unequal beam pipe radii $b_{out} \neq b_{in}$, the integral of the direct space-charge field has an additional logarithmic term, caused by different electric potentials in the two pipes. By choosing $r_1 = r_{max}$, then $z_1 = z_2$ and the first integral vanishes:

$$G_{\|}(s) = -\frac{c}{q} \int_C dr\ B_\theta^{(tot)}(r, s) + \frac{1}{\pi\varepsilon_0} \ln \frac{b_{out}}{b_{in}} \delta(s) \ , \tag{3.51}$$

where now $C = C(r, z)$ denotes the full contour of the cavity and beam pipes to ∞.

The path of integration shown in Fig. 3.7 starts at the entrance beam pipe with radius b_{in}, follows along the metallic surface up to point $z_1(r_1)$, across the cavity parallel to the axis at radius $r_1 < r_{max}$ over to $z_2(r_1)$ on the other side of the cavity wall, and then down the metallic surface to the exit beam pipe with radius b_{out}. The choice of r_1 is arbitrary as long as it lies outside the beam pipes and inside the structure. This integration contour is still limited to cavity-like structures extending to the outside of the beam tubes, although these may have different radii.

When the *radiated* azimuthal magnetic field component B_θ is substituted into the integrand, the logarithmic term has to be divided by a factor 2 due to the space charge terms left off.

3.A.3 Integration Contour for Arbitrary Structures

When the boundary of the cavity protrudes into the inside of the beam pipes, a path of integration still can be defined in such a way that the contribution of the beam pipes to the integral vanishes. It is convenient to take an integration contour consisting of a number of distinct straight pieces as shown in Fig. 3.1: start axially along the entrance pipe, proceed radially down to a radius smaller than (or equal to) the minimal radius of the structure, then axially at constant radius beyond the structure, radially up to and then axially along the exit beam pipe [Napoly93].

The EM fields are expanded into modes characterised by the *azimuthal mode number m*:

$$\begin{aligned} \{E_r, B_\theta, E_z\} &= \sum_{m=0}^{\infty} \{e_r, b_\theta, e_z\}^{(m)} \cos m\theta \ , \\ \{B_r, E_\theta, B_z\} &= \sum_{m=1}^{\infty} \{b_r, e_\theta, b_z\}^{(m)} \sin m\theta \end{aligned} \tag{3.52}$$

for both space-charge and radiated fields. Similar to the derivation of Eq. (3.45), one can derive the following relations from the homogeneous Maxwell equations with arguments $r, z, t = (s+z)/c$:

$$\begin{aligned} \left[\frac{\partial}{\partial r}(e_z) - \frac{d}{dz}(e_r + cb_\theta)\right]^{(m)} &= -\frac{m}{r}\,[cb_z]^{(m)} , \\ \left[\frac{\partial}{\partial r}(cb_z) + \frac{d}{dz}(e_\theta - cb_r)\right]^{(m)} &= -\frac{m}{r}\,[e_z]^{(m)} , \end{aligned} \tag{3.53}$$

when terms of order $1/\gamma^2$ are neglected. We rewrite the EM field components at time $t = (z+s)/c$ as functions of r, z, s, which will be indicated by a bar, e.g. $\bar{e}_z(r,z,s) = e_z(r,z,t=(z+s)/c)$. Then the total derivative with respect to z can be replaced by a partial one. Forming the sum and difference of these 2 equations, one obtains the relations

$$\begin{aligned} \left[\frac{\partial}{\partial r}S_z - \frac{\partial}{\partial z}S_r\right]^{(m)} &= 0 , \\ \left[\frac{\partial}{\partial r}D_z - \frac{\partial}{\partial z}D_r\right]^{(m)} &= 0 , \end{aligned} \tag{3.54}$$

where

$$\begin{aligned} S_z^{(m)} &= r^m[\bar{e}_z + c\,\bar{b}_z]^{(m)} , \\ S_r^{(m)} &= r^m[\bar{e}_r + c\,\bar{b}_\theta - \bar{e}_\theta + c\,\bar{b}_r]^{(m)} , \\ D_z^{(m)} &= r^{-m}[\bar{e}_z - c\,\bar{b}_z]^{(m)} , \\ D_r^{(m)} &= r^{-m}[\bar{e}_r + c\,\bar{b}_\theta + \bar{e}_\theta - c\,\bar{b}_r]^{(m)} . \end{aligned} \tag{3.55}$$

Eq. (3.54) shows that the vectors $\mathbf{S}^{(m)} = \{S_r, S_z\}$ and $\mathbf{D}^{(m)} = \{D_r, D_z\}$ are irrotational and can be *expressed as gradients of potentials*, and hence their integrals along any closed contour vanish. One may thus close the integration contour along the axis by an arbitrary path C inside the vacuum enclosure.

Longitudinal Wake Function for $m = 0$

For $m = 0$, the expressions simplify to:

$$S_z^{(0)} = \bar{e}_z \quad , \quad S_r^{(0)} = \bar{e}_r + c\,\bar{b}_\theta \tag{3.56}$$

and we find directly:

$$G_{\parallel}^{(0)}(s) = -\frac{1}{q}\int_{-\infty}^{\infty} dz\ \bar{e}_z(r=0,s,z)$$
$$= -\frac{1}{q}\int_C \Big[E_z\ dz + (E_r + cB_\theta)\ dr\Big] + \frac{\delta(s)}{2\pi\varepsilon_0}\ln\frac{b_{out}}{b_{in}}\ , \qquad (3.57)$$

where the fields are evaluated at $r, z, t = (z+s)/c$.

From the above integral, it can be seen that one has to integrate the longitudinal electric field component E_z along purely axial parts of the contour and hence the beam pipe contributions vanish. For purely radial parts the integral of the combination $E_r^{tot} + cB_\theta^{tot}$ has to be evaluated. For unequal beam pipe radii, the space-charge contribution $\ln(b_{out}/b_{in})/2\pi\varepsilon_0$ must then be subtracted. This option has been included in the computer code *ABCI* [Chin88b] which thus permits faster and more exact evaluation of the wake potentials of structures protruding into the beam pipe, such as scrapers or collimators.

Longitudinal Wake Function for $m \geq 1$

From the expressions Eqs. (3.54), one can derive the general longitudinal wake function for the m-th azimuthal mode:

$$G_{\parallel}^{(m)}(r,\theta,s) = -\frac{\cos(m\theta)}{2q}\frac{r^m}{b^m}\int_C \Big[\bar{e}_z dz + (\bar{e}_r + c\,\bar{b}_\theta)dr'\Big]\left(\frac{b^m}{r'^m} + \frac{r'^m}{b^m}\right)$$
$$-\Big[c\bar{b}_z dz - (\bar{e}_\theta - c\bar{b}_r)dr'\Big]\left(\frac{b^m}{r'^m} - \frac{r'^m}{b^m}\right)\ . \qquad (3.58)$$

The space-charge fields do not contribute to the wake function. The integral along the beam tubes vanishes since $e_z = 0$ and $r' = b$. For unequal beam tubes, only one of the integrals can be eliminated by integration at the beam tube radius, except for $m = 0$. Since the reflected EM fields in the entrance tube are usually smaller than the propagating fields in the exit tube, it is preferable to eliminate the exit tube contribution by integrating at $b_{out} < b_{in}$. If $b_{out} > b_{in}$, it is better to invert the geometry since the result is independent of the direction of beam velocity as shown in Appendix 4A.

3.A.4 Integration Contour for Transverse Wakes

Contrary to the case where $m = 0$, the direct space-charge terms for $m \geq 1$ depend on the radius of the beam tube. Hence, for transverse wakes it is no longer possible to use this approach for the case of unequal beam tube radii, and we will therefore assume $b_{in} = b_{out}$.

The asymptotic expressions for the *radiated field* components of a point charge at position $r = r_0, \theta = 0$, in a perfectly conducting tube of radius b, in the limit $z \to \pm\infty$, can be written:

$$E_r^{(m)} = -cB_\theta^{(m)} = E_\theta^{(m)} = cB_r^{m0)} = -\frac{q}{2\pi\varepsilon_0}\delta(s)\sum_{m=1}^{\infty}\frac{mr^{m-1}r_0^m}{b^{2m}}\cos(m\theta) \quad (3.59)$$

while $E_z^{(m)} = B_z^{(m)} = 0$.

The transverse wake function is in general a vector with 2 components, defined by

$$\mathbf{G}_\perp = \frac{1}{q\xi}\int_{-\infty}^{\infty} dz\ [\mathbf{E}_\perp + \mathbf{v}\times\mathbf{B}]\left(r = \xi, \theta, z, t = \frac{s+z}{c}\right) . \quad (3.60)$$

The EM fields are calculated for a beam which is displaced from the axis by a small distance $r = \xi$, and the integral in Eq. (3.60) is divided by the same quantity - hence the dimension becomes V/Cm. One can distinguish horizontal and vertical components of the transverse wake function by choosing $\theta = 0$ or $\pi/2$. However, for an axially symmetric structure it is more natural to use the radial and azimuthal components.

The transverse wake functions for mode m can be obtained from the longitudinal ones using the Maxwell equations in a form called *Panofsky-Wenzel* theorem, see Appendix to Chapter 4:

$$\frac{\partial}{\partial s}\mathbf{G}_\perp(r,\theta,s) = \nabla_\perp G_\parallel(r,\theta,s) . \quad (3.61)$$

In an axially symmetric geometry the transverse wake function can be written:

$$\mathbf{G}_\perp{}^{(m)}(r,\theta,s) = \sum_{m=1}^{\infty} r^{m-1}[\hat{r}\cos m\theta - \hat{\theta}\sin m\theta]G_\parallel^{(m)}(s) . \quad (3.62)$$

The radial and azimuthal components are equal in value, but shifted by $\pi/2m$ in azimuthal phase. The transverse wake function is in general a sum over all azimuthal modes $m \geq 1$. A pure m-th mode is excited by a thin, annular beam with intensity modulation $\propto \cos m\theta$. In practice, only the $m = 1$ or *dipole* mode is of interest, which is usually by far the strongest. It has the radial and azimuthal components:

$$G_r = -G_\perp^{(1)}\cos\theta \quad \text{and} \quad G_\theta = -G_\perp^{(1)}\sin\theta . \quad (3.63)$$

Proceeding as in the longitudinal case, one obtains the expression

$$G_{\perp}^{(m)}(s) = \frac{1}{2qb^m}\int_{\bar{C}} r'\left\{\left[e_z\, dr' - (e_r - c\, b_\theta)\, dz\right]^{(m)}\left(\frac{b^m}{r'^m} - \frac{r'^m}{b^m}\right)\right.$$
$$\left. + \left[c\, b_z\, r'dz + (e_\theta + c\, b_r)\, dr'\right]^{(m)}\left(\frac{b^m}{r'^m} + \frac{r'^m}{b^m}\right)\right\} , \quad (3.64)$$

where the fields are evaluated at r', z and $t = (z+s)/c$. The space charge fields do not contribute since their longitudinal components, as well as the relevant combinations of transverse components, are zero. The same integration contour can be used for the transverse case as the one described in the previous subsection for the longitudinal one, including the case when the structure protrudes into the beam tube.

For unequal beam tube radii, one has to integrate along the radius of the smaller beam tube. Similar to the longitudinal case it is preferable to make the exit tube the smaller one in order to eliminate its contribution to the integral.

Chapter 4

Coupling Impedances

Just as in the time domain, where the wake functions were introduced as an average over the EM forces, in the frequency domain it is also usually sufficient to study only the integrated interaction of the Fourier transforms of the EM fields with ultra-relativistic particles in order to determine their motion. The description of a problem in the frequency domain is more appropriate for analytical study, in particular since many material parameters are frequency dependent.

In general, we distinguish between longitudinal and transverse impedances according to the direction of the forces which are involved.

4.1 The Concept of Impedance

The concept of impedance is familiar mainly from the theory of electricity. There it is introduced as a generalisation of Ohm's law:

$$V = R \cdot I \;, \tag{4.1}$$

where V is the *voltage* drop across a *resistance* R traversed by a *current* I. This law holds - with certain restrictions - for direct current (DC), but it can also be used for each harmonic of an alternating current (AC) when it is written in complex notation [Jackson75]. One then has to replace the resistance R by a *complex impedance*:

$$Z = R + jX \;, \tag{4.2}$$

where now the real part R is the *resistance*, and the imaginary part X is called the *reactance*.

The impedance concept is useful for solving problems in the *frequency domain*, since it relates currents and voltages at a particular frequency. A time factor $\exp(j\omega t)$ for the current and all field components is understood to be present - but generally left off. Furthermore, the result of a calculation with complex quantities is obtained by taking only the real part of the final expression[1].

In general, an impedance is a *function of frequency* rather than just a value. The often quoted statement that *the impedance of a vacuum chamber is a certain number of Ohms* refers only to the low-frequency limit of Z/n, i.e. the impedance divided by the *harmonic number* $n = \omega/\omega_0$ (where ω_0 is the revolution frequency). At low frequencies, the imaginary part of many impedances becomes proportional to frequency - and hence to the harmonic number, while the real part vanishes more rapidly. The imaginary part of the ratio Z/n, as well as its modulus $|Z/n|$, thus tend to a single value. However, this value does not contain any information about the not less important high-frequency behaviour of the impedance function and is thus not very significant for machines with short bunches.

4.1.1 Longitudinal Impedance

In particle accelerators, the current consists of a beam of charged particles moving in high vacuum, and thus experiences no direct resistance. However, the EM fields excited by the moving charges induce currents and voltages in the surrounding vacuum chamber walls (and beyond, if the walls are transparent to field components). The concept of a *coupling impedance* is introduced to relate the beam current to the total *induced voltage* along the beam trajectory. This voltage is defined as the normalised integral over all EM force components acting on the beam [Ruggiero68]. For a beam moving with constant velocity v at an offset $\mathbf{r}_b$ from the z-axis, only the electric force is relevant in the direction of beam motion, since magnetic forces are perpendicular to it. The *longitudinal impedance* is then given by:

$$Z_{\|}(\mathbf{r}_b, \omega) = -\frac{1}{\tilde{I}} \int_{-\infty}^{\infty} dz\ \widetilde{E}_z(\mathbf{r}_b, z) \exp(jkz)\ . \tag{4.3}$$

[1] The concept of coupling impedances supersedes the *dispersion relation coefficients* U and V previously used in the literature . These coefficients are related to the longitudinal coupling impedance by $Z_{\|} = (V - jU)/e\tilde{I}$. The relation is different for the transverse coefficients and will be shown in the next section.

Here the current and field components are Fourier transforms evaluated at the (circular) frequency $\omega = 2\pi f$. Comparing this definition with that of the wake function $G_{\parallel}(\tau)$ defined in Eq. (3.1), it is clear that for $v = c$ the longitudinal coupling impedance is also the Fourier transform of the wake function $G_{\parallel}(\tau)$:

$$Z_{\parallel}(\omega) = \int_{-\infty}^{\infty} d\tau \; G_{\parallel}(\tau) \exp\left(-j\omega\tau\right) , \tag{4.4}$$

with the inverse transformation:

$$G_{\parallel}(\tau) = \int_{-\infty}^{\infty} \frac{d\omega}{2\pi} Z_{\parallel}(\omega) \exp\left(j\omega\tau\right) . \tag{4.5}$$

Since the wake function is a purely real function, the real and imaginary parts of the complex impedance can also be written as cosine and sine transforms:

$$\begin{aligned} \operatorname{Re} Z_{\parallel}(\omega) &= \int_{-\infty}^{\infty} d\tau G_{\parallel}(\tau) \cos(\omega\tau) , \\ \operatorname{Im} Z_{\parallel}(\omega) &= -\int_{-\infty}^{\infty} d\tau G_{\parallel}(\tau) \sin\left(\omega\tau\right) . \end{aligned} \tag{4.6}$$

From these equations it follows immediately that the real part of the (longitudinal) impedance is symmetric with frequency, while the imaginary part is anti-symmetric. Furthermore, the cosine and sine transforms of a real function are related by a *Hilbert transformation* [Wilson89]:

$$\operatorname{Im} Z_{\parallel}(\omega) = \frac{1}{\pi} PV \int_{-\infty}^{\infty} d\omega' \frac{\operatorname{Re} Z_{\parallel}(\omega')}{\omega - \omega'} , \tag{4.7}$$

where PV stands for the principal value, cf. Appendix 4A. This expression is also called the *Kramers-Kronig* relation [Kramers27, de Kronig28].

Although the impedance and the wake function are integral transforms of each other, in practice they contain complementary information since the impedance is usually known up to some maximum frequency, while the wake function is calculated up to some finite time or distance. Hence the impedance description is more appropriate for long range phenomena, while the wake formalism is better over short distances.

In many cases, and in particular for axially symmetric structures, the beam is assumed to move along the axis where $\mathbf{r}_b = 0$, but for ultra-relativistic beams the integral is in any case independent of the offset. This was shown in the previous Chapter for the longitudinal wake function. Since the offset $\mathbf{r}_b$ is only a parameter in the Fourier transformation, the same property holds also for the longitudinal coupling impedance.

In particle accelerators and storage rings, the most important sources of impedances are *cross section variations* of the vacuum chamber wall, as well as its *finite resistivity* (or, more generally, its finite surface impedance), where EM fields are induced by the passage of the charged particle beam. For ultra-relativistic particles, with $v/c = \beta \to 1$ and $\gamma = (1-\beta^2)^{-1/2} \to \infty$ these *wall impedances* are independent of the beam characteristics, such as bunch length and shape.

4.1.2 Electric Potential

One must distinguish the effects of the wall impedance proper, produced by alternating currents induced in the chamber walls by modulations of the beam, from effects due to differences in DC *electric potential.* Such differences are caused e.g. by single changes of the chamber cross section, and can accelerate (or decelerate) even *a coasting beam* i.e. a uniform stream of particles. Locally, the effect of the potential can be even larger than that of the coupling impedance proper. However, since the vacuum chamber is periodic in circular accelerators, the electric potential must return to its original value after at most one turn, and all DC potential differences will cancel exactly. In linacs, on the other hand, the chamber cross-section can change along the beam, and with it the electric potential. Acceleration of a charged particle beam has actually been achieved in a chamber of decreasing cross section [Friedmann79].

4.1.3 Transverse Impedance

The concept of an impedance can also be defined in the transverse directions. Both electric and magnetic field components there act on the beam through the Lorentz force $\mathbf{F}_\perp = q[\mathbf{E}_\perp + (\mathbf{v}\times\mathbf{B})_\perp]$. The integral over this EM force is usually normalised by the transverse moments of charge qr_b, rather than by the charge itself as for the longitudinal case [Hereward75]:

$$\mathbf{Z}_\perp(\mathbf{r}_b,\omega) = \frac{j}{qr_b}\int_{-\infty}^{\infty} dz \left[\widetilde{\mathbf{E}}(\mathbf{r}_b,z) + \mathbf{v}\times\widetilde{\mathbf{B}}(\mathbf{r}_b,z)\right]_\perp \exp\left(j\omega z/v\right) . \tag{4.8}$$

This changes the dimension to Ω/m, i.e. they are impedances per transverse displacement. Note that the Fourier transforms of the current and field components in this formula should be evaluated at frequency ω. In general, the transverse impedance has two spatial components, which can be chosen as horizontal and vertical, or as radial and azimuthal[2].

In many storage rings and accelerators the vacuum chambers are either circular or flat (i.e. width larger than height). Therefore the vertical impedance is either equal to or larger than the horizontal one. Hence by *transverse impedance* $Z_\perp(\omega)$ one often means the *vertical impedance* given by

$$Z_y(\omega) = \frac{j}{y_b \tilde{I}} \int_{-\infty}^{\infty} dz \left[\tilde{E}_y(z) + v_z \tilde{B}_x(z)\right] |_{x=0,y=y_b} \exp(jkz) \ . \tag{4.9}$$

The imaginary unit j is introduced in the definition in order to account for the fact that a localised transverse force does not change the transverse position of a particle, but the slope of its trajectory. After a quarter period of a transverse *betatron oscillation*, corresponding to a phase factor $\exp(j\pi/2) = j$, this is converted to a change of position.

Although less important, the *horizontal impedance* is similarly defined

$$Z_x(\omega) = \frac{j}{x_b \tilde{I}} \int_{-\infty}^{\infty} dz \left[\tilde{E}_x(z) - v_z \tilde{B}_y(z)\right] |_{x=x_b,y=0} \exp(jkz) \ . \tag{4.10}$$

For general structures with no axial symmetry, also cross terms may exist, i.e. a horizontal displacement leading to a vertical force and vice versa. However, the cross terms of the transverse impedance are usually small and therefore can be neglected.

The direct EM forces between the moving charges in a beam cause the transverse *space charge impedance.* However, in a relativistic beam on a straight trajectory a strong cancellation of electric and magnetic forces occurs and reduces the transverse impedance. This will be discussed in more detail in a later chapter.

4.1.4 Impedance Calculations

The impedance can be calculated analytically only for very few and idealised structures such as uniform, cylindrical beam pipes or closed pill-box cavities with rotational symmetry [Weiland80b]. For more complicated geometries,

[2]The real and imaginary parts of the (complex) transverse impedance replace the formerly used transverse *dispersion relation coefficients* $U_\perp$ and $V_\perp$, which are related to it by the expression $Z_\perp = jZ_0(e\nu\gamma/r_0 I)[V_\perp + j(U_\perp + V_\perp)]$.

such as periodic pill-box cavities with side tubes, one can expand the solutions into infinite series and compute the expansion coefficients numerically. Computer codes for such structures can find resonances up to arbitrary high frequencies (e.g. [Keil72a, Zotter79b]). For less symmetric structures one has to use purely numerical computer codes to calculate frequencies and strengths of resonant modes [Balakin78, Weiland83a, Weiland88]. Some of these are limited to frequencies below the beam-pipe cutoff, unless the cavity is artificially closed or the fields reflected at the cavity ports.

Most computer codes for wake potentials in the time domain also permit evaluation of impedances by including options for Fourier transformation. For good accuracy and high frequency resolution of the results one has to calculate rather long wakes of short bunches which needs considerable computer resources. To eliminate spurious oscillations introduced by the sudden cut-off due to the finite length of the computed wakes one should use *windowing functions* - but these may modify the results at other frequencies. One must also make sure that the time average over the truncated wake potential vanishes in order to avoid spurious impedances at zero frequency (when the beam pipes are equal on both sides of the structure).

4.2 Axially Symmetric Structures

For structures with axial (i.e. rotational) symmetry[3], it is most appropriate to use (circular) cylindrical coordinates. One then distinguishes longitudinal, radial and azimuthal components of the EM field. For ultra-relativistic beams in such structures, the radial and azimuthal components of the EM force have equal values and opposite signs. Then it is sufficient to calculate only a single *transverse impedance* - and the same naturally holds also for the *transverse wake function.*

In axially symmetric geometries, the EM fields and forces can be expanded into multipolar components proportional to $\exp(jm\theta)$. These components are independent of and orthogonal to each other. They are excited individually by the corresponding multipolar components of the beam current. For ultra-relativistic beams, the radial dependence of the EM fields is proportional to r^m for the longitudinal component, and to r^{m-1} for the transverse components. Hence the coupling impedances, which are integrals over the fields, have the

[3]This is often incorrectly called *cylindrical* symmetry, which term actually refers to uniform structures with arbitrary cross section and translational invariance.

same behaviour:

$$Z_{\parallel} \sim r^m \cos(m\theta) \quad \text{for} \quad m \geq 0 ,$$
$$Z_{\perp} \sim r^{m-1} \sin(m\theta) \quad \text{for} \quad m \geq 1 . \tag{4.11}$$

Higher-order multipole field components fall off more rapidly towards the axis, and thus are usually quite weak at the position of the beam. Hence for beam stability calculations, the most important multipolar components are those with the lowest mode numbers:

• the $m = 0$ or *monopole mode* in the longitudinal direction, which accelerates or decelerates the beam, and

• the $m = 1$ or *dipole mode* in one of the transverse directions, which deflects the beam.

A displaced charge of negligible transverse extent excites all multipolar field components. Beams with finite cross section excite fields up to a maximum frequency inversely proportional to the transverse beam dimensions.

4.3 Resonator Impedance

Any enlargement of the beam pipe, which does not extend over a too long distance[4], forms a *cavity.* Such a structure has an impedance function consisting of usually quite a large number of resonant peaks. These are mainly due to *trapped modes*, i.e. resonances with frequencies below the (lowest) cutoff frequency of the beam pipe. However, resonances at higher frequencies may also appear [Henke85] due to reflections of the fields on neighbouring structures, e.g. in multi-cell cavities or disc-loaded wave guides [Walkinshaw49], or due to "rattling" of fields inside the cavity before they can escape into the beam pipe.

Each of these resonances can be approximated quite well with the impedance of a *parallel resonator* circuit, as was verified by detailed field computations [Keil70]. Such a circuit consists of a resistance, an inductance, and a capacitance in parallel. The real and imaginary parts of its impedance are shown in Fig. 4.1. Other equivalent circuits have been proposed, but their frequency dependence does not fit the observed behaviour of cavity resonances so well. An obvious candidate would be a *series-parallel resonator* circuit, with resistance and inductance in series, and a capacitance in parallel. This circuit

[4] The limitations on length are given in the section on the impedance of single steps.

corresponds to a low-frequency resonator consisting of a coil and a capacitor. However, the behaviour of this impedance with increasing resistance is quite different: for a parallel resonator, the peak of the real part becomes lower and wider, while for the series-parallel resonator only the height of the peak changes, while its width remains the same. This means that in a parallel resonator circuit additional damping increases the response at off-resonance frequencies, and this was the behaviour which was observed in detailed field computations.

Using the elementary laws from circuit theory for elements in parallel, see Appendix to Chapter 1, we obtain for the *admittance*, i.e. the inverse of the impedance:

$$Y(\omega) = j\omega C + G + \frac{1}{j\omega L}. \tag{4.12}$$

where C is the capacitance, L the inductance, and G the conductance. Note that G should not be confused with the conductance of the cavity wall. The complex *impedance of a parallel resonator* circuit can thus be written:

$$Z_{\|}(\omega) = \frac{R_s}{1 + jQ_r(\omega/\omega_r - \omega_r/\omega)}, \tag{4.13}$$

where $R_s = 1/G$ is usually called the *shunt impedance.*

We have introduced the resonant frequency

$$\omega_r = \frac{1}{\sqrt{LC}}, \tag{4.14}$$

which is defined as the frequency where the real part of the impedance has its maximum $\text{Re}\, Z_{\|} = R_s$, while its imaginary part vanishes. Furthermore, we have introduced the *quality factor*:

$$Q_r = R_s\sqrt{\frac{C}{L}}\,. \tag{4.15}$$

For a real cavity, the shunt impedance is inversely proportional to the wall resistivity, and the quality factor Q_r decreases when the resistivity increases. A more elaborate model of a resonant cavity which produces the correct dependence of the shunt impedance on resistivity will be considered in Section 11.2.

The *bandwidth* of a resonance $\Delta\omega$, defined as the difference between the frequencies where the real part reaches half its maximum value, is related to

Q_r by the equation $Q_r = \omega_r/\Delta\omega$. It is interesting to note that the ratio $R_s/Q_r = \sqrt{L/C}$ does not depend on the shunt impedance of the resonator.

It is given by its reactive parts only, which are determined by the shape of the cavity, and are independent of the wall material. Therefore it is usually called the *geometrical factor* of a resonator. This factor is also related to the modal loss factors which will be discussed in the next chapter.

The resonator impedance thus has in general 3 parameters:

- the *shunt resistance* R_s,
- the *resonant frequency* ω_r,
- the *quality factor* Q_r.

At low frequencies, the impedance approaches the purely imaginary form $j\omega R_s/\omega_r Q_r$, corresponding to an inductance, while the real part vanishes quadratically as $\omega^2 R_s/\omega_r^2 Q_r^2$. At resonance, i.e. for $\omega = \omega_r$, the impedance is purely real and equals the shunt resistance. In general, the real part of the impedance is always positive.

4.3.1 Narrow-Band Resonances

For high quality factors - corresponding to narrow resonances such as found in cavities made of good metallic conductors - the expression for the impedance near the resonance can be simplified. For a small deviation from resonance frequency $\xi = \delta\omega/\omega_r << 1$, one obtains the approximation

$$Z_{\|}(\omega) \approx \frac{R_s}{1 + 2jQ_r\xi} = R_s \frac{1 - 2jQ_r\xi}{1 + (2Q_r\xi)^2} . \tag{4.16}$$

The real part has half its maximum value when the denominator equals two, i.e. when the frequency deviation becomes $\pm\omega_r/2Q_r$. The full band width at half maximum is thus again found to be $\Delta\omega = \omega_r/Q_r$.

4.3.2 Broad-Band Impedances

For an actual cavity, the impedance is a sum over a large number of resonances - often called *higher-order modes* or *HOMs* for short. These resonances may extend up to high frequencies, and get denser as the frequency increases. The finite band widths due to the wall resistivity make them overlap more and more. Above the beam pipe cutoff frequency, they are also damped by fields radiating into wave guide modes. Thus the impedance tends to decrease and becomes much smoother at very high frequencies. In other words, the impedance which acts on short bunch is *smeared out.* The narrow peaks of the cavity resonances cannot be resolved in the limited time of passage of a short bunch. In the time domain, the equivalent statement is that particles see only the wake potential over the length of a bunch and cannot be influenced by the wake potential behind it (assuming that the wakes have decayed after a full revolution).

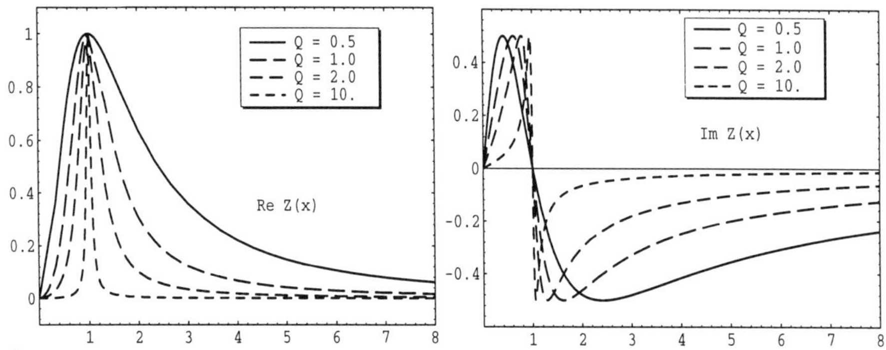

Figure 4.1: Real and imaginary parts of the longitudinal impedance versus normalised frequency ω/ω_r for resonators with various quality factors.

In this case, the actual impedance function may thus be replaced by a simpler, smoothed one. Such a *broad-band impedance* can be obtained e.g. by averaging over the resonances, or by *binning* - i.e. adding the total impedances within adjacent frequency bands [Bane80a]. Furthermore, the smoothed impedance can even be approximated by a resonator with a low quality factor - this is the so-called *broad-band resonator* model. Often this quality factor can be taken equal to one, since for most short range effects only the ratio R_s/Q_r is important. Fig. 4.1 shows the impedance of a broad-band resonator for various quality factors.

At very high frequencies, the real part of the resonator impedance falls off more rapidly ($\mathrm{Re}\, Z_{\|} \sim \omega^{-2}$) than that of a (single) cavity, $\mathrm{Re}\, Z_{\|} \sim \omega^{-1/2}$, and even faster than that of periodic cavities $\omega^{-3/2}$. However, over the limited frequency range of interest for a particular problem, e.g. over the spectrum of a bunch oscillation, the parameters of a resonator can usually be chosen to yield a satisfactory approximation. Methods to do this will be discussed in Chapter 11.

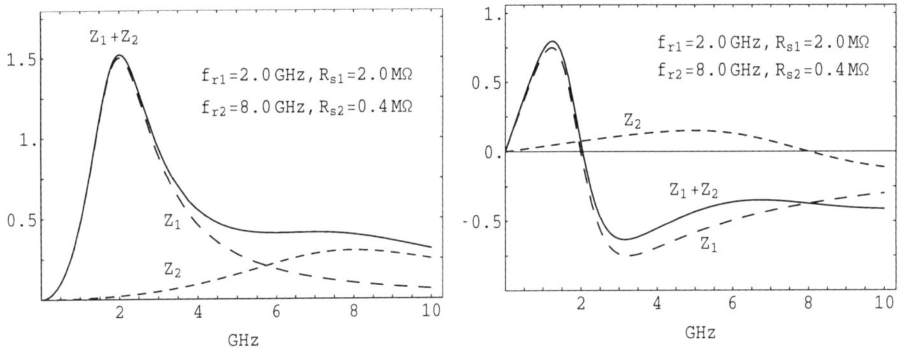

Figure 4.2: Real and imaginary parts of the longitudinal impedances of two broad-band resonators and their sum versus frequency.

The total impedance of a machine is often caused by a number of physically or electrically dissimilar structures, e.g. large cavities and many small cross section variations such as bellows. It may then become necessary to include two - or even more - broad-band resonators into the model impedance. Their resonant frequencies may extend over a large range, and thus influence the beam behaviour in rather different ways. A typical example is the combination of the broad-band impedance of an rf cavity, (with a peak near the beam pipe cut-off frequency), and another broad-band impedance of (shielded) bellows and other small, but numerous cross-section changes of the beam pipe resonating at higher frequencies due to their smaller dimensions. Fig. 4.2 shows the superposition of the impedances of two broad-band resonators with different resonant frequencies, as well as the spectra of the two lowest modes of a Gaussian bunch with rms length σ= 30 ps.

The biggest advantage of the broad-band resonator model is the fact that it permits explicit analytic evaluation of many expressions, such as wake functions, effective impedances,or loss factors, for some of the most important bunch distributions which are discussed below. Other broad-band impedance models, discussed in Chapter 11, with better high frequency behaviour have been proposed [Hofmann79b] but have not yet been used widely since they are more complicated.

4.3.3 Partial Fraction Decomposition

One can decompose the complex impedance of a resonator, given in Eq. (4.13) by a quadratic expression, into linear fractions, which then allows many subsequent mathematical operations to be performed explicitly. The longitudinal impedance can be written:

$$Z_{\|}(\omega) = \frac{\omega_r R_s}{jQ_r} \frac{\omega}{(\omega-\omega_1)(\omega-\omega_2)} = \frac{R_s}{2jQ_r'}\left[\frac{\omega_1}{\omega-\omega_1} - \frac{\omega_2}{\omega-\omega_2}\right] , \tag{4.17}$$

where $Q_r' = Q_r\sqrt{1-1/4Q_r^2}$ is a modified quality factor, and the complex frequency roots are given by

$$\omega_{1,2} = \frac{\omega_r}{Q_r}\left[\frac{j}{2} \pm Q_r'\right] . \tag{4.18}$$

For the usual case of quality factors larger than 1/2, the modified quality factor Q_r' is real, and thus $\omega_2 = -\omega_1^*$ which permits simplification of many expressions. For smaller quality factors, Q_r' is purely imaginary, and with $Q_r'' = jQ_r' = Q_r\sqrt{1/4Q_r^2 - 1}$ one can write

$$\omega_{1,2} = \frac{j\omega_r}{Q_r}\left[\frac{1}{2} \pm Q_r''\right] . \tag{4.19}$$

In this case there is no simple relation between the two roots, and both terms need to be written explicitly.

4.3.4 Reduced and Transverse Impedances

In circular accelerators with revolution frequency ω_0, one finds that for beam stability the quantity $Z_{\|}/n$, i.e. the impedance divided by the harmonic number $n = \omega/\omega_0$, is actually of more basic importance than the impedance itself. Due to the extra frequency factor, the symmetry properties of this *reduced impedance* $Z_{\|}/n$ are changed compared to $Z_{\|}$: the real part is now odd in frequency, while its imaginary part is even. Its frequency decomposition actually is even simpler, and can be written:

$$\frac{Z_{\|}}{n}(\omega) = \frac{\omega_0 R_s}{jQ_r} \frac{\omega_r}{(\omega-\omega_1)(\omega-\omega_2)} = \frac{\omega_0 R_s}{2jQ_r'}\left[\frac{1}{\omega-\omega_1} - \frac{1}{\omega-\omega_2}\right] . \tag{4.20}$$

Simply by redefining the parameters in this expression, the same formula can be used also for the *transverse resonator impedance.* The shunt resistance R_s

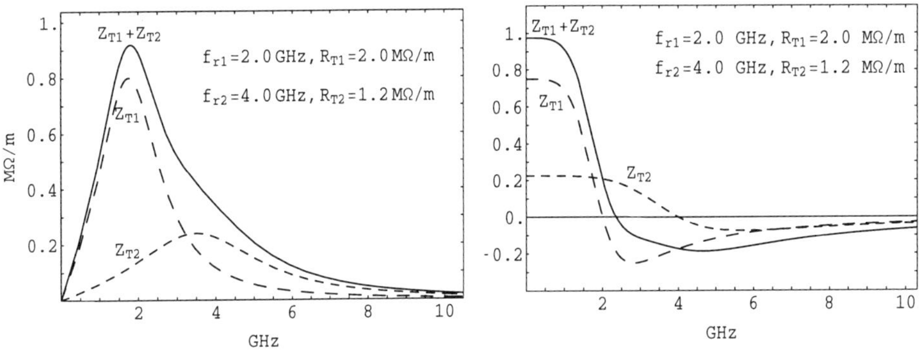

Figure 4.3: Real and Imaginary parts of transverse impedances of two broad-band resonators and their sum versus frequency.

is replaced by the *transverse resistance* $R_\perp$, with dimensions Ω/m since it is measured per unit transverse displacement, and the resonant frequencies and quality factors are changed to those of the transverse resonant modes. The transverse impedance of two rather unequal resonators is shown in Fig. 4.3.

The expression Eq. (4.20) can be used for both $Z_\parallel/n$ and $Z_\perp$, depending on the interpretation of the resistance R as the longitudinal shunt impedance R_s, or the transverse one $R_\perp$.

Appendix 4.A General Properties of Impedances

The coupling impedances have important general properties which will be discussed in this Appendix. Certain properties arise as consequences of the definition of the impedance. These properties are always valid. For example, the relation between the real and the imaginary parts of the impedance or symmetry in respect to the change of the sign of its argument, follow from its definition. Others applying to specific cases have a more limited range of validity. For example, an expression for the radial dependence of the impedance has only been derived for axially symmetric structures excited by a ultra-relativistic particles. Nevertheless, all these properties are useful, e.g. they provide powerful methods to simplify impedance calculations and to verify their correctness.

4.A.1 Symmetry with Respect to Frequency

The symmetry properties of the real and imaginary parts of the impedance with respect to frequency were mentioned in Chapter 1:

$$Z_{\|}(-\omega) = Z_{\|}^{*}(\omega) , \qquad (4.21)$$

where the *asterisk* indicates the complex conjugate value. Hence the real part $Z_R = \text{Re } Z_{\|}(\omega)$ is a symmetric, while the imaginary part $Z_I = \text{Im } Z_{\|}(\omega)$ is an anti-symmetric function of frequency. It should be emphasised that this property is due to the fact that the impedance is the Fourier transform of the purely real wake function, cf. Eq. (4.4).

As we will see throughout this book calculations of expressions are greatly facilitated by an analytical continuation of the impedance function given in Eq. (4.3) into the complex frequency plane $\omega_c = \omega + j\omega_I$ In this domain it has a more general symmetry property:

$$Z_{\|}(-\omega_c^{*}) = Z_{\|}^{*}(\omega_c) . \qquad (4.22)$$

For purely imaginary ω_c that gives $Z_{\|}(\omega_I) = Z_{\|}^{*}(\omega_I)$. This means that the impedance is purely real on the imaginary axis: $Z_I(\omega_I) = 0$ for Re $\omega_c = 0$. In particular, the DC impedance is always real, $Z_I(0) = 0$.

Similarly, from the definition of the transverse impedance Eq. (4.8) one obtains:

$$Z_{\perp}(-\omega_c^*) = -Z_{\perp}^*(\omega_c) \ . \tag{4.23}$$

For a real frequency ω, the symmetry property of the transverse impedance with respect to frequency is reversed in comparison to that of the longitudinal impedance: Re $Z_{\perp}(\omega)$ is an anti-symmetric, while Im $Z_{\perp}(\omega)$ is a symmetric function of frequency.

4.A.2 Relation between Real and Imaginary Parts

Several other properties of the impedance can be derived from the analytic continuation of the impedance into the complex ω_c plane.

Since the wake function cannot travel faster than light, it must vanish in front of an exciting particle moving with (almost) that velocity. This property is loosely called *causality*:

$$G_{\parallel}(s) \ = \ 0 \quad \text{for} \quad s = c\tau < 0 \ . \tag{4.24}$$

Using this fact, one can show that the longitudinal impedance has no singularities in the lower half-plane of the complex frequency.

The wake function is given by Eq. (4.5). The integral in this equation can be evaluated with the help of the residue theorem:

$$\oint d\omega Z_{\parallel}(\omega) \exp(j\omega\tau) = 2\pi j \sum_k Z_{\parallel}(\omega_k) \exp(j\omega_k\tau) \ . \tag{4.25}$$

If the complex poles ω_k of the impedance function are known, one can evaluate the integral from $-\infty$ to $+\infty$ by closing the integration contour over a path on which the contribution vanishes, e.g. with a semi-circle whose radius tends to infinity as shown in Fig. 4.4. The real part of the exponent is Re $j\omega_c\tau = -\tau\omega_I$. For $\tau < 0$, it is negative for a negative imaginary part of the complex frequency, and one must therefore close the contour in the lower half-plane $\omega_I < 0$. If there are no poles of the impedance with negative imaginary part, the integral will be zero as required by Eq. (4.24).

On the other hand, when $\tau > 0$, the contour of integration should be closed in the upper half-plane $\omega_I > 0$ resulting in the expression Eq. (4.25).

Now we are ready to derive the relation between the real and imaginary parts of the longitudinal impedance. The derivation follows closely a similar derivation of the *dispersion relation* connecting the real and imaginary parts of

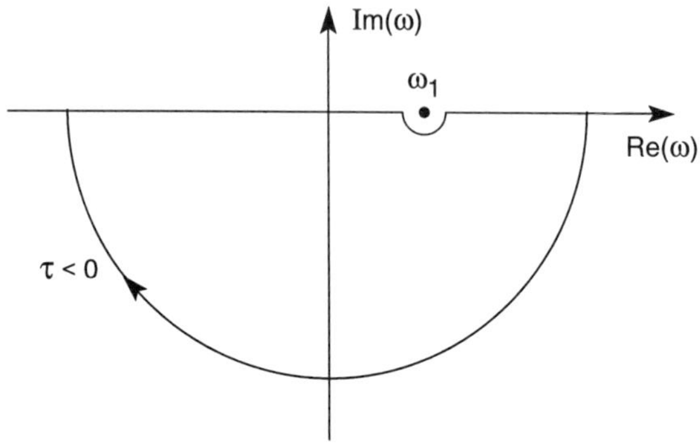

Figure 4.4: Integration path in the complex frequency plane.

the dielectric permittivity [Kramers27, de Kronig28]. To derive this relation we perform the integration of the expression $Z_{\parallel}(\omega_c)/(\omega_c - \omega_1)$, where ω_1 is real, over the contour shown in Fig. 4.4. Since the contribution of the integral over the semi-circle of a large radius vanishes in this case, the integral can be written as the sum of its *principal value* [Morse53] and the residue times $j\pi$ (since the pole lies on the real axis, its contribution is halved), which must be equal to zero since there are no other poles inside the contour :

$$PV \int_{-\infty}^{\infty} d\omega \frac{Z_{\parallel}(\omega)}{\omega - \omega_1} + j\pi Z_{\parallel}(\omega_1) = 0 \ . \tag{4.26}$$

The first term in this expression is the result of the integration along the real axis. The second term is the contribution of a infinitesimal semi-circle above the pole at $\omega_c = \omega_1$. The integral over a semi-circle of a radius $R \to \infty$ does not contribute to the value of the integral.

Separating the real and imaginary parts in Eq. (4.26) and interchanging the names of ω and ω_1, one obtains the following two relations between the real and the imaginary parts of the longitudinal impedance:

$$\begin{aligned} Z_R(\omega) &= \frac{1}{\pi} PV \int_{-\infty}^{\infty} Z_I(\omega_1) \, \frac{d\omega_1}{\omega_1 - \omega} \ , \\ Z_I(\omega) &= -\frac{1}{\pi} PV \int_{-\infty}^{\infty} Z_R(\omega_1) \, \frac{d\omega_1}{\omega_1 - \omega} \ . \end{aligned} \tag{4.27}$$

One can recognise these expressions as *the Hilbert transformations.* These relations allow to calculate the complex impedance from either the real or imaginary part alone.

Similar relations hold also for the transverse impedance. Since the transverse wake in front of an exciting charge also vanishes in the ultra-relativistic limit, the same Eqs. (4.27) are valid for the real and imaginary parts of the transverse impedance as well.

4.A.3 The Panofsky-Wenzel Theorem

There exist relations between the transverse derivative of the longitudinal wake function and the longitudinal derivative of the transverse wake function. Similar relations hold for the impedances. These relations are valid for general conditions and are referred to as the *Panofsky-Wenzel theorem* [Panofsky56], although the authors stated it only for the special case of an axially symmetric cavity. We present here a general derivation of this relation which shows that the theorem is valid also for non-symmetric structures. A similar proof has recently been published [Henke95].

The only condition necessary is that the transverse components of the particle velocity can be neglected, and that its longitudinal component is constant. The trajectory of the exciting point charge is assumed to be parallel to the z axis, but may be displaced from it by $\mathbf{r}_q$. Following behind at a distance s, a test particle of charge q moves on the trajectory $z = v(t - t_0) - s$ parallel to the first one, but displaced from the axis by $\mathbf{r}_e$.

To derive the desired relation we start with the Maxwell equation: $\nabla \times \mathbf{E} = -\partial \mathbf{B}/\partial t$. From this equation the magnetic induction $\mathbf{B}$ can be found:

$$\mathbf{B} = -\int_{t_1}^{t} dt \, \nabla \times \mathbf{E} \, . \tag{4.28}$$

The lower limit of integration t_1 is arbitrary provided that it is far enough in the past to ensure that $\nabla \times \mathbf{E}$ is negligible.

The Lorentz force $\mathbf{F}$ acting on the trailing particle can then be written:

$$\mathbf{F} = q \left[\mathbf{E} - \mathbf{v} \times \int_{t_1}^{t} dt' \nabla \times \mathbf{E} \right] \, . \tag{4.29}$$

Since we assumed that the particle velocity is constant it can be brought inside the integral:

$$\mathbf{F} = q \left[\mathbf{E} - \int_{t_1}^{t} dt' \left[\nabla(\mathbf{v} \cdot \mathbf{E}) - \mathbf{v}(\nabla \cdot \mathbf{E}) \right] \right] \, . \tag{4.30}$$

We separate the longitudinal and transverse components:

$$F_{\parallel} = q\, E_z\ , \qquad \mathbf{F}_{\perp} = q\left[\mathbf{E}_{\perp} - v\int_{t_1}^{t} dt'\,\left[\nabla_{\perp} E_z - \frac{\partial \mathbf{E}_{\perp}}{\partial z}\right]\right]\ . \tag{4.31}$$

The differential operator $\nabla_{\perp}$ acts on the transverse coordinates $\mathbf{r}_e$ only.

Now we use the definitions of the longitudinal and the transverse wake functions Eqs. (3.1) and (3.17) rewritten as integrals over time:

$$G_{\parallel}(\mathbf{r}_q, \mathbf{r}_e, s) \ = \ -\frac{1}{q}\int_{-\infty}^{\infty} dt'\ E_z|_{(\mathbf{r}_q,\mathbf{r}_e,z=vt'-s,t')}\ , \tag{4.32}$$

$$\mathbf{G}_{\perp}(\mathbf{r}_q, \mathbf{r}_{\perp}, s) \ = \ \frac{v}{q}\int_{-\infty}^{\infty} dt\ \left[\mathbf{E}_{\perp} - v\int_{t_1}^{t} dt'\ (\nabla_{\perp} E_z - \frac{\partial \mathbf{E}_{\perp}}{\partial z})\right]_{(\mathbf{r}_q,\mathbf{r}_e,z=vt'-s,t')} \tag{4.33}$$

To obtain the desired relation, we need to calculate the derivative of the transverse wake function with respect to s. The derivative of the first term, which depends on s through $z = vt - s$, may be replaced by $\partial/\partial s = -\partial/\partial z$. The second and third terms depend on s only through the time $t = (z+s)/v$ in the upper limit of the integral. Then the derivative with respect to s may be replaced by $\partial/\partial s = \partial/v\partial t$ with the result being that the value of the integral equals the integrand taken at the upper limit. As a result, the first and the third terms cancel each other, and the gradient factors out to operate on the expression of the longitudinal wake function Eq. (3.1):

$$\frac{\partial \mathbf{G}_{\perp}}{\partial s}|_{(\mathbf{r}_q,\mathbf{r}_e,s)} = \nabla_{\perp} G_{\parallel}(\mathbf{r}_q, \mathbf{r}_e, s)\ . \tag{4.34}$$

By applying the Fourier transformation in the coordinate s to both sides of this equation we obtain a similar relation for the longitudinal and transverse impedances:

$$\mathbf{Z}_{\perp}(\omega, \mathbf{r}_e) = \frac{v}{\omega}\nabla_{\perp} Z_{\parallel}(\omega, \mathbf{r}_e)\ . \tag{4.35}$$

That is what we intended to prove, and which relation has become known as the *Panofsky-Wenzel theorem*.

4.A.4 The Radial Dependence of the Impedance

For *axially symmetric* structures, the radial dependence of the coupling impedance in the *ultra-relativistic case* can be found explicitly [Bane80b, Weiland83a]. To obtain this dependence, note that the radiated part of the EM field satisfies the homogeneous wave equation. For the Fourier harmonic $\widetilde{\mathbf{E}}_\omega$ it is:

$$\nabla^2\widetilde{\mathbf{E}}_\omega + k^2\widetilde{\mathbf{E}}_\omega = 0 \, , \quad k = \omega/c \, . \tag{4.36}$$

The synchronous component of the field, i.e., the component whose phase velocity equals the velocity of the particle, is then found as follows:

$$\widetilde{\mathbf{E}}_k^{(s)}(r,\theta) = \int_{-\infty}^{\infty} dz\, \widetilde{\mathbf{E}}_\omega(r,\theta,z)\, e^{-jkz} \, . \tag{4.37}$$

In general, the boundary conditions for $\widetilde{\mathbf{E}}_\omega(r,\theta,z)$ mix components of $\widetilde{\mathbf{E}}_k(r,\theta)$ with different wave numbers k. However, for axially symmetric structures, the azimuthal harmonics $\widetilde{\mathbf{E}}_m$ which are the coefficients in the expansion of the synchronous component

$$\widetilde{\mathbf{E}}_k^{(s)}(r,\theta) = \sum_m \widetilde{\mathbf{E}}_m(k,r)\, e^{jm\theta} \tag{4.38}$$

do not mix. Hence, these coefficients can be treated independently of each other.

The equations for the projections of the azimuthal harmonics can be obtained with the Eqs. (4.36) and (4.37).

$$\begin{aligned} \left[\frac{1}{r}\frac{\partial}{\partial r}\left(r\frac{\partial}{\partial r}\right) - \frac{m^2}{r^2}\right] \widetilde{E}_m^{(z)}(k,r) &= 0 \, , \\ \left[\frac{1}{r}\frac{\partial}{\partial r}\left(r\frac{\partial}{\partial r}\right) - \frac{(m\pm 1)^2}{r^2}\right] \widetilde{E}_m^{(\pm)}(k,r) &= 0 \, , \end{aligned} \tag{4.39}$$

where $\widetilde{E}_m^{(\pm)} \equiv \widetilde{E}_m^{(r)} \pm j\widetilde{E}_m^{(\theta)}$. The solutions of Eqs. (4.39) which are finite on the axis $r = 0$ are:

$$\begin{aligned} \widetilde{E}_m^{(z)}(k,r) &= F_m^{(z)}(k) r^m \, , \quad m \ge 0 \, , \\ \widetilde{E}_m^{(\pm)}(k,r) &= F_m^{(\pm)}(k) r^{m\pm 1} \, , \quad m \pm 1 \ge 0 \, . \end{aligned} \tag{4.40}$$

The functions $F_m^{(z)}(k)$ and $F_m^{(\pm)}(k)$ are defined by the boundary condition for $\widetilde{\mathbf{E}}_k^{(s)}(r,\theta)$. The longitudinal impedances for each mode m can then be expressed in terms of $\widetilde{E}_m^{(z)}(k,r)$:

$$Z_{\|m} = -\frac{\widetilde{E}_m^{(z)}(k,r)}{q} = -\frac{F_m^{(z)}(k)}{q}\, r^m \;, \quad m \geq 0 \;. \tag{4.41}$$

The transverse impedances can be derived from the longitudinal component with the help of the Panofsky-Wenzel theorem Eq. (4.35). Using expressions for the gradient in cylindrical coordinates, we obtain:

$$\begin{aligned} Z_{rm} &= -\frac{1}{qk}\frac{\partial \widetilde{E}_m^{(z)}(k,r)}{\partial r} = -\frac{mF_m^{(z)}(k)}{qk}\, r^{m-1} \;, \quad m \geq 1 \;, \\ Z_{\theta m} &= \frac{jm}{qk}\frac{\widetilde{E}_m^{(z)}(k,r)}{r} = \frac{jmF_m^{(z)}(k)}{qk}\, r^{m-1} \;, \quad m \geq 1 \;. \end{aligned} \tag{4.42}$$

These formulae give the scaling of the impedance with the offset $r = r_e$ of the test particle. From the last two equations one can see that in rotationally symmetric structures $Z_{\theta m} = -jZ_{rm}$.

These equations can also be written in dimensionless form, $Z_{\|m} \sim (r/b)^m$, and Z_{rm} and $Z_{\theta m} \sim (r/b)^{m-1}$. Here b is a characteristic transverse dimension, e.g. the pipe radius. Since the transverse size of the bunch is usually much smaller than b, mainly to prevent particle losses, the monopole mode ($m = 0$) dominates the longitudinal impedance, while the dipole mode ($m = 1$) is the most important transverse one.

4.A.5 Independence of Impedance on the Integration Path

In Chapter 3 it was shown that the integral for the wake function for axially symmetric structures is independent of the path of integration as long as it is parallel to the axis. The impedance is the Fourier transform of the wake function thus has the same property. For the monopole mode $m = 0$ this can also be seen directly from Eq. (4.41).

The longitudinal impedance can therefore be calculated by integrating the field over z at any value of the coordinate r. In particular, it is convenient to integrate the field along the pipe wall and the line which is its imaginary continuation inside the structure. Since the longitudinal field on the wall is zero, the integration for the case $b_1 = b_2$ can be performed over the line inside the structure only. By rescaling the respective impedance with the offset r, this procedure can be applied to any other mode m as well.

Using Eq. (4.42) one obtains for *the dipole* mode $m = 1$ the relation

$$Z_{r1} = \frac{1}{kb} Z_{\|1}|_{r=b} \ . \tag{4.43}$$

In many cases it is sufficient to find the longitudinal impedance. The transverse impedance can then be obtained by applying Eq. (4.43) for $m = 1$, or the Panofsky-Wenzel theorem for any value of m.

4.A.6 Directional Symmetry of Impedance

The independence of the impedance on the direction of the bunch motion is an important feature, in particular for colliders with beams circulating in opposite directions. This property was first investigated for *longitudinal* impedance of an axially symmetric, but longitudinally asymmetric cavity with side pipes of equal radii [Gluckstern88]. This result corroborates observation that the wake potentials obtained numerically, e.g. with the code *ABCI*, are independent on the direction of the beam motion.

Another proof of the same statement is based on the reciprocity relation applied to the tensor Green function [Bisognano89]. Here we present a more general, and physically more transparent derivation [Heifets90b] of this property for *longitudinal* and *transverse* impedances. The result is valid for cavities without azimuthal symmetry, and also for arbitrary particle velocity as long as it is constant. The result shows that the impedance is directionally symmetric only if the entrance and exit beam pipes have equal cross sections.

We consider a cavity of arbitrary shape, traversed by a bunch travelling parallel to the coordinate axis z with offset $\mathbf{r}_b$. We attach the superscripts + or - to all quantities pertaining to the positive or negative direction of bunch motion.

The current densities of the two beams of opposite charges have only z-components. Their Fourier harmonics are:

$$\begin{aligned} \tilde{J}_z^+ &= q\delta(\mathbf{r}-\mathbf{r}_b)\exp(-j\omega z/v) \ , \qquad & \tilde{\rho}^+ &= \tilde{J}_z^+/v \ , \\ \tilde{J}_z^- &= -q\delta(\mathbf{r}-\mathbf{r}_b)\exp(j\omega z/v) \ , \qquad & \tilde{\rho}^- &= -\tilde{J}_z^-/v \ . \end{aligned} \tag{4.44}$$

Note that

$$\tilde{\mathbf{J}}^- = -\tilde{\mathbf{J}}^{+*} \ , \qquad \tilde{\rho}^- = \tilde{\rho}^{+*} \ . \tag{4.45}$$

The fields $\tilde{\mathbf{E}}^{\pm}$, excited by these sources, satisfy the following wave equations:

$$\left(\nabla^2 + \frac{\omega^2}{c^2}\right) \tilde{\mathbf{E}}^{\pm} = \frac{\nabla \tilde{\rho}^{\pm}}{\varepsilon_0} + j\omega\mu_0 \tilde{\mathbf{J}}^{\pm} . \tag{4.46}$$

The boundary conditions for the tangential to the wall field components are:

$$\tilde{\mathbf{E}}^{\pm}_{tan}(\mathbf{r}, z) = 0 , \tag{4.47}$$

and the radiation conditions for the radiated part of the EM field:

$$\lim \tilde{\mathbf{E}}^{\pm}_{rad}(z, \mathbf{r}, t) = 0 \quad \text{for} \quad z \to \pm\infty . \tag{4.48}$$

Note that the fields $\tilde{\mathbf{E}}^{+}$ and $\tilde{\mathbf{E}}^{-}$ satisfy the same wave equation. Also the boundary conditions, Eq. (4.47) are also valid for $\tilde{\mathbf{E}}^{\pm *}$. We further need the additional assumption that there are no incident waves accompanying the particles. Then we may conclude that the electric fields can be written as complex conjugates of one another:

$$\tilde{\mathbf{E}}^{-*} = \tilde{\mathbf{E}}^{+} . \tag{4.49}$$

From the Maxwell equation $j(\omega/c)\widetilde{\mathbf{H}}^{\pm} = -\nabla \times \tilde{\mathbf{E}}^{\pm}$, it also follows that:

$$\widetilde{\mathbf{H}}^{+*} = -\widetilde{\mathbf{H}}^{-} . \tag{4.50}$$

We now multiply the Eqs. (4.46) by $\tilde{\mathbf{E}}^{\mp}$ subtract the results and integrate the difference over the volume V of the cavity and its beam pipes up to $z = \pm\xi, \xi \to \infty$. We then obtain the *Lorentz reciprocity theorem* [Collin50]:

$$\int dV \; [\tilde{\mathbf{E}}^{-} \cdot \tilde{\mathbf{J}}^{+} - \tilde{\mathbf{E}}^{+} \cdot \tilde{\mathbf{J}}^{-}] = \int d\mathbf{S} \cdot [\tilde{\mathbf{E}}^{+} \times \widetilde{\mathbf{H}}^{-} - \tilde{\mathbf{E}}^{-} \times \widetilde{\mathbf{H}}^{+}] . \tag{4.51}$$

The integration on the right-hand side is performed over the surface enclosing the volume V. Since the vector product of the electric and the magnetic fields is tangential to a perfectly conducting wall, its scalar product with the (normal) surface element vector $d\mathbf{S}$ vanishes. Hence it is sufficient to perform the integration only over the areas of the beam pipe openings $S_{1,2}$.

The exciting beam currents have only z components. Their scalar products with the electric field of a beam moving in the opposite directions become $\pm q\delta(\mathbf{r} - \mathbf{r}_b) \exp(\mp jkz) E_z^{\mp}$. The integration over the transverse coordinates on the l.h.s. of Eq. (4.51) then simply replaces $\mathbf{r}$ by $\mathbf{r}_b$, while the integral over z gives the longitudinal impedance.

From Eq. (4.3) we define the longitudinal impedances $Z^{\pm}$:

$$Z^{\pm}(\omega, \mathbf{r}) = \mp\frac{1}{q}\int_{-\infty}^{\infty} dz\ \widetilde{E}_z^{\pm}(\omega, \mathbf{r}, z)\ \exp(\pm j\omega z/v)\ . \tag{4.52}$$

We thus obtain the following expression for the difference between the impedances for two opposite directions:

$$\begin{aligned} Z^-(\omega, \mathbf{r}_b) - Z^+(\omega, \mathbf{r}_b) &= \int_{S_1} d\mathbf{S}\cdot[\widetilde{\mathbf{E}}^+ \times \mathbf{H}^- - \widetilde{\mathbf{E}}^- \times \widetilde{\mathbf{H}}^+] \\ &- \int_{S_2} d\mathbf{S}\cdot[\widetilde{\mathbf{E}}^+ \times \widetilde{\mathbf{H}}^- - \widetilde{\mathbf{E}}^- \times \widetilde{\mathbf{H}}^+]\ , \end{aligned} \tag{4.53}$$

where the subscripts 1 and 2 refer to the beam pipe openings at $z = \pm\xi$, respectively. Using Eqs. (4.49) and (4.50), this equation can be rewritten in the form:

$$\begin{aligned} Z^-(\omega, \mathbf{r}_b) - Z^+(\omega, \mathbf{r}_b) &= \int_{S_1} d\mathbf{S}\cdot[\widetilde{\mathbf{E}}^+ \times \widetilde{\mathbf{H}}^{+*} + \widetilde{\mathbf{E}}^{+*} \times \widetilde{\mathbf{H}}^+] \\ &- \int_{S_2} d\mathbf{S}\cdot[\widetilde{\mathbf{E}}^+ \times \widetilde{\mathbf{H}}^{+*} + \widetilde{\mathbf{E}}^{+*} \times \widetilde{\mathbf{H}}^+]\ . \end{aligned} \tag{4.54}$$

The r.h.s. of this equation being a sum of two complex conjugate quantities is real. Hence, the imaginary parts of the two impedances must be equal:

$$\operatorname{Im} Z^+(\omega, \mathbf{r}_b) = \operatorname{Im} Z^-(\omega, \mathbf{r}_b)\ . \tag{4.55}$$

The integrals on the r.h.s. of Eq. (4.54) have a simple physical meaning. They give the energy flow of the EM field in the beam pipes. As a consequence of the radiation condition, Eq. (4.48), the only part of the EM field remaining far from the cavity is the direct space-charge field of the bunch. For the case when both side pipes have similar and equal cross sections, these space-charge fields are the same. From that it follows that the real parts of the longitudinal impedances are also equal. Applying the *Panofsky-Wenzel theorem*, Eq. (4.35), we see that the same is true for the transverse impedances.

However, for unequal beam pipe cross sections, the space-charge fields of the two beams are different, even at $z = \pm\infty$. In this case the real parts of the impedances for the two directions may differ by a constant.

In the ultra-relativistic case, $\gamma \to \infty$, for the beam pipes with circular cross sections, the difference of the energy of the space-charge fields in the beam pipes with radii b_1 and b_2 is proportional to the constant $\ln(b_1/b_2)$. Hence the difference of the real parts of the impedances is also proportional to this value [Balakin83], cf. also Appendix to Chapter 3.

Chapter 5

Loss Factors and Effective Impedances

The (longitudinal) loss factor was introduced to simplify the calculation of energy loss of bunched beams passing through vacuum chambers of high energy particle accelerators or storage rings.

The energy lost by the beam, including the synchrotron radiation which is often the dominant loss in electron machines - has to be restored by the rf system which therefore must have enough power. Furthermore, it is important to make sure that the energy released will not overheat sensitive elements which could be deformed or even destroyed by too high temperatures. A word of caution is necessary here: the fact that the beam loses energy inside a certain structure does not mean that it is also absorbed there. The part of the energy spectrum with frequencies above the beam pipe cutoff will in general propagate out of a cavity or a cavity-like object and travel along the beam pipe. In the same way some of the energy of neighbouring structures may leak back into the structure under consideration. This happens in particular in long arrays of cavities, such as disk-loaded waveguides, where the energy flowing in through the irises and the energy flowing out of them tend to become equal. In this case fields at frequencies above the iris cutoff must also be included in the estimate of energy deposition.

The concept of a loss factor has been generalised to include the change of energy loss with displacement of the beam trajectory which is usually assumed to remain parallel to the axis. Its dependence on displacement is related to the transverse loss factor, now often called *kick factor*, since it essentially describes a localised deflection or "kick" exerted on a beam. For structures with several resonances, one can define *modal loss factors* for each of these resonances.

The loss factors are generally defined for ultra-relativistic bunch motion $v \approx c$, and corrections would have to be applied for beams with lower velocities. Also the vacuum chamber walls are usually assumed to be perfectly conducting, except for the discussion on the resistive wall loss factor in Section 5.1.

5.1 The Longitudinal Loss Factor

The energy change of a bunch passing through a structure is the product of its charge q and of the induced voltage $V_{ind} = -Z_{\|} I_b$. The voltage is proportional to the current $I_b = q\omega_0/2\pi$, and thus also to the charge q. Hence the total energy change is proportional to the square of the charge and can be written simply as

$$\Delta\mathcal{E} = -\kappa_{\|} q^2 \ . \tag{5.1}$$

The proportionality factor $\kappa_{\|}$ is called the total (longitudinal) *loss factor*. Note that the energy change is always negative, corresponding to an energy loss, while the loss factor will in general be positive.

In the frequency domain, the energy loss is given by the integral over the product of the impedance $Z_{\|}(\omega)$ and the power density of the particle distribution $h(\omega) = \tilde{\lambda}(\omega)\tilde{\lambda}^*(\omega)$. For circular machines, the power density spectrum consists of discrete lines at multiples of the revolution frequency ω_0. The loss factor is then given by the infinite sum:

$$\kappa_{\|} = \frac{\omega_0}{2\pi} \sum_{p=-\infty}^{\infty} Z_{\|}(p\omega_0)\, h(p\omega_0) \ . \tag{5.2}$$

For equally spaced (and populated) bunches, the spectrum envelope depends simply on the particle distribution - in particular the bunch length and separation. For non-uniform distributions of bunches, modulations of the spectrum envelope will occur.

Infinite sums are often difficult to evaluate, hence one tends to replace them by integrals. This is a good approximation as long as the spacing between the spectral lines is small compared to the scale of variations of the impedance. For short bunches, a slowly varying, *broad-band* impedance is sufficient since single resonances cannot be resolved in the time of the bunch passage.

However, for long bunches or *continuous (or "coasting") beams*, narrow resonant peaks can become important, in particular for multi-bunch or multi-turn oscillations. A beam consisting of a number of bunches may cause *coherent* excitation of *EM* fields if one of its spectral lines falls close to a structure resonance. In this case, the *EM* fields of all bunches are in phase and add linearly, hence the energy or power increase quadratically. This may lead to strongly enhanced energy loss and has to be taken into account for multi-bunch operation.

The loss factor thus depends not only on the structure, but also on the bunch shape, in particular on its rms length σ (in time) which is often written as its argument. Under the conditions discussed above, it can be expressed by the integral:

$$\kappa_{\parallel}(\sigma) = \frac{1}{2\pi} \int_{-\infty}^{\infty} d\omega\, Z_{\parallel}(\omega)\, h(\omega, \sigma) \,. \tag{5.3}$$

The power density is always an even function of frequency, while the imaginary part of an impedance is odd. Hence their product does not contribute to the integral which extends over both positive and negative frequencies. Therefore it is sufficient to integrate only over the even real part of the impedance. One may further limit the integration to positive frequencies and double the result to get:

$$\kappa_{\parallel}(\sigma) = \frac{1}{\pi} \int_{0}^{\infty} d\omega\, \mathrm{Re}[Z_{\parallel}(\omega)]\, h(\omega, \sigma) \,. \tag{5.4}$$

In the time domain, the loss factor can be obtained directly by integrating the product of the wake potential $W_{\parallel}(\tau)$ and the line density of the particle distribution $\lambda(\tau)$:

$$\kappa_{\parallel}(\sigma) = \int_{-\infty}^{\infty} d\tau\, W_{\parallel}(\tau)\, \lambda(\tau). \tag{5.5}$$

Sometimes this integral is difficult to evaluate, and often an alternate method is more appropriate. First express the wake potential as a convolution with the wake function, $G_{\parallel}(\tau)$; inverting the order of the two integrations over t and τ, one gets the expression:

$$\kappa_{\parallel}(\sigma) = \int_{0}^{\infty} dt\; G_{\parallel}(t)\; S(t) \,, \tag{5.6}$$

where

$$S(t) = \int_{-\infty}^{\infty} d\tau\; \lambda(\tau)\; \lambda(\tau - t) \tag{5.7}$$

is the *auto-correlation function* of the line density. Since $G_{\parallel}(t) = 0$ for $t < 0$, the integration may actually be limited to positive values of the delay.

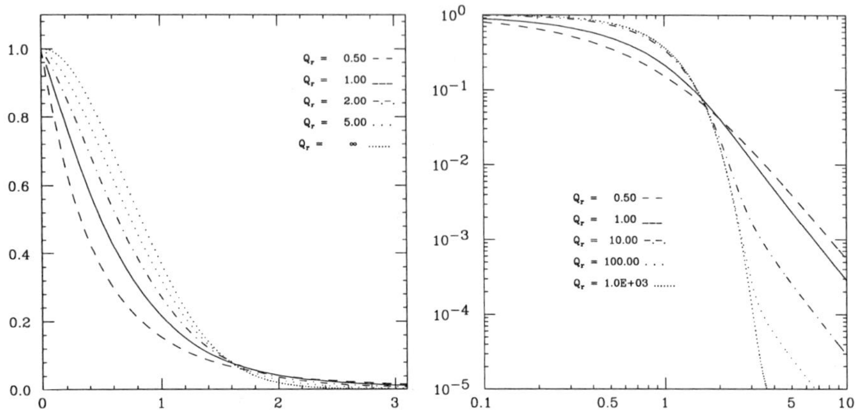

Figure 5.1: Loss factor of a Gaussian bunch in a resonator versus resonant frequency times bunch length $\omega_r\,\sigma$: a) linear scale, b) logarithmic scale.

5.1.1 Gaussian Bunch in a Resonator

For a Gaussian distribution with rms length σ_s, or $\sigma = \sigma_s/\beta c$ in time, the power density $h(\omega) = \exp(-\omega^2\sigma^2)$ can be obtained from the Fourier transform of the normalised line density $\tilde{\lambda}(\omega) = \exp[-\omega^2\sigma^2/2]$. Hence the loss factor is given by

$$\kappa_{\scriptscriptstyle\parallel}(\sigma) = \frac{1}{2\pi}\int_{-\infty}^{\infty} d\omega\, Z_{\scriptscriptstyle\parallel}(\omega)\exp\left(-\omega^2\sigma^2\right)\,. \tag{5.8}$$

For a resonator impedance, after partial fraction decomposition, this integral can be evaluated analytically in terms of the complex error function defined by [Abramowitz65]:

$$\mathrm{w}(z) = \mathrm{e}^{-z^2}\left[1 + \frac{j}{\sqrt{\pi}}\int_0^{z^2}\frac{\mathrm{e}^t}{\sqrt{t}}dt\right]. \tag{5.9}$$

and can be written:

$$\kappa_{\scriptscriptstyle\parallel}(\sigma) = \frac{R_s}{4Q_r'}\Big[\omega_1\mathrm{w}(\omega_1\sigma) - \omega_2\mathrm{w}(\omega_2\sigma)\Big]\,, \tag{5.10}$$

where $\omega_{1,2} = (\omega_r/Q_r)(j/2 \pm Q_r')$ and $Q_r' = \sqrt{Q_r^2 - 1/4}$. Usually $Q_r \geq 1/2$, then $\omega_2 = -\omega_1^*$ and $\mathrm{w}(\omega_2\sigma) = \mathrm{w}^*(\omega_1\sigma)$, and the loss factor can be written in the simpler form:

$$\kappa_{\scriptscriptstyle\parallel}(\sigma) = \frac{R_s}{2Q_r'}\mathrm{Re}\Big[\omega_1\mathrm{w}(\omega_1\sigma)\Big]\,. \tag{5.11}$$

Fig. 5.1 shows the loss factors for a Gaussian bunch in a resonator.

The direct calculation of the loss factor in the time domain leads to an integral of the form $\int_0^\infty d\tau \; \mathrm{w}(a + b\tau) \exp(-\tau^2)$, which cannot be performed easily. However, using the auto-correlation function, which for a Gaussian is:

$$S(t) = \frac{1}{2\sqrt{\pi}\sigma} \exp\left(-\frac{t^2}{4\sigma^2}\right) , \tag{5.12}$$

the integrand becomes again an exponential divided by a linear term, and can be evaluated with the same result as obtained in the frequency domain.

For short bunches, for which $|\omega_1|\sigma = \omega_r \sigma \ll 1$, one can use the first two terms of the power series expansion $\mathrm{w}(z) = 1 + 2jz/\sqrt{\pi}$ to get:

$$\kappa_{\parallel}(\sigma) \approx \frac{\omega_r R_s}{2Q_r} \left[1 - \frac{2}{\pi} \frac{\omega_r \sigma}{Q_r}\right] . \tag{5.13}$$

The leading term[1] depends only on the ratio R_s/Q_r, and is independent of the shunt impedance and quality factor alone.

For long bunches $\omega_r \sigma \gg 1$, one needs the first two terms of the asymptotic series $\mathrm{w}(z) = j/(\sqrt{\pi}z) + j/(2\sqrt{\pi}z^3)$ to obtain $\kappa_{\parallel}(\sigma) = \omega_r R_s/(4\sqrt{\pi} Q_r^2 \omega_r^3 \sigma^3)$. Since R_s/Q_r does not depend on the quality factor, the loss factor decreases as $1/Q_r$ for long bunches.

5.1.2 Resistive Wall Loss Factor

The more complicated dependence of the resistive wall impedance, which is proportional to the square root of frequency, at least in the usual range of interest (see next Chapter) makes the evaluation of infinite sums and integrals more difficult than for resonator impedances. However, the loss factor can be found also in this case analytically except for a constant given by an integral which is easily evaluated numerically.

For a circular cylindrical chamber with non-zero resistivity ρ_c and radius b, the real part of the impedance can be written (see Section 6.4 below)

$$\mathrm{Re} Z_{\parallel}(\omega) = \frac{Z_0 \delta_0}{2b} \sqrt{\frac{|\omega|}{\omega_0}}, \tag{5.14}$$

[1] In the US literature, the *Linac Ohm* is often used as unit of the shunt impedance. It equals half a conventional Ohm, and thus $\kappa_{\parallel} = (\omega/4)(R_s^{linac}/Q_r)$.

where $\delta_0 = \sqrt{2\rho_c/\omega_0\mu}$ is the skin depth at the revolution frequency ω_0, μ is the permeability of the wall material, and $Z_0 = 120\pi\Omega$ the free space impedance. The loss factor for a Gaussian bunch is then found to be

$$\kappa_{\parallel}(\sigma) = \frac{C_L}{2\pi} \frac{\omega_0 Z_0}{(\omega_0\sigma)^{3/2}} \frac{\delta_0}{b} , \tag{5.15}$$

where the constant $C_L = \int_0^\infty e^{-x^2}\sqrt{x}dx = \Gamma(3/4)/2 = 0.613$. For a flat chamber of half-height h and half-width w, one gets an estimate of the loss factor by replacing the radius b with its averaged value $\langle b \rangle = 2/(1/h + 1/w)$.

The $\sigma^{-3/2}$ dependence of the loss factor can lead to considerable power dissipation and heating of the vacuum chamber walls for short bunches with high charge. However, for cryogenically cooled (e.g. superconducting) walls, for which it is rather costly to remove heat, the resistivity is also much lower which in turn reduces the power dissipation.

5.2 The Kick Factor

The *transverse loss factor* is also called *kick factor*, which name better describes the physical action of the *EM* fields on the particles and is therefore adopted here. Its definition is analogous to that of the longitudinal loss factor:

$$\kappa_{\perp}(\sigma) = \frac{1}{2\pi}\int_{-\infty}^{\infty} d\omega \; Z_{\perp}(\omega)\, h(\omega,\sigma) = \frac{1}{\pi}\int_0^{\infty} d\omega \; \mathrm{Im}\Big[Z_{\perp}(\omega)\Big]\, h(\omega,\sigma) \tag{5.16}$$

and can be calculated in the same way as the longitudinal one. However, since $Z_{\perp}$ is an impedance per unit (transverse) displacement, also the dimensions of the kick factor differ from that of the loss factor by m^{-1} and becomes V/C m. Again, one can also calculate it in the time domain:

$$\kappa_{\perp}(\sigma) = \int_{-\infty}^{\infty} d\tau \; W_{\perp}(\tau)\; \lambda(\tau) = \int_0^{\infty} dt \; G_{\perp}(t)\; S(t) , \tag{5.17}$$

where the second equation expresses the kick factor in terms of the transverse Green function $G_{\perp}$ and the auto-correlation function $S(t)$ defined in Eq. (5.7).

5.2.1 Gaussian Bunch in a Resonator

As shown before, the transverse (dipole) impedance of a resonator can be expanded into partial fractions to get:

$$Z_\perp(\omega) = \frac{jR_\perp}{2Q'_r}\left[\frac{1}{\omega-\omega_2} - \frac{1}{\omega-\omega_1}\right] , \tag{5.18}$$

while the spectral power density for a Gaussian is $\exp\left(-\omega^2\sigma^2\right)$. Hence the loss factor is given by

$$\kappa_\perp(\sigma) = \frac{j\omega_r R_\perp}{2Q'_r}\int_{-\infty}^{\infty}\frac{d\omega}{2\pi}\left[\frac{1}{\omega-\omega_2} - \frac{1}{\omega-\omega_1}\right]\exp(-\omega^2\sigma^2) \,. \tag{5.19}$$

Like other integrals before, this one can also be expressed by the complex error function $\mathrm{w}(z)$ to yield

$$\kappa_\perp(\sigma) = \frac{\omega_r R_\perp}{4Q'_r}\left[\mathrm{w}(\omega_1\sigma) - \mathrm{w}(\omega_2\sigma)\right] . \tag{5.20}$$

For the usual case $Q_r \geq 1/2$, for which $\omega_2 = -\omega_1^*$, one can simplify this expression to get:

$$\kappa_\perp(\sigma) = \frac{\omega_r R_\perp}{2Q'_r}\mathrm{Im}\Big[\mathrm{w}(\omega_1\sigma)\Big] , \tag{5.21}$$

where $\omega_1 = \omega_r Q'_r/Q_r + j(\omega_r/2Q_r)$ as given after Eq. (5.10). The same result can be obtained in the time domain by evaluating the convolution of the wake function and the auto-correlation function.

For short bunches, when the argument $|\omega_1\sigma| = \omega_r\sigma$ is small compared to unity, the complex error function can be approximated by the first two terms of the power series $\mathrm{w}(z) \approx 1 + 2jz/\sqrt{\pi}$ to get

$$\kappa_\perp(\sigma) = \frac{\omega_r R_\perp}{Q_r}\frac{\omega_r\sigma}{\sqrt{\pi}} . \tag{5.22}$$

For long bunches, the first two terms of the asymptotic series are $\mathrm{w}(z) \approx (j/\sqrt{\pi}z)(1 + 1/2z^2)$, and are sufficient to obtain the expression

$$\kappa_\perp(\sigma) = \frac{R_\perp}{Q_r}\frac{1}{2\sqrt{\pi}\sigma} . \tag{5.23}$$

In these approximations, the kick factor is proportional to σ for short bunches, and to σ^{-1} for long ones, hence it must have a maximum in between. Both

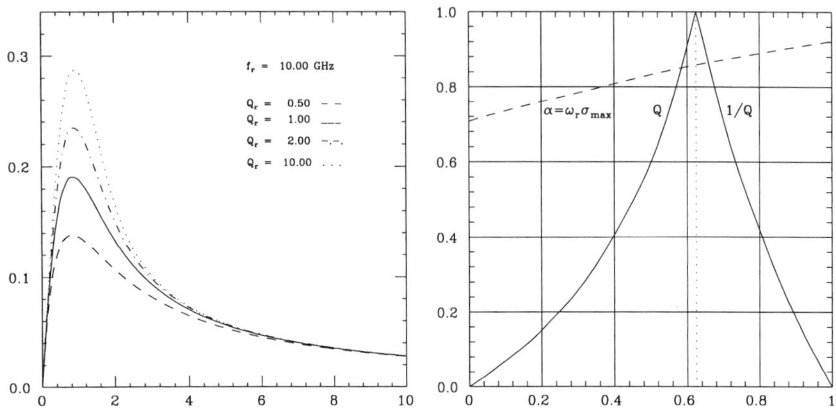

Figure 5.2: Transverse loss factor of Gaussian bunch in a resonator versus $\omega_r \, \sigma$ and broad-band parameters Q_r and α_L versus the ratio parameter r_L.

approximations depend only on the ratio $R_\perp/Q_r$, but the location and the value of the maximum depend also on the resonant frequency and, to a lesser degree, on the quality factor Q_r as shown in Fig.5.2a.

It is thus possible to estimate the parameters of an equivalent transverse broad-band impedance from the dependence of the kick factor on bunch length. In particular, $R_\perp/Q_r$ is simply given by the asymptotic value of the quantity $2\sqrt{\pi}\kappa_\perp\omega_r\sigma$. The resonant frequency is given by $f_r = \alpha_L c/2\pi\hat{\sigma}$, where $\hat{\sigma}$ is the location of the maximum of the kick factor, and α_L is a numerical factor. This factor, and the quality factor Q_r, can be obtained from the ratio of the product of the bunch length and the kick factor at the maximum and at infinity, i.e. from $r_L = \hat{\sigma}\hat{\kappa}_\perp/(\sigma\kappa_\perp)_\infty$ as shown in Fig.5.2b.

However, for the determination of equivalent broad-band impedances by means of the kick factors as described above, one should not mix dissimilar elements such as e.g. large rf cavities (with correspondingly low resonant frequencies) and small bellows (with much higher ones). Their contributions tend to cancel, and un-physically low quality factors may be obtained. Furthermore, it must be pointed out that a similar procedure is less satisfactory in the longitudinal direction, since the asymptotic loss factors do not become independent of the quality factors for long bunches as in the transverse case, as can be seen in Fig. 5.1.

5.3 The Modal Loss and Kick Factors

The *total (longitudinal) loss factor* of a structure must be distinguished from the *modal loss factor* caused by a single resonant mode. This is usually defined for vanishing bunch length and is thus simply the integral over the real part of the impedance $Z_{\parallel n}(\omega)$ of the n-th resonant mode:

$$\kappa_{\parallel n} = \frac{1}{\pi} \int d\omega \ \mathrm{Re}[Z_{\parallel n}(\omega)] \ , \tag{5.24}$$

where the integration extends over the region near the resonant frequency ω_n. The modal loss factor for a cavity resonance is proportional to the *geometric factors* $(R_s/Q_r)_n$ i.e. the shunt impedance divided by the quality factor of each mode. These ratios are independent of the wall resistivity to first order:

$$\kappa_{\parallel n} = \frac{\omega_n}{2} \left(\frac{R_s}{Q_r} \right)_n \ . \tag{5.25}$$

Several computer programs for *EM* fields in the frequency domain calculate the *geometric factors* in addition to the resonant frequencies for closed or for periodic cavities. These programs evaluate the induced voltages $V_n = \int dz \ E_{zn} \exp(jkz)$ by integrating the longitudinal component of the electric field of the n-th mode. By dividing with the stored energy U_n of the same mode one obtains the *modal loss factors*:

$$\kappa_{\parallel n} = \frac{V_n V_n^*}{4U_n} \ . \tag{5.26}$$

Also the wake function can be expressed in terms of the modal loss factors. For $\tau > 0$ one finds [Bane80b]

$$W_{\parallel}(\tau) = 2 \sum_{n=1}^{\infty} \kappa_{\parallel n} \cos(\omega_n \tau) \exp(-\frac{\omega_n \tau}{2Q_{rn}}) \ , \tag{5.27}$$

while $W_{\parallel} = 0$ in front of the bunch[2]. This expression is particularly useful when the resonances are determined by measurements or by performing computations in the frequency domain. At large distances behind a bunch, only a few dominant modes contribute significantly to the wake function, while the

[2] We have changed the sign of the original expression since we count a decelerating wake field as positive, and we have added a damping term for generality.

large number of smaller ones approximately cancel each other by adding with arbitrary phases. For short-range wakes, however, most modes add in phase and one needs to compute a rather large number of resonances to get the correct result. It is then easier to replace the large number of discrete modes by a continuous high-frequency tail of the impedance. For periodic structures this can be obtained from the so-called *optical resonator model* [Bane80b, Brandt85], cf. Chapter 9.

Integration of the product of the line density and the wake function, expressed as an infinite series over all modes, yields the expression

$$\kappa_{\parallel} = \sum_{n=1}^{\infty} \kappa_{\parallel n} \int_{-\infty}^{\infty} dt S(t) \cos \omega_n t \ . \tag{5.28}$$

The cosine transform of the auto-correlation function $S(t)$ is the power density $h(\omega)$ of the line density λ, and the total loss factor can be found by summing over all resonances:

$$\kappa_{\parallel}(\lambda) = \sum_{n=1}^{\infty} \kappa_{\parallel n} h(\omega_n, \lambda) \ . \tag{5.29}$$

In particular, for a Gaussian bunch with rms length σ

$$\kappa_{\parallel}(\sigma) = \sum_{n=1}^{\infty} \kappa_{\parallel n} \exp(-\omega_n^2 \sigma^2) \ . \tag{5.30}$$

In practice, the upper limit of the summation index is the highest mode for which the modal loss factor is still significant. For single cavities with smooth beam pipes, the modal loss factors become small when the resonant frequency of a mode exceeds the beam pipe cutoff frequency. Then the *EM* fields can propagate out of the cavity and deposit the energy somewhere else. However, if the *EM* fields are partially reflected back, also modes above cutoff may have significant loss factors.

With the advent of better time-domain programs, it is nowadays more efficient to compute the short-range wake directly. However, only wake potentials of finite distributions can be found with these programs. In addition, one needs rather long computing times to find to good accuracy the long-range wake potential far behind a bunch. There are similar limitations in the frequency domain, since the reconstructed wake potential far behind a bunch depends rather critically on the *exact* values of the resonant frequencies of the strongest modes. This limitation is sometimes expressed by the *sampling theorem* $\Delta t \propto 1/\Delta f$. The finite accuracy of numerical calculations or measurements thus limits the precise knowledge of long-range wakes. In addition, resonant frequencies of cavities or cavity-like objects may vary in time due to changes in dimensions such as caused by fluctuations of temperature or pressure. The effects of long-range wakes should then be treated in statistical manner.

In the case when the wake functions are excited by a train of equidistant bunches with spacing s_b, the interference of the fields in the same structure due to different bunches may have to be taken into account [Wilson91] if they build up coherently. In the ultra-relativistic limit, the loss factor of a mode with wave number $k_n = \omega_n/c$ and quality factor Q_{rn} should be multiplied by the quantity $F(k_n s_b/2Q_{rn}, k_n s_b)$, where the function F of two arguments is given by:

$$F(x,y) \equiv \frac{\sinh x}{\cosh x - \cos y} \,. \tag{5.31}$$

If $Q_{rn} \gg 1$ and the condition for the resonance excitation $k_n s_b = 2\pi m$, m an integer, is fulfilled, the loss factor can be substantially enhanced, i.e.,

$$F \sim 2Q_{rn}/\pi m \,. \tag{5.32}$$

The possibility of increased, coherent energy loss must be taken into account in the design of cooling systems. There may also be coherent cancellation of fields if their phases happen to be opposite, but since the frequencies of higher modes in cavities are likely to change under various external conditions, it is usually necessary to count with the worst case.

5.3.1 The Modal Kick Factor

For modes propagating with light velocity in an axially symmetric structure, the longitudinal electric fields can be written [Bane80b] $E_{zmn} = E_{mn}(r/b)^m$, where b is the radius of the beam pipes, while E_{mn} is the amplitude of the m-th *azimuthal* and n-th *radial mode.* We have left off the factor $\cos(m\theta)$ which is to be understood for the m-th azimuthal mode. The modal loss factors for any $m \geq 0$ can be defined by replacing the induced voltage in the expression Eq. (5.26) for the $m = 0$ case by the voltage defined as the integral of the synchronous component of the electric field $V_{mn} = \int dz\, E_{mn} \exp(jkz)$ over the length of the cavity:

$$\kappa_{\parallel mn} = \frac{V_{mn} V_{mn}^*}{4U_{mn}} \,. \tag{5.33}$$

The transverse momentum "kick" experienced by a particle [Bane84] is given by $j\omega\Delta P = \partial E_{zmn}/\partial r = mE_{mn}r^{m-1}/b^m$. The radial component of the wake function for the m-th azimuthal mode is obtained by integrating the kick (per unit length) over $dz = cdt$:

$$G_{\perp m}(\tau) = 2mc\left(\frac{r}{b}\right)^{2m-1}\sum_{n=1}^{\infty}\frac{\kappa_{\parallel mn}}{\omega_{mn}}\sin(\omega_{mn}\tau)\exp\left(-\frac{\omega_{mn}\tau}{2Q_{rmn}}\right) . \tag{5.34}$$

The finite width of the resonances is taken into account by exponential damping factors of each term in the sum.

The dipole mode, with azimuthal mode number $m = 1$, is thus the lowest one with a non-vanishing transverse wake function. It is usually the dominant contribution, and often referred to simply as *the transverse wake function.* Its gradient (with respect to radius) then is given by the infinite sum

$$G'_{\perp}(\tau) = \frac{2c}{b^2}\sum_{n=1}^{\infty}\frac{\kappa_{\parallel 1n}}{\omega_{1n}}\sin(\omega_{1n}\tau) , \tag{5.35}$$

where we have left off the exponential damping factors for simplicity. In practice, the summation needs to be truncated at some finite mode number, above which the loss factors become negligible.

The total kick factor is obtained by integration of the product of the transverse wake function with the auto-correlation function as in Eq. (5.17):

$$\kappa_{\perp}(\sigma) = \frac{2c}{b^2}\sum_{n=1}^{\infty}\frac{\kappa_{\parallel 1n}}{\omega_{1n}}\int_0^{\infty} dt S(t)\sin\omega_{1n}t . \tag{5.36}$$

The weighting factors in the sum for the total loss factor are different from the longitudinal case. This is discussed below for a Gaussian distribution.

5.3.2 Gaussian Distribution

For a Gaussian distribution with line density $\lambda(\tau) = \exp(-\tau^2/2\sigma^2)/\sqrt{2\pi}\sigma$, the auto-correlation function is given by Eq. (5.12). Substituting this expression into Eq. (5.36) yields [Zotter80]

$$\kappa_{\perp}(\sigma) = \frac{c}{\sqrt{\pi}\sigma b^2}\sum_{n=1}^{\infty}\frac{\kappa_{\parallel 1n}}{\omega_{ln}}\int_0^{\infty} dt\sin(\omega_{1n}t)\exp\left(-\frac{t^2}{4\sigma^2}\right) . \tag{5.37}$$

This integral can be expressed by the complex error function of purely imaginary argument, which is better known under the name of *Dawson integral* [Abramowitz65] $D(z) = \exp(-z^2) \int_0^z dt \exp(t^2)$. One then gets

$$\kappa_\perp(\sigma) = \frac{2c}{\sqrt{\pi} b^2} \sum_{n=1}^{\infty} \frac{\kappa_{\parallel 1n}}{\omega_{1n}} \mathrm{D}(\omega_{1n}\sigma) \; . \tag{5.38}$$

Since the Dawson integral is available in most computer libraries, the evaluation of this expression is just as easy as that for the longitudinal loss factor Eq. (5.30).

5.4 Resonant Cavities

A single cavity with side tubes has an impedance whose real part decreases as $\omega^{-1/2}$ at high frequencies, and hence the loss factor diverges for zero bunch length, where it is given by Eq. (5.4) with a spectral power density equal to unity. This is no longer the case for periodic arrays of cavities, for which the impedance decreases as $\omega^{-3/2}$ at high frequencies, and hence the integral remains finite. The transition from the first to the second situation will be discussed in Chapter 9.

For perfectly conducting cavity (and tube) walls, the real part of the impedance is zero below the first cavity resonance situated at $\omega_{01} = j_{01}c/d$, where j_{01} is the first zero of the Bessel function J_0, and d is the outer radius of the cavity. The loss factor is then given by

$$\kappa_{\parallel}(\sigma) \rightarrow \frac{1}{\pi} \int_{\omega_{01}}^{\infty} d\omega \mathrm{Re} Z_{\parallel}(\omega) \approx \mathrm{Re} Z_{\parallel}(\omega_{01}) \frac{1}{2\sqrt{\pi}\sigma} \, \mathrm{erfc}(\omega_{01}\sigma) \; , \tag{5.39}$$

where $\mathrm{erfc}(z) = 1\text{-}\,\mathrm{erf}\,(z)$ is the complementary error function [Abramowitz65], which approaches $\exp(-z^2)/\sqrt{\pi}z$ for large arguments. Hence in this case the loss factor decreases exponentially for long bunches, faster than σ^{-3} as is the case for a resonator impedance, cf. section 5.1.1. However, the finite resistivity of the walls adds a real part of the impedance at low frequencies, and the decrease of the loss factor comes closer to that of the resistive wall which is only proportional to $\sigma^{-3/2}$ as shown above.

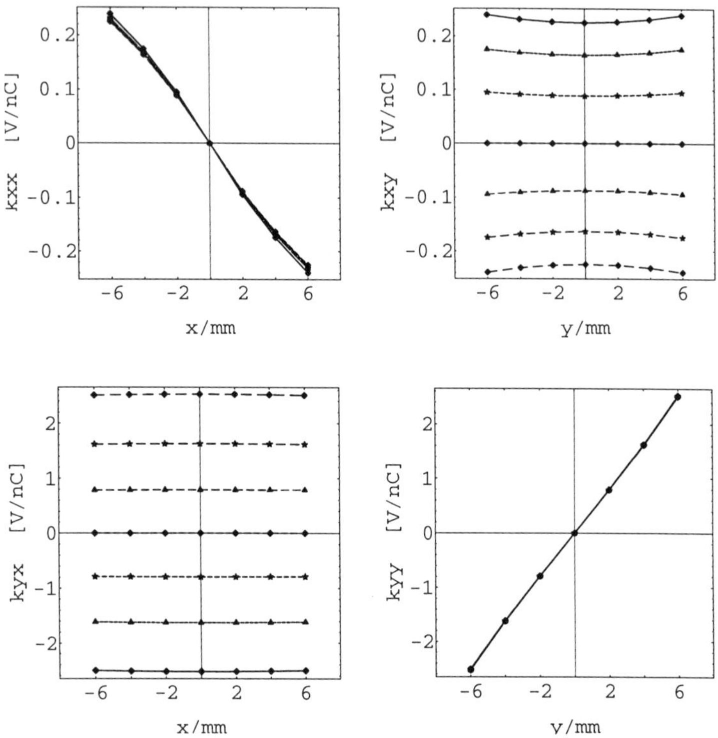

Figure 5.3: Kick factors as function of both horizontal and vertical beam displacement in the shielded *LEP* bellows.

5.5 Structures Without Axial Symmetry

Many vacuum chambers have elliptic or oval cross section, sometimes even with holes or slots distributed around the circumference. Besides, a number of structures like separators, pickup plates, collimators etc. have often mirror symmetry about two orthogonal planes, but are not rotationally symmetric. The loss factors depend in general on the displacement of the beam, and do not necessarily vanish at the beam axis, which is usually still assumed to be parallel to the axis formed by the intersection of symmetry planes. It is then no longer possible to divide the energy loss by the displacement, and we therefore use not normalised kick factors $k_{xx}, k_{y}y$ and the cross terms $k_{xy}, k_{x}y$ causing *wake field coupling* into the other plane.

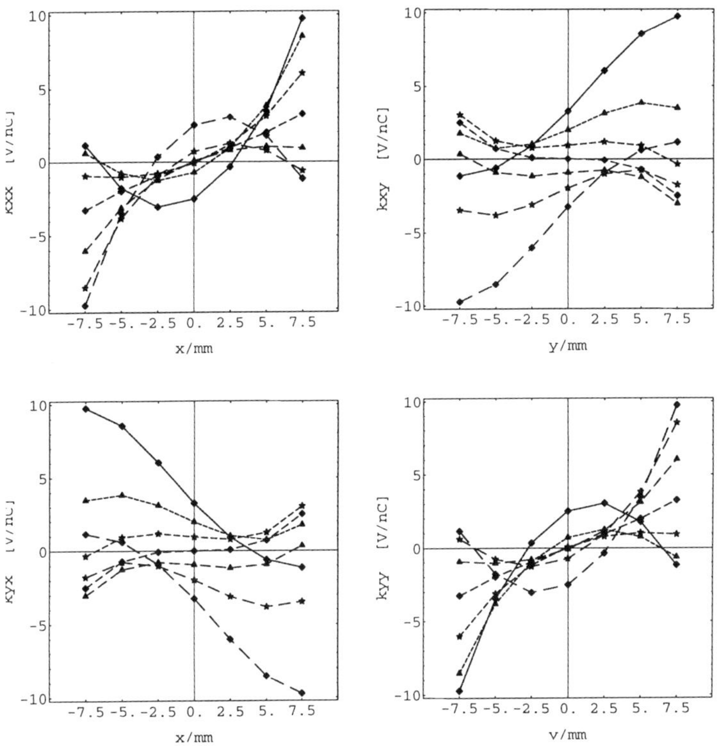

Figure 5.4: Kick factors as function of both horizontal and vertical beam displacement for beam position monitor cavities.

For off-axis beams, the gradient of the transverse loss factors can become negative, i.e. particles following the exciting ones are focused toward the beam axis rather than deflected to the outside as they invariably are for rotationally symmetric structures. This fact was recognized many years ago, e.g. for the *rf quadrupole accelerator (RFQ)* [Puglisi87], and is also the principle of *rf focusing* by oval beam holes in cavities with circular cross section (or equivalently by circular beam holes in oval cavities) such as proposed for e.g. the *Compact Linear Collider (CLIC)* which is presently studied at *CERN* [Schnell91].

As can be seen in Fig. 5.3, the horizontal loss factor is small, but has a negative gradient for the *LEP shielded bellows*. For computational reasons, the oval beam chamber and large number of narrow slots between the sliding-contact fingers were approximated here by a rectangular beam pipe and a small number of wide slots.

Focusing happens only in one plane, and is always smaller than defocusing in the orthogonal direction. There is no violation of energy conservation, since the longitudinal loss factor remains always positive. Nevertheless, the fact that kick factors may be negative is not commonly known. Recently it has been proposed [Wagner96] to use this fact to compensate large kick factors in storage rings, which often lead to bunch current limitations by the *transverse mode coupling instability*; however, this effect has not yet been tested in practice.

The horizontal and vertical kick factors for the *electro-static separators* of the *LEP* storage ring at *CERN* have been computed with the code *MAFIA* [Weiland88] and are shown in Fig.5.4. A similar picture is shown for a beam pipe with four elongated recessions for a stripline pickup structure, Fig.5.5. Also shown are the *cross-coupling kick factors*, leading to a vertical kick for a horizontal displacement and vice versa.

5.5.1 Fully Asymmetric Structures

When structures have only one, or even no symmetry plane, there is no obvious axis. Since the dependence of the kick factors on beam displacement is no longer linear, it does not make sense in this case to divide them (or the transverse wake function) by it.

The wake potentials and the corresponding loss factors are defined as integrals along a straight line assumed to be the beam trajectory, usually parallel to the beam pipes attached to either end of the structure. It is also no longer useful to expand the *EM* fields into azimuthal modes. However, most structures used in actual accelerators have at least one plane of symmetry, as e.g. a row of pumping holes connecting a beam chamber to a pump chamber such as often used in electron storage rings.

General 3-D codes have been developed in recent years [Weiland88, Nelson96] to compute short range wake potentials or resonances of structures of arbitrary shape. However, for the computation of fields of short bunches, the required time and memory often still exceed the available resources even with the fastest computers existing today.

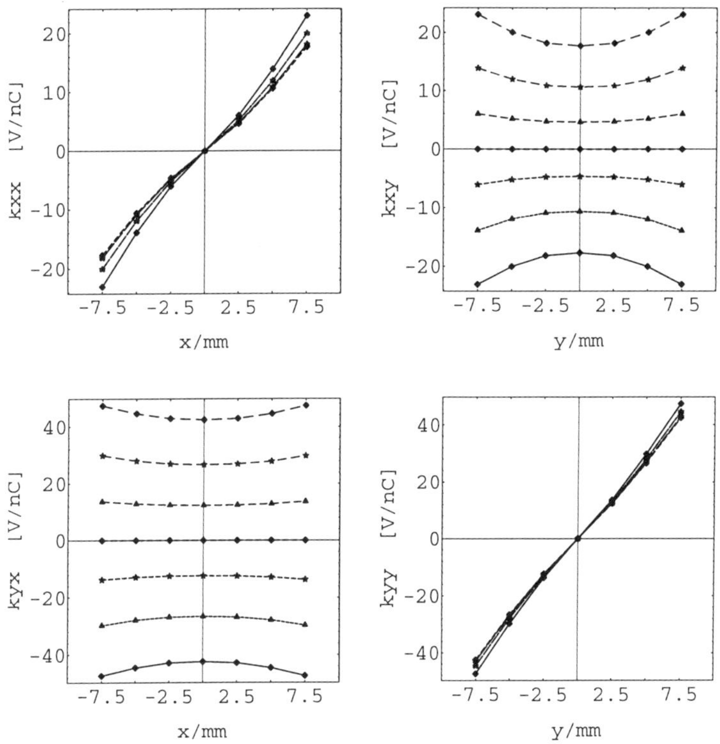

Figure 5.5: Kick factors as function of both horizontal and vertical beam displacement in *LEP* electro-static separators.

5.6 Effective Impedances

The coherent stability of bunched beams is in general determined not by the coupling impedance itself, but by its convolution with the bunch oscillation spectrum, which is often called *effective impedance.* Apart from the normalisation, this definition agrees completely with that of the loss factors - except that the lowest longitudinal shape oscillation is a dipole mode with $\bar{m}=1$. The mode $\bar{m}=0$ cannot be excited coherently, but is the dominant contribution to the energy loss. In circular accelerators, the bunch oscillation spectrum consists of a large number of discrete lines, spaced by multiples of the revolution frequency ω_0, with a *spectrum envelope* whose width is inversely proportional to the bunch length. For M equally spaced bunches, only each M-th line is excited, and the spectral lines are spaced by the *bunch frequency* $M\omega_0$. If the bunches perform coherent shape oscillations, the spectral envelopes are shifted to higher frequencies. The *effective impedance* is defined as the convolution of the impedance and the spectrum of the oscillations considered.

In general, an effective impedance - similar to the loss factor - depends not only on the coupling impedance, but also on the bunch distribution and its oscillation mode. Arbitrary distributions are possible - at least in principle - for heavy particles like protons or ions, for which synchrotron radiation is usually negligible. Such bunches are often nearly as long as the rf wave length, and thus experience quite strong deformations due to non-linearities of the applied rf fields. Also the potential well effect may alter the line density. Nevertheless, at higher current levels, local instabilities often lead to an approximately parabolic line density.

5.6.1 Longitudinal Oscillation Modes

Due to the combined action of the applied rf voltage and the differential drift of particles of different energy, the particles in bunch perform longitudinal oscillations. These *synchrotron oscillations* consist of a periodic oscillations of the time delay from the *synchronous particle* (or the rf phase or relative position), and of energy deviation about its *nominal energy*. However, these incoherent oscillations occur with arbitrary phases, their external fields essentially cancel each other, and cannot be seen from the outside.

On the other hand, *coherent oscillations* of the bunch as a whole will induce external *EM* fields which interact with the vacuum chamber wall and also produce signals which can be measured with beam diagnostics instruments. The bunch shape oscillations in longitudinal *phase space*, i.e. energy deviation against time (or rf phase), are characterised by the *bunch shape mode number* m. The same letter is also often used to describe the azimuthal field dependence $\sim \exp(jm\theta)$. In order to avoid confusion, we use in this book the symbol $\bar{m}$ for the bunch shape oscillation mode.

The *stationary mode* $\bar{m} = 0$ is the bunch distribution itself, while oscillations are described by mode numbers $\bar{m} \geq 1$: $\bar{m} = 1$ dipole, $\bar{m} = 2$ quadrupole, $\bar{m} = 3$ sextupole mode etc.). The mode number $\bar{m}$ also equals the number of zero crossings of the line density modes, which are the projections of the phase-space oscillations onto the time axis.

The *effective longitudinal impedance* for the $\bar{m}-$th bunch-shape mode of oscillation can be written [Sacherer73]:

$$Z_{eff}^{\bar{m}} = \sum_{p=-\infty}^{\infty} \frac{Z_{\|}}{n}(\omega_p) H_{\bar{m}}(\omega_p) \; . \tag{5.40}$$

The *spectral frequencies* of a beam consisting of k_b uniformly spaced, equal bunches are given by

$$\omega_p = \omega_0(k_b p + n_b + \bar{m} Q_s) \; , \tag{5.41}$$

where ω_0 is the revolution frequency, and $Q_s = \omega_s/\omega_0$ the synchrotron tune. The *coupled-bunch mode number* $0 \le n_b < k_b$ describes the phase shift between bunches ($\Delta\phi = 2\pi n_b/k_b$). For a single bunch $k_b = 1$, hence $n_b = 0$, and the spectral frequencies are simply $\omega_p = \omega_0(p + \bar{m} Q_s)$.

The quantity $H_{\bar{m}}(\omega)$ is the normalised *spectral power density*[3] of the $\bar{m}$-th mode of oscillation, which is given by

$$H_{\bar{m}}(\omega) = C_{\bar{m}} \tilde{\lambda}_{\bar{m}}(\omega) \tilde{\lambda}^*_{\bar{m}}(\omega). \tag{5.42}$$

where $\tilde{\lambda}_{\bar{m}}(\omega)$ are the Fourier transforms of the line density modes $\lambda_{\bar{m}}(\tau)$, and the normalisation constants $C_{\bar{m}}$ are determined by the condition:

$$\sum_{p=-\infty}^{\infty} H_{\bar{m}}(\omega_p) = 1. \tag{5.43}$$

An alternative way of writing the expression for the effective impedance is $Z_{eff}^{\bar{m}} = \sum Z_{\|}(\omega_p) H_{0m}(\omega_p)$, where $H_{0\bar{m}}$ is the (normalised) cross power density of the stationary ($\bar{m} = 0$) mode, acting on the $\bar{m}$-th longitudinal mode of oscillation. These two definitions are essentially the same, since in general $\tilde{\lambda}_{0\bar{m}}(\omega_p) = \lambda_{\bar{m}}(\omega_p)(\omega_0/\omega_n)$.

In the longitudinal case, only oscillation modes with $\bar{m} \ge 1$ exist and may lead to collective instability of a beam. The threshold currents are proportional to the effective impedances which therefore should be kept sufficiently small in order to keep the bunches stable. The *stationary mode* with $\bar{m} = 0$, defined as $\sum_p Z_{\|}(\omega_p) H_0(\omega, p)$, is relevant for the *potential well effect*, which may lead to bunch lengthening and to changes of the distribution function. The energy dissipation in the surrounding structures may lead to excessive heating and has to be taken into account in the design of the cooling system.

[3]More exactly, $H_{\bar{m}}(\omega)$ is the *self power density* $H_{\bar{m},\bar{m}}$, while the *cross power density* $H_{\bar{m},\bar{m}'}(\omega)$ refers to the interaction of two different modes of oscillation.

For a purely inductive impedance, usually appropriate for bunches which are long compared to the vacuum chamber radius, the coupling impedance divided by the mode number $Z_{\parallel}/n$ is constant and thus can be taken out of the sum in Eq. (5.40). In this case, the effective impedance equals the coupling impedance.

5.6.2 Spectral Power Density of Parabolic Bunches

The best way to describe bunch oscillations are the *eigenmodes* of a distribution, which depend also on the particular impedance considered. For long proton bunches of full length $\tau_L = L_b/c$, a parabolic line density is usually quite appropriate:

$$\lambda(\tau) \propto \begin{cases} 1 - \left[\dfrac{2\tau}{\tau_L}\right]^2 & \text{for } |\tau| < \dfrac{\tau_L}{2} \\ 0 & \text{otherwise .} \end{cases} \tag{5.44}$$

Its eigenmodes are given by Legendre polynomials [Besnier78]. However, it was found in practice that it is quite adequate and easier to use *sinusoidal modes* [Sacherer73]. These modes are given by:

$$\lambda_{\bar{m}}(\tau) = \frac{2}{\tau_L} \begin{cases} \cos\left[(\bar{m}+1)\pi\tau/\tau_L\right] & \bar{m} \text{ even} \\ \sin\left[(\bar{m}+1)\pi\tau/\tau_L\right] & \bar{m} \text{ odd} \end{cases} \tag{5.45}$$

while $\lambda_{\bar{m}}(\tau) = 0$ for $|\tau| > \tau_L/2$. From the Fourier transforms of these modes one can find the spectral power densities. Apart from a constant factor, which is determined by the normalisation, they are given by:

$$h_{\bar{m}}(\omega) = (\bar{m}+1)^2 \frac{1 + (-)^{\bar{m}} \cos(\omega\tau_L)}{\left[(\omega\tau_L)^2 - \pi^2(\bar{m}+1)^2\right]^2} . \tag{5.46}$$

The normalisation factor is given by $C_{\bar{m}} = 1/\sum_p h_m(\omega_p)$. We define $\omega_p\tau_L = \omega_0\tau_L(pk_b + n_b + \bar{m}Q_s) = b(p+d)$, where we introduced the *bunching factor* b and the *fractional detuning* d:

$$b = k_b\omega_0\tau_L = k_b\frac{L_b}{R} , \qquad d = \frac{n_b + \bar{m}Q_s}{k_b} , \tag{5.47}$$

where R is the average radius of the accelerator. Then the power density at a spectral frequency can be written:

$$h_{\bar{m}}(\omega_p) = \frac{(\bar{m}+1)^2}{b^4} \sum_{i=1}^{2} \frac{1+(-)^{\bar{m}} \cos b(p_i+d)}{(p-q_1)^2(p-q_2)^2}, \tag{5.48}$$

where the complex poles are given by:

$$p_{1,2} = \frac{\omega_{1,2}}{k_b \omega_0} - d = \frac{h_r}{k_b Q_r}\left(\frac{j}{2} \pm Q_r'\right) - \frac{n+\bar{m}Q_s}{k_b}, \tag{5.49}$$

and the real poles are located at

$$q_{1,2} = \frac{\pm\pi(\bar{m}+1)}{b} - d\,. \tag{5.50}$$

By partial fraction decomposition and applying the appropriate summation formulae [Zotter78b], one finds the surprisingly simple result $C_{\bar{m}} = 2\pi b$, independent of the mode number $\bar{m}$, and thus

$$H_{\bar{m}}(\omega_p) = 2\pi \frac{(\bar{m}+1)^2}{b^3} \sum_{i=1}^{2} \frac{1+(-)^{\bar{m}} \cos b(p_i+d)}{(p-q_1)^2(p-q_2)^2} \tag{5.51}$$

for the normalised spectral power densities of sinusoidal modes.

5.6.3 Effective Impedance of Parabolic Bunches

We evaluate the effective longitudinal impedance for a parallel resonator, for which the infinite series can again be summed analytically. The coupling impedance divided by the harmonic number, evaluated at the spectral frequencies, can be written, cf. Eq. (5.40):

$$\frac{Z_{\parallel}}{n}(\omega_p) = -j\frac{Z_1}{(p-p_1)(p-p_2)}, \tag{5.52}$$

with the coefficient:

$$Z_1 = \frac{h_r R_s}{k_b^2 Q_r}, \tag{5.53}$$

where $h_r = \omega_r/\omega_0$ is the harmonic number of the resonant frequency. The effective impedance can thus be found from the expression:

$$Z_{eff}^{\bar{m}} = -jZ_1 2\pi \frac{(\bar{m}+1)^2}{b^3} \sum_{p=-\infty}^{\infty} \frac{1+(-)^{\bar{m}} \cos b(p+d)}{(p-p_1)(p-p_2)(p-q_1)^2(p-q_2)^2}. \tag{5.54}$$

Again the infinite series can be summed after partial fraction decomposition (see Appendix A). The result[Zotter78a] can be written in compact form:

$$Z_{eff}^{\bar{m}} = -j\Big[D_1 Z_1 + D_2 Z_2\Big] \,, \tag{5.55}$$

where Z_1 is given by Eq.(5.53), and

$$Z_2 = \frac{2\pi^2(\bar{m}+1)^2}{k_b^2 b^3}\frac{R_s}{Q_r'} . \tag{5.56}$$

In terms of $z_i = b(p_i + d)$, the two (complex) factors are

$$\begin{aligned} D_1 &= \tfrac{1}{2}\sum_{i=1}^{2} \frac{1}{(p_1 - q_i)(p_2 - q_i)} \,, \\ D_2 &= \sum_{i=1}^{2} (-)^i \frac{\cot(\pi p_i)[1 + (-)^{\bar{m}} \cos z_i] + (-)^{\bar{m}} \sin z_i}{(p_i - q_1)^2 (p_i - q_2)^2} \,, \end{aligned} \tag{5.57}$$

where all parameters have been defined in Eqs.(5.47, 5.49, 5.50). These expressions are readily evaluated by computer, and have been compared with direct, term-by-term summation of the series in Eq.(5.40). Because the effective impedance can be computed much faster using Eq.(5.55) than by direct summation, these expressions have been incorporated into several computer codes which evaluate the stability of bunched beams such as *BBI* [Hofmann79a] and *ZAP* [Zisman86].

In particular, to evaluate the effective impedance of large machines with short bunches, the direct summation of the series requires a large number of terms for good accuracy. The summation over p should be performed up to an upper limit larger than $h_r = \omega_r/\omega_0$ or R/L_b - whichever is bigger - before the series may be truncated. For a machine as large as *LEP*, one would need millions of terms. In addition, the sums sometimes first seem to converge to an intermediate, incorrect value before reaching the correct one [Zotter78a]. Therefore one cannot use the standard criterion for terminating summation when the change of the sum becomes small.

The effective impedances of the multi-bunch modes for the case of 4 bunches as function of the resonator frequency are shown in Fig. 5.6. One can see that the real part of the impedance of the modes $n_b = 0$ or 2, i.e. with zero or π phase shift between bunch oscillations, remains relatively small, while the intermediate modes $n_b = 1$ and 3, with $\pm\pi/2$ phase shift, oscillate strongly

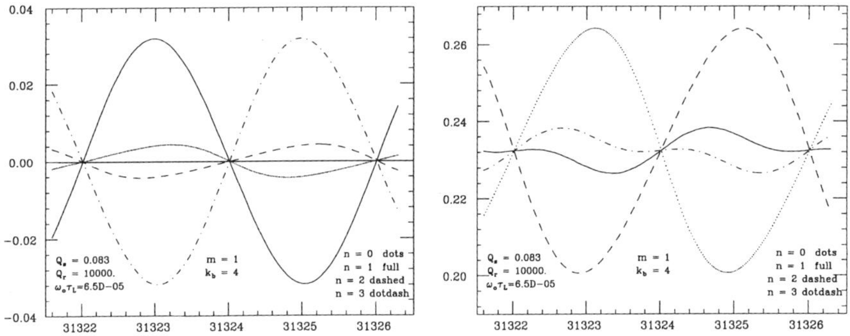

Figure 5.6: Real and imaginary parts of effective impedance for sinusoidal modes of 4 equidistant bunches in a narrow-band cavity versus $h_r = \omega_r/\omega_0$.

between positive or negative values. The imaginary part of the effective impedance oscillates about a much larger average value, hence its relative change is quite small.

For stability calculations, the effective impedances of all resonances of possibly many cavities have to be added. Their impedances oscillate with different amplitudes and phases, and thus their sum can only be found numerically. Their maxima depend rather critically on the exact values of the resonant frequencies, and if these cannot be determined to the required precision, only *worst case* estimates can be given. These may be exaggeratedly pessimistic, but the values are often needed to determine the required damping of higher modes in cavities.

The problem of determining the total impedance is aggravated by the fact that the resonant mode frequencies may vary from cell to cell in multi-cell cavities, or from cavity to cavity, due to small differences in their dimensions which are permitted by fabrication tolerances, and due to different positions of the *tuners* used to keep the fundamental frequency constant. The positions of variable tuners furthermore change with operating temperature and ambient pressure, and the higher mode frequencies may thereby be influenced in an unpredictable way.

5.6.4 Effective Impedance of Gaussian Bunches

Electron bunches in circular accelerators usually have a nearly Gaussian distribution, caused by the strong emission of synchrotron radiation in bending (and other) magnets. The equilibrium between quantum excitation and radiation damping leads to a Gaussian distribution in energy. In addition, the bunches have to be short compared to the rf wave length in order to get a sufficient quantum life time [Sands70]. Then the rf voltage is usually quite linear over the length of the bunch, and the line density distribution becomes also nearly Gaussian:

$$\lambda(\tau) = \frac{1}{\sqrt{2\pi}\sigma} \exp\left(-\frac{\tau^2}{2\sigma^2}\right) , \qquad (5.58)$$

where $\sigma = \sigma_s/c$ is the rms bunch length in units of time.

Suitable line density modes are products of Hermite polynomials of order $\bar{m}$ with a Gaussian factors [Besnier78]. From their Fourier transforms one obtains the spectral power densities

$$h_{\bar{m}}(\omega) = (\omega\sigma)^{2\bar{m}} \exp\left(-\omega^2\sigma^2\right) \qquad (5.59)$$

and the normalised ones $H_{\bar{m}}(\omega_p) = C_{\bar{m}} h_{\bar{m}}(\omega_p)$. The normalisation constants can be found by evaluating the expression $C_{\bar{m}} = 1/\sum_p h_{\bar{m}}(\omega_p)$. One can sum the required infinite series using the expression:

$$\sum_{p=-\infty}^{\infty} \mathrm{e}^{-(xp+y)^2} = \frac{\Gamma(\bar{m}+1/2)}{x} + \ldots \qquad (5.60)$$

For small values of the parameter x, this expression is valid to an extremely good accuracy. This parameter for our problem is equal to the bunching factor $b = k_b\sigma_s/R$, which is always small compared to unity. The next term in the expansion is multiplied by the factor $\exp\left(-\pi^2/b^2\right)$. For a single bunch in *LEP*, e.g., $b \approx 10^{-5}$, and hence the neglected term is of the order of $\exp\left(-10^{10}\right)$. Including only the first term of the expansion one gets $C_{\bar{m}} = b/\Gamma(\bar{m}+1/2)$. The same result is obtained by replacing the infinite sum by an integral.

The effective impedances for a Gaussian bunch in a resonator is given by

$$Z_{eff}^{\bar{m}} = -jZ_{\bar{m}} \sum_{p=-\infty}^{\infty} (\omega_p\sigma)^{2\bar{m}} \mathrm{e}^{-(\omega_p\sigma)^2} \left[\frac{\omega_1}{\omega_p-\omega_1} - \frac{\omega_2}{\omega_p-\omega_2}\right] , \qquad (5.61)$$

where

$$Z_{\bar{m}} = \frac{\omega_0\sigma R_s}{2Q_r' \, \Gamma(\bar{m}+1/2)} . \qquad (5.62)$$

These infinite series can be summed analytically [Zotter80] in terms of the complex error function $\mathrm{w}(z)$. Defining $z = b\ (p+d)$ and $z_{1,2} = \omega_{1,2}\sigma =$

$b\ (p_{1,2} + d)$ one finds for $i = 1, 2$:

$$S_0(p_i) = \sum_{p=-\infty}^{\infty} \frac{\mathrm{e}^{-z^2}}{p - p_i} = \pi \left[j\mathrm{w}(z_i) - (\cot \pi p_i + j)\mathrm{e}^{-z_i^2} \right] \tag{5.63}$$

and

$$S_n(p_i) = \sum_{p=-\infty}^{\infty} \frac{z^n \mathrm{e}^{-z^2}}{p - p_i} = z_i^n\, S_0(p_i) + \sum_{\ell=0}^{[\frac{n-1}{2}]} \Gamma(\ell + \tfrac{1}{2}) z_i^{n-1} . \tag{5.64}$$

Here $[a]$ means the integer part of a. Like the sums in Eq.(5.60), these expressions are a very good approximation for large rings with short bunches, where the bunching factor $b = k_b \omega_0 \sigma << 1$. Thus one finds the effective impedance for Gaussian bunches in a resonator

$$\begin{aligned} Z_{eff}^{\bar{m}} &= -jZ_{\bar{m}} \left\{ \sum_{k=0}^{\bar{m}-1} \Gamma(k + \tfrac{1}{2}) \left(z_1^{\bar{m}_k} - z_2^{\bar{m}_k} \right) - j\pi \left[z_1^{2\bar{m}} \mathrm{w}(z_1) - z_2^{2\bar{m}} \mathrm{w}(z_2) \right] \right. \\ &- \left. \pi \left[\mathrm{e}^{-z_1^2} (j + \cot \pi p_1) - \mathrm{e}^{-z_2^2} (j + \cot \pi p_2) \right] \right\} , \end{aligned} \tag{5.65}$$

where we wrote for brevity $\bar{m}_k = 2\bar{m} - 2k - 1$.

When the resonance is not too narrow, i.e. the quality factor $Q_r \ll \omega_r/\omega_0$, the imaginary part of the arguments of the *cotangent* is large: Im $[\pi p_{1,2}] = \pi\omega_r/(2Q_r\omega_0) \gg 1$. Then the *cotangent* tends exponentially towards $(-j)$, the two terms in the square brackets approach zero very quickly, and thus can be neglected with very small error. The effective impedance of Hermitian modes in wide- and narrow-band resonators is shown in Fig. 5.7.

Short Gaussian Bunches

For bunches short compared to the resonator wavelength, $|z_i| = \omega_r \sigma \ll 1$ since $\sigma \ll 1/\omega_r$. Then in Eq.(5.65) only the $k = \bar{m} - 1$ term in the sum over k is important, since all other terms are proportional to higher powers of $z_{1,2}$ and thus are negligible. Hence the effective impedance for the $\bar{m}$-th mode of oscillation of a short bunch can be approximated by the expression:

$$Z_{eff}^{\bar{m}} = -j \frac{\omega_0 R_s}{\omega_r Q_r} \frac{(\omega_r \sigma)^2}{\bar{m} - 1/2} . \tag{5.66}$$

The first factor is just the low frequency impedance of a resonator, divided by the harmonic number:

$$\frac{Z_{\|}}{n}(\omega = 0) = j \frac{\omega_0 R_s}{\omega_r Q_r} . \tag{5.67}$$

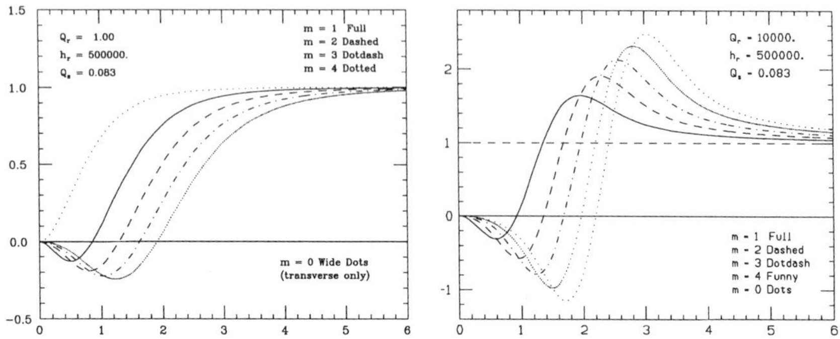

Figure 5.7: Real and imaginary parts of effective impedance versus bunch length for a Gaussian bunch in a broad - and a narrow-band resonator.

The effective impedances of the oscillation modes with $\bar{m} \geq 1$, which are important for (longitudinal) instabilities, are purely imaginary and negative. For that reason they are often called *capacitive*, which is misleading as the effective impedance is still proportional to frequency and hence *inductive*, although with a negative value of the inductance.

Long Gaussian Bunches

For bunches which are long compared to the resonant wavelength, one can evaluate Eq.(5.65) by the asymptotic expression for the complex error function [Abramowitz65] $\mathrm{w}(z) \approx (j/\pi z) \sum_k \Gamma(k+1/2)/z^{2k}$. Then all terms of the finite sum over k are exactly cancelled by the first $\bar{m}-1$ terms of this expansion, and the remaining series starts with $k = \bar{m}$. Keeping only the two lowest terms, one gets for long bunches:

$$Z_{eff}^{\bar{m}} = j\frac{Z_{\|}}{n}\left[1 + \frac{1}{(\omega_r\sigma)^2}\left(1 - \frac{1}{Q_r^2}\right)\right], \tag{5.68}$$

i.e. the effective impedance approaches the low frequency inductance.

5.7 Effective Transverse Impedance

The effective transverse impedance for the $\bar{m}$-th longitudinal mode of a dipole oscillation of a bunched beam can be defined [Sacherer74]

$$Z_{\perp\bar{m}}^{eff} = \sum_{p=-\infty}^{\infty} Z_{\perp}(\omega_p) H_{\bar{m}}(\omega_p - \omega_\xi) , \tag{5.69}$$

where $\omega_p = \omega_0(p + n_b + \bar{m}Q_s + Q_\beta)$ are the spectral frequencies for transverse oscillations. They are shifted also by the betatron tune Q_β in the plane of oscillation investigated. Furthermore, $\omega_\xi = \omega_0\xi/\eta$ is the *chromatic frequency*, where $\xi = PdQ_\beta/Q_\beta dP$ is the chromaticity (P is the longitudinal particle momentum). Except for the shift of the bunch spectrum by the chromatic frequency ω_ξ, the effective transverse impedance is thus identical to the transverse loss (or kick-) factor. For high-energy electron machines, one can usually replace $\eta = \alpha - 1/\gamma^2$ by the momentum compaction factor α when $\gamma \gg \alpha^{-1/2}$.

The transverse impedance is proportional to the longitudinal one divided by the harmonic number n. Hence the effective transverse impedances of a resonator are given by the same expressions as the longitudinal ones. However, the shunt resistance R_s has to be replaced by the transverse resistance (per unit displacement) $R_\perp$, and the resonant frequency and quality factor are those of a transverse resonance. Furthermore, the transverse spectral frequencies must be used, and the chromatic frequency subtracted - which may make a considerable difference when the chromaticity is large. The main difference is due to the fact that the rigid dipole mode with $\bar{m} = 0$, which does not contribute to a longitudinal mode, usually dominates the effective transverse impedance.

5.7.1 Transverse Resonator Impedance

The effective *transverse impedance* of a resonator, with resonant frequency ω_r, the quality factor Q_r and the transverse resistance $R_\perp$, can be obtained from Eq.(5.61). The argument of the complex error function then is

$$z_{1,2} = (\omega_{1,2} - \omega_\xi)\sigma , \tag{5.70}$$

with $\omega_{1,2} = \frac{\omega_r}{Q_r}(\frac{j}{2} \pm Q_r'), Q_r' = \sqrt{Q_r^2 - 1/4}$ as before. For a not too high quality factor, i.e. $Q_r << \omega_r/\omega_0$, one can again neglect terms containing factors $[\cot(\pi z_i) + j]$, to get approximately:

$$Z^{eff}_{\perp\bar{m}} = -2jZ_{\bar{m}}\left\{\sum_{k=0}^{\bar{m}-1}\Gamma(k+1/2)\mathrm{Re}\Big[z_1^{2\bar{m}-2k-1}\Big] + \mathrm{Im}\Big[\pi(z_1^{2m}\mathrm{w}(z_1)\Big]\right\}. \quad (5.71)$$

We rewrite the equation as $Z^{eff}_{\perp\bar{m}} = Z_\perp(0)F_{\bar{m}}(\omega_r\sigma)$, where $Z_\perp(0) = jR_\perp/Q_r$ is the low frequency transverse impedance of the resonator. The most important contribution to the impedance for transverse stability considerations is produced by the rigid dipole mode with $\bar{m} = 0$. For low chromaticity ($\xi \approx 0$), the form factor $F_0(x)$ is given by

$$F_0(x) = x\sqrt{\pi}\frac{Q_r}{2jQ'_r}[\mathrm{w}(z_1) - \mathrm{w}(z_2)] = \frac{\sqrt{\pi}Q_r}{Q'_r}x\ \mathrm{Im}\left\{\mathrm{w}\left(\frac{x}{Q_r}\left[\frac{j}{2} + Q'_r\right]\right)\right\}, \quad (5.72)$$

where the second equation is valid for the usual case $Q_r \geq 1/2$.

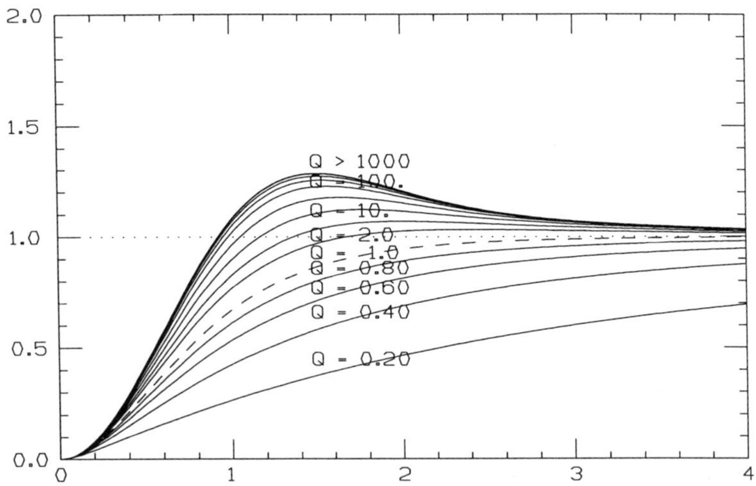

Figure 5.8: Form factor $F_0(x)$ for effective transverse impedance $\bar{m} = 0$ versus product of resonator frequency and bunch length $x = \omega_r\sigma$.

When the bunch length is short compared to the wave length of the resonance, i.e. when $|z_{1,2}| = \omega_r\sigma \ll 1$, one may use the lowest terms of the power series expansion of the complex error function $\mathrm{w}(z) \approx 1 + 2jz/\sqrt{\pi} + ...$ to get the same expressions as for the longitudinal case. In particular for $\bar{m} = 0$, the function $F_0(x)$ can be approximated by $2x^2$ for very short bunches. On the other hand, for long bunches when $x \gg 1$, the asymptotic expansion of the complex error function $\mathrm{w}(z) = j/(\sqrt{\pi}z)$ can be used to show that $F_0(x)$ approaches unity. Both limits are independent of the quality factor Q_r as can be seen in Fig. 5.8.

In the following two appendices we show the summation of the infinite series for the effective impedances of *sinusoidal modes* for parabolic bunch distributions, and of *Hermitian modes* applicable to Gaussian bunches.

Appendix 5.A Sinusoidal modes

In this Appendix we present mathematical details of the derivation of some expressions used in the subsection on the effective impedance of parabolic bunches.

5.A.1 Fourier Transforms of Line Density Modes

By evaluating the integral

$$\int_{-\infty}^{\infty} d\tau \lambda(\tau) \exp(-j\omega\tau) = \tilde{\lambda}(\omega) \tag{5.73}$$

one finds for sinusoidal modes [Sacherer76]

$$\tilde{\lambda}(\omega) = \frac{4\pi(\bar{m}+1)}{(\omega\tau_L)^2 - \pi^2(\bar{m}+1)^2} \begin{cases} (-)^{\bar{m}/2}\cos(\omega\tau_L) & \bar{m} \quad \text{even} \\ (-)^{(\bar{m}+1)/2}\sin(\omega\tau_L) & \bar{m} \quad \text{odd}\ . \end{cases} \tag{5.74}$$

Hence the power density at frequency ω becomes

$$h_{\bar{m}}(\omega) = (\bar{m}+1)^2 \frac{1+(-)^{\bar{m}}\cos(\omega\tau_L)}{[(\omega\tau_L)^2 - \pi^2(\bar{m}+1)^2]^2}\ , \tag{5.75}$$

which has to be evaluated at the spectral frequencies $\omega_p = \omega_0(pk_b + n_b + \bar{m}Q_s)$. We express the argument $\omega_p\tau_L = b\ (p+d)$ in terms of the *bunching factor* b and the *relative detuning* d with the definitions

$$b = k_b\omega_0\tau_L = k_b\frac{L_b}{R}\ , \qquad d = \frac{n_b + \bar{m}Q_s}{k_b}. \tag{5.76}$$

With the abbreviation

$$N_p = 1 + (-)^{\bar{m}}\cos b(p+d) \tag{5.77}$$

the expression for the spectral power density can be rewritten

$$h_{\bar{m}}(\omega_p) = \frac{(\bar{m}+1)^2 N_p}{[b(p+d)^2 - \pi^2(\bar{m}+1)^2]^2} = \frac{(\bar{m}+1)^2 N_p/b^4}{(p-q_1)^2(p-q_2)^2}\ , \tag{5.78}$$

where the (second order) poles are located at

$$q_{1,2} = \pm\pi\frac{(\bar{m}+1)}{b} - d. \tag{5.79}$$

5.A.2 Normalisation Factors

To find the normalisation factors $C_{\bar{m}} = 1/\sum_p h_{\bar{m}}(\omega_p)$ one needs to sum the series over the spectral power densities. One can perform a partial fraction decomposition of h_m, taking into account the double roots:

$$\frac{1}{(p-q_1)^2(p-q_2)^2} = \sum_{i=1}^{2}\left[\frac{A_i}{p-q_i} + \frac{B_i}{(p-q_i)^2}\right]. \tag{5.80}$$

Comparison of coefficients of equal powers yields:

$$A_1 = -A_2 = -\frac{2}{(q_1-q_2)^3} \quad = \quad \frac{2b^3}{\pi^3(\bar{m}+1)^3} \tag{5.81}$$

and

$$B_1 = B_2 = \frac{1}{(q_1-q_2)^2} \quad = \quad \frac{b^2}{\pi^2(\bar{m}+1)^2}. \tag{5.82}$$

The summation over p can be performed [Zotter78b] using the expressions:

$$\begin{aligned}\sum_{p=-\infty}^{\infty}\frac{\cos(xp+y)}{(p-q)^2} &= \pi^2\left[\frac{\cos(xq+y)}{\sin^2(\pi q)} - \frac{x}{\pi}\frac{\sin[\pi q-(xq+y)]}{\sin(\pi q)}\right], \\ \sum_{p=-\infty}^{\infty}\frac{\cos(xp+y)}{p-q} &= -\pi\frac{\cos[\pi q-(xq+y)]}{\sin(\pi q)},\end{aligned} \tag{5.83}$$

which are also valid for $x = y = 0$:

$$\sum_{p=-\infty}^{\infty}\frac{1}{p-q} = -\pi\cot(\pi q) \quad , \quad \sum_{p=-\infty}^{\infty}\frac{1}{(p-q)^2} = \frac{\pi^2}{\sin^2(\pi q)}. \tag{5.84}$$

With the abbreviation $u_{1,2} = b(q_{1,2}+d) = \pm\pi(\bar{m}+1)$, one finds $\cos(u_{1,2}) = (-)^{\bar{m}+1}$, while $\sin(u_{1,2}) = 0$. Hence

$$\begin{aligned}\sum_{p=-\infty}^{\infty}\frac{N_p}{p-q_{1,2}} &= -\pi\left[\cot(\pi q_{1,2}) + (-)^{\bar{m}}\frac{\cos(\pi q_{1,2}-u_{1,2})}{\sin(\pi q_{1,2})}\right] = 0\,, \\ \sum_{p=-\infty}^{\infty}\frac{N_p}{(p-q_{1,2})^2} &= \pi^2\left[\frac{\cos u_{1,2}}{\sin^2(\pi q_{1,2})} + (-)^m\frac{b}{\pi}\frac{\sin(\pi q_{1,2}-u_{1,2})}{\sin(\pi q_{1,2})}\right] = \pi b.\end{aligned} \tag{5.85}$$

Therefore one gets simply

$$\sum_{p=-\infty}^{\infty} h_{\bar{m}}(\omega_p) = \pi b(B_1+B_2). \tag{5.86}$$

From Eq. (5.82), one obtains the normalisation constant $C_{\bar{m}} = \pi b/2$ and

$$H_{\bar{m}}(\omega_p) = \frac{\pi(\bar{m}+1)^2}{2b^3} \frac{1+(-)^{\bar{m}}\cos b(p+d)}{(p-q_1)^2(p-q_2)^2} . \qquad (5.87)$$

5.A.3 Effective Impedance

The coupling impedance of a parallel resonator, evaluated at the spectral frequencies $\omega_p = b(p+d)$, can be written:

$$\frac{Z_{\parallel}}{n} = \frac{-jZ_1}{(p-p_1)(p-p_2)} , \qquad (5.88)$$

where $Z_1 = h_r R_s / k_b^2 Q_r$ and

$$p_{1,2} = \frac{\omega_{1,2}}{k_b \omega_0} - d, \quad \omega_{1,2} = \frac{\omega_r}{Q_r}\left[\pm\frac{j}{2} + Q'_r\right] . \qquad (5.89)$$

One can again perform a partial fraction decomposition of the product of the impedance and the power density, which is only slightly more complicated than before. Apart from the constant factor $C_1 = -jZ_1 2\pi(\bar{m}+1)^2/b^3$, the impedance contains an infinite sum over p of the fractions:

$$\frac{1}{(p-p_1)(p-p_2)(p-q_1)^2(p-q_2)^2} = \sum_{i=1}^{2}\left[\frac{A'_i}{p-p_i} + \frac{A''_i}{p-q_i} + \frac{B'_i}{(p-q_i)^2}\right] . \qquad (5.90)$$

The coefficients can again be determined by comparison of equal powers:

$$\begin{aligned} A'_{1,2} &= \frac{\pm 1}{(p_1-p_2)(p_{1,2}-q_1)^2(p_{1,2}-q_2)^2} = \frac{A_{1,2}}{p_1-p_2} \\ B'_{1,2} &= \frac{1}{(q_1-q_2)^2(p_1-q_{1,2})(p_2-q_{1,2})} = \frac{B_{1,2}}{q_1-q_2} , \end{aligned} \qquad (5.91)$$

while $A''_{1,2}$ are not needed since they are multiplied by the sums $\sum_p N_p/(p-q_{1,2})$ which equal zero, cf. Eq. (5.85). The effective impedance can thus be written

$$Z_{eff}^{\bar{m}} = C_1 \sum_{i=1}^{2}\left[A'_i \sum_{p=-\infty}^{\infty} \frac{N_p}{p-p_i} + B'_i \sum_{p=-\infty}^{\infty} \frac{N_p}{(p-q_i)^2}\right] = C_1 \sum_{i=1}^{2}\left[A'_i N_i + B'_i\right] , \qquad (5.92)$$

with the numerator

$$N_p = \frac{\cos(\pi p) + (-)^{\bar{m}} \sin(\pi p - z)}{\sin(\pi p)} = \cot(\pi p)\,[1 + (-)^{\bar{m}} \cos z] + (-)^{\bar{m}} \sin z \ . \tag{5.93}$$

For $N_{1,2}$ the same expression holds, with all p, z replaced by $p_{1,2}, u_{1,2}$. One should be careful in the numerical evaluation of trigonometric functions of complex argument, since they diverge exponentially when the imaginary part of the argument becomes large. Substituting the expressions for the coefficients $A'_{1,2}$ and $B'_{1,2}$, and using $p_1 - p_2 = (2h_r Q'_r)/(k_b Q)$, $q_1 - q_2 = 2\pi(\bar{m}+1)$, yields finally:

$$Z_{eff}^{\bar{m}} = -j\Big[D_1 Z_1 + D_2 Z_2\Big] \ , \tag{5.94}$$

with $2D_1 = \sum_i \bar{B}_i$ and $D_2 = \sum_i \bar{A}_i N_i$. The coefficient Z_1 has been defined in Eq. (5.53), while Z_2 is given in Eq. (5.56).

Appendix 5.B Hermitian Modes

In this appendix we present mathematical details of the derivation of expressions used in the subsection on the effective impedance of Gaussian bunches.

5.B.1 Fourier Transforms of Line Density Modes

The line density modes of Gaussian bunches can be written in terms of Hermite polynomials $H_{\bar{m}}$ of order $\bar{m}$: $\lambda^{(\bar{m})}(\tau) = H_{\bar{m}}(\tau/2\sqrt{\sigma})\exp(-\tau^2/2\sigma^2)$. They have the Fourier transforms [Abramowitz65]

$$\tilde{\lambda}^{(\bar{m})}(\omega) = (\omega\sigma)^{\bar{m}} \exp(-\frac{\omega^2\sigma^2}{2}), \tag{5.95}$$

from which one obtains the spectral power density, cf. Eq. (5.59):

$$h_{\bar{m}}(\omega_p) = (\omega_p\sigma)^{2\bar{m}} e^{-\omega_p^2\sigma^2} . \tag{5.96}$$

The normalised spectral power density H_m has the additional factor $C_{\bar{m}} = 1/\sum_p h_{\bar{m}}(\omega_p)$. Also these infinite series can be summed analytically [Zotter80].

5.B.2 Normalisation Factors

The (elliptic) *theta function* of the third kind ϑ_3 has the series representation [Abramowitz65]:

$$\vartheta_3(z,q) = 1 + 2\sum_{n=0}^{\infty} q^{n^2}\cos(2nz) = \sum_{p=-\infty}^{\infty} \exp\left(2jzp + p^2 \ln q\right), \qquad (5.97)$$

where q is called the *nome*. By substituting $q = \mathrm{e}^{-x^2}$ and $z = jxy$ one can deduce the expression

$$\sum_{p=-\infty}^{\infty} \mathrm{e}^{-(xp+y)^2} = \mathrm{e}^{-y^2}\vartheta_3\left(jxy, \mathrm{e}^{-x^2}\right) . \qquad (5.98)$$

The *Jacobi transformation* [Whittaker69] for theta function of complex argument yields:

$$\vartheta_3(v, \mathrm{e}^{j\pi u}) = \mathrm{e}^{\frac{v^2}{j\pi u}}\vartheta_3(v/u, \mathrm{e}^{-j\frac{\pi}{u}}) . \qquad (5.99)$$

Substituting $v = jx^2y$ and $u = x^2/j\pi$ one obtains

$$\begin{aligned}\sum_{p=-\infty}^{\infty} \mathrm{e}^{-(xp+y)^2} &= \frac{\sqrt{\pi}}{x}\vartheta_3\left(\frac{\pi y}{x}, \mathrm{e}^{-\frac{\pi^2}{x^2}}\right) \\ &= \frac{\sqrt{\pi}}{x}\left[1 + 2\mathrm{e}^{-\frac{\pi^2}{x^2}}\cos\frac{2\pi y}{x} + \ldots\right] \end{aligned} \qquad (5.100)$$

up to terms of second order.

The series converges very rapidly for $x \ll 1$, since powers of π/x appear in the negative exponents of all but the first term of the expansion. Already the second term is completely negligible for $x \ll 1$. This is always the case when x is the bunching factor of an electron accelerator. Hence the summation formula

$$\sum_{p=-\infty}^{\infty} \mathrm{e}^{-(xp+y)^2} = \frac{\sqrt{\pi}}{x} \qquad (5.101)$$

gives an extremely good approximation.

By differentiation with respect to the parameter y one obtains the other sums required:

$$\sum_{p=-\infty}^{\infty} (xp+y)^{2m}\, e^{-(xp+y)^2} = \frac{1}{x}\Gamma\left(\bar{m}+\tfrac{1}{2}\right) ,$$

$$\sum_{p=-\infty}^{\infty} (xp+y)^{k}\, e^{-(xp+y)^2} = \begin{cases} \dfrac{1}{x}\Gamma\left(\dfrac{k+1}{2}\right) & \text{for } k=2\bar{m} , \\ 0 & \text{for } k=2\bar{m}+1 , \end{cases} \tag{5.102}$$

where $\Gamma(x)$ is the Gamma function and $\Gamma(1/2)=\sqrt{\pi}$. Since only even powers $2\bar{m}$ appear in the sum over p, the normalisation constant becomes

$$C_{\bar{m}} = \frac{b}{\Gamma(\bar{m}+1/2)}. \tag{5.103}$$

The accuracy of the approximation can be judged from the example of a single bunch with rms length $\sigma = 30$ ps ($\sigma_s = 1$ cm) in *LEP*. With a revolution frequency of $\omega_0 = 2\pi 11$ KHz $= 7.10^4 s^{-1}$, the bunching factor b is about 8.10^{-6} and $\exp\left(-\pi^2/b^2\right) \approx \exp\left(-10^{11}\right)$ which is indeed negligible.

5.B.3 Effective Impedance

To evaluate the effective impedance, one needs to sum series of the form $\sum_p x^k e^{-(xp+y)^2}/(p-p_i)$, cf. Eq. (5.64), which can be found by integration [Zotter80] of Eq. (5.101) with respect to the parameter ξ:

$$\sum_p e^{-\xi(p^2+2a_1p+a_2^2)} = \sqrt{\frac{\pi}{\xi}} e^{\xi(a_1^2-a_2^2)}. \tag{5.104}$$

The integral over the right hand side of this equation can again be expressed in terms of the *complex error function* [Abramowitz65].

One then obtains the sum

$$S_0(p_{1,2}) = \sum_{p=-\infty}^{\infty} \frac{e^{-(xp+y)^2}}{p-p_i} = \pi\left[j\mathrm{w}(u_{1,2}) - e^{-u_{1,2}^2}(\cot\pi p_i + j)\right] , \tag{5.105}$$

where $u_{1,2} = xp_{1,2}+y$. When the imaginary part of the argument is large compared to unity, the cotangent tends exponentially towards $-j$, and thus the second term in the square brackets can be neglected. This is the case for not too narrow, high frequency resonances, for which $Q_r \ll \omega_r/\omega_0$. In particular, it is always true for broad-band resonators, where the sum thus becomes simply $S_0(p_{1,2}) = j\pi \mathrm{w}(u_{1,2})$.

Finally, by differentiation with respect to the parameter x one finds also the other required sums

$$\begin{aligned} S_k(p_{1,2}) &= \sum_{p=-\infty}^{\infty} (xp+y)^k \frac{e^{-(xp+y)^2}}{(p-p_{1,2})} \\ &= u_{1,2}^k S_0(p_{1,2}) + \sum_{\ell=0}^{[\frac{k-1}{2}]} \Gamma(\ell+1/2) u_{1,2}^{k-2\ell-1} , \end{aligned} \tag{5.106}$$

again with $u_{1,2} = xp_{1,2} + y$. The square bracket in the upper limit of the sum on the r.h.s. stands for the integer part of the expression. All these summation formulae are excellent approximations for small bunching factors $b \ll 1$, which is always the case in electron machines. They can therefore be used to evaluate the effective impedance for Gaussian bunches in resonator impedances much faster than by direct summation.

Chapter 6

Uniform Cylindrical Pipes

We now turn to the calculation of the coupling impedances for various structures which are used in modern particle accelerators. The simplest one is a smooth vacuum chamber consisting of a straight cylindrical pipe. In the following Chapters we will consider some more complicated elements.

The *EM* fields excited in a smooth, perfectly conducting pipe can have two distinct polarisations:

(1) *transverse magnetic fields*, usually called *TM* (or E) modes, for which the longitudinal component of the magnetic field vanishes, i.e. $H_z = 0$, and

(2) *transverse electric fields*, referred to as *TE* (or H) modes, for which the longitudinal component of the electric field is zero ($E_z = 0$). The propagation constant of these fields is real at least over a finite frequency range. Therefore they can propagate undamped, with a phase velocity which is always larger than the speed of light [Jackson75]. Since particles can only move with velocities smaller than the speed of light $v \leq c$, they cannot excite wake fields in such pipes. Nevertheless, the study of the propagation of rf fields in such pipes is important in several aspects. In particular, it clarifies certain features of the *EM* fields which one has to keep in mind when considering more complicated cases.

The situation is quite different when the pipe wall is not perfectly conducting. Strictly speaking, the *EM* fields no longer consist of pure *TM* or *TE* modes but are a mixture of both, in which E_z as well as H_z components are different from zero. The only exception are *monopole* ($m = 0$) modes in axially symmetric pipes. In this case the *TM* and *TE* modes do not mix even when conductivity of the wall is finite.

Due to dissipation of field energy in the wall resistance, the (longitudinal) propagation constant is no longer real, and the amplitude of the *EM* wave decreases along the pipe. The phase velocity of some field harmonics may become lower than the speed of light, thus permitting an effective transfer of particle energy to the *EM* field. If these frequencies lie inside the spectrum of a bunch, wake fields are excited. This in turn may influence its motion, and should therefore be taken into account. For a point charge, this is always the case since its spectrum contains all frequencies.

Certain elements of the vacuum chamber, such as beam-position monitors, cavity windows, kickers, etc., need to be made of insulating (e.g. ceramic) materials. Furthermore, in fast-cycling accelerators, where the guiding magnetic field changes rapidly, the whole vacuum chamber is made of non-conducting materials in order to avoid excessive eddy currents. In some cases also materials with high magnetic permeability, such as pole pieces of bending or focusing magnets are exposed to the beam or at least its *EM* fields. Even when the chamber is uniform over a sufficient length, it can then still produce wake fields which need to be understood in order to avoid excessive limitations of the beam current. Such problems constitute the subject of this Chapter.

6.1 Maxwell Equations in Cylindrical Coordinates

To study the excitation and propagation of *EM* fields in uniform, cylindrical pipes it is most appropriate to start from the Maxwell equations for the *Fourier components of the EM fields*, which are functions of (angular) frequency $\omega = 2\pi f$, and are identified here by a tilde over the symbol. If the beam pipe has circular cross section, we call it *axially or rotationally symmetric*[1]. In this case it is appropriate to use (circular) cylindrical coordinates r, θ, z.

We furthermore assume that the material properties do not change with time or in space, and thus the coefficients of the Maxwell equations as well as the boundary conditions are constant. All field and source quantities involved may therefore be expanded into double Fourier integrals in t and θ:

$$\widetilde{\mathbf{E}}_m(r, z) = \int_{-\infty}^{\infty} d\theta \ \mathrm{e}^{jm\theta} \int_{-\infty}^{\infty} dt \ \mathbf{E}(\mathbf{r}, z, t) \ \mathrm{e}^{j\omega t} \ . \tag{6.1}$$

[1]The often used designation *cylindrically symmetric* is not quite correct, as it actually applies to cylinders of arbitrary cross section.

Similar relations hold for the magnetic field strength **H**, the current density **J**, and the charge density ρ.

Since the azimuthal angle is cyclic, i.e. all quantities remain unchanged when 2π is added to θ, only integer values $m = 0, \pm 1, \pm 2 \ldots$ are allowed. Hence the Fourier integral over θ degenerates into a Fourier series.

In components, the Maxwell equations for the field harmonics become:

$$\begin{aligned}
\left[\frac{jm}{r}\widetilde{E}_{zm} - \frac{\partial \widetilde{E}_{\theta m}}{\partial z}\right] + j\omega\mu\widetilde{H}_{rm} &= 0 \, , \\
\left[\frac{\partial \widetilde{E}_{rm}}{\partial z} - \frac{\partial \widetilde{E}_{zm}}{\partial r}\right] + j\omega\mu\widetilde{H}_{\theta m} &= 0 \, , \\
\left[\frac{1}{r}\frac{\partial}{\partial r}\left(r\widetilde{E}_{\theta m}\right) - \frac{jm}{r}\widetilde{E}_{rm}\right] + j\omega\mu\widetilde{H}_{zm} &= 0 \, ;
\end{aligned} \tag{6.2}$$

$$\begin{aligned}
\left[\frac{jm}{r}\widetilde{H}_{zm} - \frac{\partial \widetilde{H}_{\theta m}}{\partial z}\right] - (j\omega\varepsilon + \sigma_c)\widetilde{E}_{rm} &= \tilde{J}_{rm} \, , \\
\left[\frac{\partial \widetilde{H}_{rm}}{\partial z} - \frac{\partial \widetilde{H}_{z\omega}}{\partial r}\right] - (j\omega\varepsilon + \sigma_c)\widetilde{E}_{\theta m} &= \tilde{J}_{\theta m} \, , \\
\left[\frac{1}{r}\frac{\partial}{\partial r}\left(r\widetilde{H}_{\theta m}\right) - \frac{jm}{r}\widetilde{H}_{rm}\right] - (j\omega\varepsilon + \sigma_c)\widetilde{E}_{zm} &= \tilde{J}_{zm} \, ;
\end{aligned} \tag{6.3}$$

$$\frac{1}{r}\frac{\partial}{\partial r}\left(r\widetilde{E}_{rm}\right) + \frac{jm}{r}\widetilde{E}_{\theta m} + \frac{\partial \widetilde{E}_{zm}}{\partial z} = \frac{\tilde{\rho}_m}{\varepsilon} \, ; \tag{6.4}$$

$$\frac{1}{r}\frac{\partial}{\partial r}\left(r\widetilde{H}_{rm}\right) + \frac{jm}{r}\widetilde{H}_{\theta m} + \frac{\partial \widetilde{H}_{zm}}{\partial z} = 0 \, . \tag{6.5}$$

The permittivity ε, the permeability μ, and the conductivity σ_c of the media have been assumed to be independent of both spatial coordinates and time, but they may depend on frequency ω.

The electric field of an ultra-relativistic particle is uniquely distributed in space: its longitudinal component is reduced by a factor γ compared to its magnitude in the rest frame. In particular for light particles, such as electrons or positrons, with very large energy factors $\gamma \gg 1$, the longitudinal electric field is therefore extremely small. At the same time, the magnetic field has only an azimuthal component. Such a field distribution is sometimes referred to as "pancake field" [Chao93]. The charge and current densities of a bunch must furthermore satisfy the *continuity equation*:

$$\frac{1}{r}\frac{\partial}{\partial r}(r\widetilde{J}_{rm}) + \frac{jm}{r}\widetilde{J}_{\theta m} + \frac{\partial \widetilde{J}_{zm}}{\partial z} + j\omega\widetilde{\rho}_m = 0 \,. \qquad (6.6)$$

This completes the set of equations which - together with the appropriate boundary conditions - are required to determine the *EM* fields.

6.2 Eigenmodes of *EM* Field in a Pipe

Before calculating the *EM* field excited by a bunch, it is instructive to study the field propagation in a uniform pipe. For this we need to solve only the homogeneous Maxwell equations, since no free charges or currents are present inside the pipe ($\rho = \mathbf{J} = 0$). The fields are assumed to have been excited by an unspecified external source.

The uniformity of the geometry allows further simplification of the equations by applying an additional Fourier transformation in the longitudinal coordinate z to the field harmonics:

$$\begin{aligned}\widetilde{\mathbf{E}}_{mp}(r) &= \int_{-\infty}^{\infty} dz \widetilde{\mathbf{E}}_m(r,z)\, e^{jpz} \,, \\ \widetilde{\mathbf{H}}_{mp}(r) &= \int_{-\infty}^{\infty} dz \widetilde{\mathbf{H}}_m(r,z)\, e^{jpz} \,. \end{aligned} \qquad (6.7)$$

Here the longitudinal propagation constant p should be understood as a free (complex) parameter. The solution of the Maxwell equations for $\widetilde{\mathbf{E}}\,(r)$ and $\widetilde{\mathbf{H}}\,(r)$ can then be represented in terms of elementary cylindrical waves $\mathcal{Z}_m(\nu r)$ and their derivatives $\mathcal{Z}'_m(\nu r)$ with respect to the argument νr, and an exponential factor $F_m = \exp[j(m\theta + pz + \omega t)]$. The electric field components can be written:

$$\begin{aligned}\widetilde{E}_r &= \sum_{m=-\infty}^{\infty} \left[A_m \frac{jp}{\nu}\mathcal{Z}'_m(\nu r) - B_m \frac{m\mu\omega}{\nu^2 r}\mathcal{Z}(\nu r)\right] F_m \,, \\ \widetilde{E}_\theta &= -\sum_{m=-\infty}^{\infty} \left[A_m \frac{mp}{\nu^2 r}\mathcal{Z}_m(\nu r) + B_m \frac{j\mu\omega}{\nu}\mathcal{Z}'(\nu r)\right] F_m \,, \\ \widetilde{E}_z &= \sum_{m=-\infty}^{\infty} A_m \mathcal{Z}(\nu r) F_m \,, \end{aligned} \qquad (6.8)$$

and the magnetic field components become:

$$\begin{aligned}
\widetilde{H}_r &= \sum_{m=-\infty}^{\infty} \left[A_m \frac{mk_c^2}{\mu\omega\nu^2 r} \mathcal{Z}_m(\nu r) + B_m \frac{jp}{\nu} \mathcal{Z}'(\nu r) \right] F_m , \\
\widetilde{H}_\theta &= - \sum_{m=-\infty}^{\infty} \left[A_m \frac{jk_c^2}{\mu\omega\nu} \mathcal{Z}'_m(\nu r) - B_m \frac{mp}{\nu^2 r} \mathcal{Z}(\nu r) \right] F_m , \\
\widetilde{H}_z &= \sum_{m=-\infty}^{\infty} B_m \mathcal{Z}(\nu r) F_m , \qquad (6.9)
\end{aligned}$$

where the radial propagation constant is given by

$$\nu^2 = k_c^2 - p^2 , \qquad k_c^2 = \omega^2 \varepsilon\mu - j\mu\omega\sigma_c . \qquad (6.10)$$

Both ε and μ are in general complex quantities (cf. Appendix).

The radial function $\mathcal{Z}_m(\nu r)$ is an appropriately chosen Bessel function. For the internal region of the pipe, which contains the axis $r = 0$, it must be a Bessel function of the first kind $\mathcal{Z}_m(\nu r) = J_m(\nu r)$ in order for the fields to remain finite. In the outermost region stretching to infinity, it should be a Hankel function of the first kind $\mathcal{Z}_m(\nu r) = H_m^{(1)}(\nu r)$ so the fields tend towards zero for $r \to \infty$. The unknown coefficients A_m and B_m, as well as the longitudinal propagation constant p can be determined from the source terms which excite the *EM* fields, and from the conditions of continuity of the field components at the interfaces between regions.

For the case of a metallic pipe of inner radius b, we usually need to distinguish only two regions: $r < b$ (vacuum) and $r > b$ (metal). Except for a very thin wall (or very low frequencies), the outer radius of the pipe can be taken as infinite since the fields do not penetrate very far into the metal. The expansions Eqs. (6.8) and (6.9) are valid in both regions. In the inner region, where $k = k_0 = \omega/c$, the coefficients will be identified with the superscript i. In the outer region they are characterised by the superscript e.

The continuity condition for the tangential components [Jackson75] $\widetilde{E}_z$ and $\widetilde{E}_\theta$ at the surface of the metal wall $r = b$ give the following system of four equations for the four unknown coefficients A_m^i, B_m^i, A_m^e and B_m^e. Introducing the notations:

$$u = b\sqrt{k_0^2 - p^2} , \qquad v = b\sqrt{k_c^2 - p^2} \qquad (6.11)$$

for the arguments of the Bessel functions, we write for short J_m for $J_m(u)$ and H_m for $H_m^{(1)}(v)$ to get:

$$\frac{mp}{u^2} J_m A_m^i - \frac{j\mu_0\omega}{u} J'_m B_m^i - \frac{mp}{v^2} H_m A_m^e + \frac{j\mu\omega}{v} H'_m B_m^e = 0 ,$$

$$
\begin{aligned}
J_m A^i_m - H_m A^e_m &= 0 \,, \\
-\frac{jk_0^2}{\mu_0 \omega u} J'_m A^i_m - \frac{mp}{u^2} J_m B^i_m + \frac{jk_c^2}{\mu \omega v} H'_m A^e_m + \frac{mp}{v^2} H_m B^e_m &= 0 \,, \\
J_m B^i_m - H_m B^e_m &= 0 \,. \qquad (6.12)
\end{aligned}
$$

This constitutes a homogeneous system of linear equations for the coefficients. In order for non-trivial solutions to exist, its determinant must vanish, which yields an equation for the propagation constants p:

$$
J_m^2 \left\{ \left[\frac{\mu_0}{u} \frac{J'_m}{J_m} - \frac{\mu}{v} \frac{H'_m}{H_m} \right] \left[\frac{k_0^2}{\mu_0 u} \frac{J'_m}{J_m} - \frac{k_c^2}{\mu v} \frac{H'_m}{H_m} \right] - m^2 p^2 \left[\frac{1}{v^2} - \frac{1}{u^2} \right]^2 \right\} = 0 \,. \qquad (6.13)
$$

An infinite number of discrete values of possible propagation constants p_{mn} satisfy this equation where $n = 1, 2, \ldots, \infty$ is the number of its root. Eq. (6.13) cannot be solved analytically, and approximate solutions must be found by numerical methods.

In order to satisfy the continuity conditions at the surface of a metal with finite conductivity, in general a superposition of fields of both polarisations *TM* and *TE* is needed. The only exceptions are axially symmetric modes with $m = 0$, which can satisfy the continuity conditions across the boundary separately. For TM modes, the propagation constants are then given by the equation:

$$
\frac{k_0^2}{\mu_0 u} \frac{J'_0}{J_0} = \frac{k_c^2}{\mu v} \frac{H_0^{(1)\prime}}{H_0^{(1)}} \,, \qquad (6.14)
$$

while for TE modes one gets:

$$
\frac{\mu_0}{u} \frac{J'_0}{J_0} = \frac{\mu}{v} \frac{H_0^{(1)\prime}}{H_0^{(1)}} \,. \qquad (6.15)
$$

The inequality $\omega \ll \sigma_c/|\varepsilon|$ holds very well, in the frequency range of interest, for practically all metals used for vacuum chambers. For copper, e.g. , $\sigma_c/|\varepsilon| \approx 10^{17}$ Hz, which exceeds by far all frequencies excited even by the shortest bunches occurring in accelerators. Since the propagation constant p is smaller or of the same order of magnitude as ω, its inclusion does not change this estimate. Under these conditions, the radial propagation constant given in Eq. (6.10) is quite well approximated by the expression:

$$
\nu = \sqrt{-j\omega\mu\sigma_c} \,, \qquad \text{Im } \nu > 0 \,. \qquad (6.16)
$$

The sign of its imaginary part is chosen in such a way as to ensure that the fields are decreasing with r when penetrating into the metal for all wave numbers. The imaginary part Im ν is the inverse of the *skin depth* δ, which defines the effective length of high frequency field penetration into a metal.

Usually the skin depth is much smaller than the pipe radius b. In this case, Eq. (6.13) can be substantially simplified by replacing the Hankel functions with their asymptotic expressions [Abramowitz65]

$$H_m^{(1)}(v) \approx \sqrt{\frac{2}{\pi v}} \exp\left(jv - j\pi(\frac{m}{2} + \frac{1}{4})\right) \qquad \text{for} \quad |v| \gg 1 \; . \tag{6.17}$$

Consequently, $H_m^{(1)\prime}(v)/H_m^{(1)}(v) \approx j$, and Eq. (6.13) can be written:

$$J_m^2 \left\{ \left[\frac{\mu_0}{u}\frac{J_m'}{J_m} - \frac{j\mu}{v}\right]\left[\frac{k_0^2}{\mu_0 u}\frac{J_m'}{J_m} - \frac{jk_c^2}{\mu v}\right] - m^2p^2\left[\frac{1}{v^2} - \frac{1}{u^2}\right]^2 \right\} = 0 \; . \tag{6.18}$$

This equation is still too complicated to be solved analytically. But for the $m = 0$ modes, Eqs. (6.14) and (6.15) are now much simpler:

$$\begin{aligned} \frac{k_0^2}{\mu_0 u}\frac{J_1(u)}{J_0(u)} - \frac{jk_c^2}{\mu v} &= 0 \qquad \text{for TM fields} \; , \\ \frac{\mu_0}{u}\frac{J_1(u)}{J_0(u)} + \frac{j\mu}{v} &= 0 \qquad \text{for TE fields} \; , \end{aligned} \tag{6.19}$$

where $u = b\sqrt{k_0^2 - p^2}$ and $v = b\sqrt{k_c^2 - p^2}$ were given by Eq. (6.11), and are quite easily solved for p by computer or by approximate (perturbation) methods.

6.2.1 Infinite Conductivity

We first consider a pipe with infinite wall conductivity $\sigma_c \to \infty$ ($|v| \to \infty$). In this case Eq. (6.18) can be satisfied if in two ways:

$$J_m(u) = 0 \qquad \text{or} \qquad J_m'(u) = 0 \; . \tag{6.20}$$

Physically, the first of these conditions means simply that both tangential components of the electric (E_z and E_θ) and the normal component of magnetic (H_r) field must vanish on a perfectly conducting surface for the *TM* mode. Similarly, the second condition means the same for the relevant components

(E_θ and H_r) of the TE mode. Since in each case these two boundary conditions constitute a full set, each mode can be excited separately. Hence, in a smooth, perfectly conducting pipe, the two polarisations TM and TE do not mix for any m. Which one of them is excited in a particular case depends on the coupling to the external rf source.

Eqs. (6.20) are satisfied by the infinite number of roots of the Bessel functions $J_m(j_{mn}) = 0$ and $J'_m(j'_{mn}) = 0$, respectively[2]. Consequently, the propagation constants p^E of the TM modes, and p^M of the TE modes, have an infinite number of values:

$$\begin{aligned} p^E_{mn} &= \sqrt{\left(\frac{\omega}{c}\right)^2 - \left(\frac{j_{mn}}{b}\right)^2}\,, \\ p^M_{mn} &= \sqrt{\left(\frac{\omega}{c}\right)^2 - \left(\frac{j'_{mn}}{b}\right)^2}\,. \end{aligned} \tag{6.21}$$

For frequencies $\omega > cj_{mn}/b$ or $\omega > cj'_{mn}/b$ the propagation constant is real, and such harmonics propagate without damping. In the opposite case, p is purely imaginary and such harmonics are damped or *evanescent*.

When the propagation constants are found as roots of a determinant, only the ratio of the unknown coefficients is defined. The magnitude of these harmonic amplitudes can be determined if the power of the external rf source is known. Consider e.g. the case when an rf harmonic of frequency ω with power P is injected into the beam pipe. The Poynting vector $\mathbf{S} = \mathbf{E} \times \mathbf{H}^\star$ is real and has only a longitudinal component

$$S_z = E_r H_\theta^\star\,, \tag{6.22}$$

where the superscript $\star$ denotes the complex conjugate value. The power flow of the EM field can be evaluated as the integral of S_z over the cross section of the vacuum chamber:

$$P = \int_0^b r dr \int_0^{2\pi} d\theta S_z \tag{6.23}$$

In particular, for an axially symmetric TM field ($m = 0$), the integral yields for the n-th harmonic:

$$P_n^E = A_n^2 \frac{\omega k_0^2 b^4 J_1^2(j_{0n})}{2\mu_0 j_{0n}^2} p_n^E\,, \tag{6.24}$$

[2]For $m = 0$, sometimes the first index is left off for brevity.

from which the harmonic amplitude A_n can be determined.

Similarly, the power flow for the n-th harmonic of a TE field is:

$$P_n^M = B_n^2 \frac{\mu_0 b^4 J_1^2(j'_{0n})}{2\omega j'^2_n} p_n^M \ . \tag{6.25}$$

One can then also evaluate the stored energy of the EM field in a beam pipe of length L:

$$U = \int_0^L dz \int_S d\theta r dr \Big(\varepsilon|\mathbf{E}|^2 + \mu|\mathbf{H}|^2\Big) \ . \tag{6.26}$$

6.2.2 Finite Conductivity

The conductivity of most metals is very large but finite. Usually it is a function of temperature, and may become infinite at very low temperatures for DC if the metal becomes super-conducting. However, it always remains finite for AC. The most interesting frequencies in accelerators are those excited by particle bunches, and can reach quite high values for very short bunches. The shortest ones currently used are of the order of 1 mm, hence frequencies up to a few 100 GHz may be important. On the lower frequency end, particle oscillations occur at frequencies of the order of KHz in electron accelerators, and even lower in proton accelerators.

At these frequencies, the current flows mainly on the surface of highly-conducting metals. The characteristic depth of penetration is called the *skin depth* and is given by

$$\delta_s \equiv \sqrt{\frac{2}{\mu\sigma_c\omega}} \ . \tag{6.27}$$

In most cases it is smaller than the thickness of the vacuum chamber wall. The penetration of EM fields into the metal is characterised by the same quantity which is usually much smaller than the radius b of the beam pipe. The fields close to the surface therefore behave like plane waves. The tangential components of the electric and magnetic fields have to satisfy the boundary condition (cf. Appendix to Chapter 1):

$$\widetilde{\mathbf{E}}_t = \zeta \cdot \widetilde{\mathbf{H}}_t \times \mathbf{n} \ , \tag{6.28}$$

where $\mathbf{n}$ is the unit vector normal to the surface, pointing into the metal. The quantity

$$\zeta \equiv \sqrt{\frac{j\mu\omega}{\sigma_c}} = (1+j)\sqrt{\frac{\mu\omega}{2\sigma_c}} \tag{6.29}$$

is called the (complex) *surface impedance* of a metal. Since the tangential field components must be continuous at the surface, the same condition Eq. (6.28) must also hold for the field components in the vacuum region next to the surface [Leontovich48].

The fact that the metal conductivity is very large allows us to calculate the propagation constants using perturbative methods. One possible approach is to find correction terms due to the finite conductivity to the solutions for infinite conductivity. We call the propagation constant for perfect conductivity p_0 and put $p = p_0 + \delta p$, and make the assumption that $\delta p \ll p_0$.

For an axially symmetric (m=0) TM mode, one gets from Eq. (6.11) approximately $u \approx j_{0n} - p_0 \delta p^E b^2 / j_{0n}$, and hence $J_0(u) \approx J_1(j_{0n}) p_0 \delta p^E b^2 / j_{0n}$. Substituting these approximations into the first of Eqs. (6.19), the correction term due to finite resistivity to the n-th propagation constant becomes:

$$\delta p_n^E = \frac{\omega \varepsilon_0 \zeta}{p_n^E b} \tag{6.30}$$

A similar calculation can be done for an axially symmetric (m=0) TE mode. In this case, the correction term to the propagation constant is found from the second of Eqs. (6.19):

$$\delta p_n^M = \frac{j'^2_n \zeta}{\mu_0 \omega p_n^M b^3 [1 - J_2(j'_{0n}) / J_0(j'_{0n})]} \, . \tag{6.31}$$

The real and imaginary parts of the correction terms are equal and positive. The conductivity appears in the denominators of Eqs. (6.30) and (6.31), which justifies our original assumption of small corrections. The (small) imaginary part of the correction term becomes the *damping decrement* of the waves in the beam pipe due to the energy losses in the wall. The (small) real part increases the propagation constant, and thereby decreases the phase velocity slightly.

The same result can be obtained by considering the energy flow into the metal wall [Landau60]. The damping decrement of the field amplitude is the ratio of this quantity to the energy flow through the pipe given in Eq. (6.24).

6.3 *EM* Fields Excited by a Charge

We turn now to the evaluation of the wake function due to the finite resistivity of the wall. We consider a bunch of charge q with rotationally symmetric distribution ρ and with infinitesimal extent in the longitudinal direction, moving on the axis of a pipe of radius b nearly with the speed of light[3]. The charge and the current densities of such a bunch are:

$$\rho(\mathbf{r}) = \rho(r)\delta(z - ct) \ , \qquad \mathbf{J}(\mathbf{r}) = \hat{\mathbf{z}} J_z \ , \tag{6.32}$$

where $\hat{\mathbf{z}}$ is the unit vector in the z direction, and $J_z(\mathbf{r}) = c\rho(\mathbf{r})$. The transverse distribution function $\rho(r)$ has the following normalisation:

$$\int r dr \rho(r) = \frac{q}{2\pi} \ . \tag{6.33}$$

The Fourier transforms of ρ and J_z are

$$\widetilde{\rho}(r, z) = \frac{\rho(r)}{c} \exp(-jk_0 z) \ , \qquad \widetilde{J}_z(r, z) = c\widetilde{\rho}(r) \tag{6.34}$$

where the propagation constant $k_0 = \omega/c$.

The field excited by such a bunch is also axially symmetric, i.e. it has only harmonics with azimuthal mode number $m = 0$. Only the components $\widetilde{E}_r$, $\widetilde{E}_z$ and $\widetilde{H}_\theta$ are coupled to the bunch and are excited in the pipe. They are all independent of the azimuth θ. The only non-trivial equations of the system of Eqs. (6.2)-(6.5) for these components are:

$$\begin{aligned}
\frac{\partial \widetilde{E}_r}{\partial z} - \frac{\partial \widetilde{E}_z}{\partial r} + j\omega\mu_0 \widetilde{H}_\theta &= 0 \ , \\
\frac{\partial \widetilde{H}_\theta}{\partial z} + j\omega\varepsilon_0 \widetilde{E}_r &= 0 \ , \\
\frac{1}{r}\frac{\partial}{\partial r}\left(r\widetilde{H}_\theta\right) - j\omega\varepsilon_0 \widetilde{E}_z &= \tilde{J}_z \ , \\
\frac{1}{r}\frac{\partial}{\partial r}\left(r\widetilde{E}_r\right) + \frac{\partial \widetilde{E}_z}{\partial z} &= \frac{\tilde{\rho}}{\varepsilon} \ .
\end{aligned} \tag{6.35}$$

For azimuthally *non-symmetric fields* which are proportional to $\exp(jm\theta)$ with $m \neq 0$, all six *EM* field components may be different from zero and have to be be calculated from the full Maxwell equations, cf. Eqs. (6.66).

[3]The general case for arbitrary charge velocity and walls consisting of many concentric layers will be treated in the Appendix.

6.3.1 *EM* Fields in Vacuum

Taking into account the dependence of the charge density on the coordinates, we search for a solution in the following form:

$$\begin{aligned} \widetilde{E}_z(r,z) &= F_z \exp(-jk_0 z) \ , \\ \widetilde{E}_r(r,z) &= F_r(r) \exp(-jk_0 z) \ , \\ \widetilde{H}_\theta(r,z) &= F_\theta(r) \exp(-jk_0 z) \ , \end{aligned} \tag{6.36}$$

where F_z is constant. Since all field components have the factor $\exp(-jk_0 z)$, we shall omit this factor in the following.

Direct integration of Eqs. (6.35) yields the following solution for the amplitudes F_r and F_θ in terms of the constant $\widetilde{E}_z = F_z$ and the charge density $\tilde{\rho} = \tilde{J}_z/c$:

$$F_\theta(r) = j\frac{\omega\varepsilon_0}{2} r F_z + \frac{1}{r}\int r dr \rho(r) \ , \tag{6.37}$$

while $F_r(r) = F_\theta(r)/c\varepsilon_0$. To proceed we need to specify the charge density distribution. As an example we consider a bunch with a uniform density[4] for $r < a$ and zero for $r > a$.

We distinguish two regions: (1) $0 < r < a$ occupied by the bunch, and (2) $a < r < b$ outside the bunch but inside the beam pipe. The transverse components of the *EM* fields excited by the bunch then are:

$$\widetilde{E}_r(r) = \begin{cases} \dfrac{jkF_z r}{2} + \dfrac{qr}{2\pi a^2 \varepsilon_0 c} & \text{for} \quad 0 < r < a \ , \\ \dfrac{jkF_z r}{2} + \dfrac{q}{2\pi r \varepsilon_0 c} & \text{for} \quad a < r < b \ , \end{cases} \tag{6.38}$$

$$\widetilde{H}_\theta(r) = \begin{cases} \dfrac{j\omega\varepsilon_0 F_z r}{2} + \dfrac{qr}{2\pi a^2} & \text{for} \quad 0 < r < a \ , \\ \dfrac{j\omega\varepsilon_0 F_z r}{2} + \dfrac{q}{2\pi r} & \text{for} \quad a < r < b \ . \end{cases} \tag{6.39}$$

The constant F_z will be determined below. These fields remain finite at $r = 0$.

The second (space charge) term in each of this equations is the same as in free space. The first term (proportional to the constant F_z) appears due

[4] A solution for a different charge distribution can be found in ref.[Chao93].

to the presence of the beam pipe. It vanishes when $b \to \infty$, see below Eq. (6.48). The radial component of the Lorentz force $\widetilde{E}_r - Z_0\widetilde{H}_\theta$ is zero in the approximation used (when $\gamma \to \infty$). For relative velocities which are smaller then 1, the Lorentz force is not zero, but is decreased by a factor γ^{-2}, which is small for high-energy accelerators. Such *space charge force compensation* in a pipe holds only for the monopole $m = 0$ mode.

6.3.2 *EM* Fields inside the Metal Wall

Next we want to find the field inside the metal wall for $r > b$. For this we need to solve Eqs. (6.35), substituting $\tilde{\rho} = \tilde{J}_z = 0$, since there are no free charges or currents inside the metal, and using the effective *complex permittivity* of the metal instead of ε_0

$$\varepsilon_c = \varepsilon - \frac{j\sigma_c}{\omega}. \tag{6.40}$$

We search again field components of the form $\widetilde{E}_z(r,z) = G_z(r)$, $\widetilde{E}_r(r,z) = G_r(r)$, and $\widetilde{H}_\theta(r,z) = G_\theta(r)$. Note that G_z is no longer constant as the field should go to zero at $r \to \infty$. The system of equation for the functions $G_{z,r,\theta}$ is:

$$\begin{aligned}
\frac{\partial G_z}{\partial r} + jk_0 G_r - j\omega\mu G_\theta &= 0 \,, \\
G_\theta + c\varepsilon_c G_r &= 0 \,, \\
\frac{1}{r}\frac{\partial}{\partial r}(rG_\theta) - j\omega\varepsilon_c G_z &= 0 \,, \\
\frac{1}{r}\frac{\partial}{\partial r}(rG_r) - jk_0 G_z &= 0 \,.
\end{aligned} \tag{6.41}$$

In terms of the Hankel function of the first kind and order zero $H_0^{(1)}(x)$, which vanishes for $x \to \infty$, the solutions can be written $G_z(r) = \alpha H_0^{(1)}(\nu r)$ where α is a proportionality factor, and

$$\begin{aligned}
G_r(r) &= -\frac{jk_0}{\nu^2}\frac{\partial G_z}{\partial r} \,, \\
G_\theta(r) &= -\frac{j\omega\varepsilon_c}{\nu^2}\frac{\partial G_z}{\partial r} \,.
\end{aligned} \tag{6.42}$$

The propagation constant ν is defined by

$$\nu^2 = \omega^2(\varepsilon_c \mu - \varepsilon_0 \mu_0) \; . \tag{6.43}$$

Using the expressions for the electric field in the metal wall, one can find the energy loss per unit length, dissipated as heat:

$$\frac{d\mathcal{E}}{dz} = -\sigma_c \int_V dV \mathbf{E}^2(\mathbf{r}, t) \; . \tag{6.44}$$

The integration here should be performed over the total volume V of the metal wall. Using the *Parceval theorem* of the Fourier analysis, which states

$$\int_{-\infty}^{\infty} dz \; \mathbf{E}^2(r, z) = \int_{-\infty}^{\infty} dk_0 \; |\widetilde{\mathbf{E}}(r)|^2 \; , \tag{6.45}$$

one finds

$$\frac{d\mathcal{E}}{dz} = -\sigma_c \int_b^{\infty} r dr \int_{-\infty}^{\infty} dk_0 |\widetilde{\mathbf{E}}(r)|^2 \; . \tag{6.46}$$

Calculation for the case $\nu b \gg 1$ gives [Chao93]:

$$\frac{d\mathcal{E}}{dz} = -\frac{q^2}{2\pi\varepsilon_0 b^2} \; . \tag{6.47}$$

The two remaining unknown constants α, defined before Eq. (6.42), and F_z in Eq. (6.36), can be determined by matching the *EM* field at the boundary $r = b$ between the metal and the vacuum regions. Since $\widetilde{E}_r$ is discontinuous there, due to induced charges on the metal surface, one should choose the components $\widetilde{E}_z$ and $\widetilde{H}_\theta$. This yields $\alpha = F_z / H_0^{(1)}(\nu b)$ with

$$F_z = \frac{jZ_0 q}{\pi b^2 k_0} \left[1 - \frac{2\nu\varepsilon_c H_1^{(1)}(\nu b)}{b k_0^2 (\varepsilon_c - \varepsilon_0) H_0^{(1)}(\nu b)} \right]^{-1} \; . \tag{6.48}$$

6.4 Finite Wall Conductivity: Longitudinal

Now we turn to the calculation of the longitudinal *coupling impedance* and *wake function* for a uniform cylindrical pipe with finite wall resistivity.

6.4.1 Longitudinal Impedance

The constant F_z given above is related to the longitudinal impedance per length L. From the definition

$$\frac{Z(\omega)}{L} = -\frac{1}{|\tilde{I}|L}\int_0^L dz\ \tilde{E}_z(z)\ e^{jk_0 z} = -\frac{F_z}{q} \tag{6.49}$$

one finds the following expression:

$$\frac{Z(\omega)}{L} = -\frac{jZ_0}{\pi k_0 b^2}\left[1 - \frac{2(\nu b)}{(k_0 b)^2}\ \frac{\varepsilon_c}{\varepsilon_c - \varepsilon_0}\ \frac{H_1^{(1)}(\nu b)}{H_0^{(1)}(\nu b)}\right]^{-1}, \tag{6.50}$$

where $\varepsilon_c \neq \varepsilon_0$ is the permittivity of the metal, and $Z_0 \equiv \sqrt{\mu_0/\varepsilon_0} = 377$ Ohm the free-space impedance.

For the high conductivity of most metals used for vacuum chambers, and for frequencies relevant to bunch lengths in accelerators and storage rings, the radial propagation constant ν is very well approximated by Eq. (6.16). In this case $\nu b \gg 1$, and expression Eq. (6.50) can be substantially simplified by replacing the Hankel functions by their asymptotic values:$H_1^{(1)}(\nu b)/H_0^{(1)}(\nu b) \approx -j[1 - j/2\nu b]$. With $\varepsilon_c/\varepsilon_0 \approx \nu^2/k_0^2$, we can write it [Chao82] as:

$$\frac{Z(k)}{L} = -\frac{Z_0}{2\pi b}\left[\left(\frac{\nu}{k_0} + \frac{k_0}{\nu}\right)\left(1 - \frac{j}{2\nu b}\right) + j\frac{k_0 b}{2}\right]^{-1}, \tag{6.51}$$

The term $j/(2\nu b)$ is a small correction and usually can be neglected. At not too high frequencies, and for good metal conductivity, the term k_0/ν is also small compared to ν/k_0. Under these conditions the expression for the resistive wall impedance simplifies:

$$\frac{Z(k)}{L} = -\frac{Z_0}{2\pi b}\left[\frac{\nu}{k_0} + j\frac{k_0 b}{2}\right]^{-1}, \tag{6.52}$$

As we will see shortly, the behaviour of the longitudinal wake is quite different at small and large distances s behind the bunch. We define a *characteristic distance* s_c by the following expression [Bane95]:

$$s_c \equiv \left(\frac{2cb^2\varepsilon_0}{\sigma_c}\right)^{1/3}. \tag{6.53}$$

(Note, that we use MKS system, while in the ref.[Bane95] the CGS system had been used). For most cases, the characteristic distance is quite small: e.g. for a copper pipe with radius $b = 10^{-2}$ m and $\sigma_c = 6.4 \cdot 10^7 \mathrm{Sm}^{-1}$, it is $s_c = 20\mu$m.

Introducing dimensionless wave number $\kappa \equiv k_0 s_c$ Eq. (6.52) becomes:

$$\frac{Z(\kappa)}{L} = \frac{Z_0 s_c}{2\pi b^2} \left[\frac{1 - j\mathrm{sgn}(\kappa)}{|\kappa|^{1/2}} + j\frac{\kappa}{2} \right]^{-1} . \tag{6.54}$$

6.4.2 Longitudinal Wake

The Fourier transform of the impedance

$$G_{\parallel}(s) = \frac{c}{2\pi s_c} \int_{-\infty}^{\infty} d\kappa Z(\kappa) \exp(j\kappa \frac{s}{s_c}) \tag{6.55}$$

gives the wake function $G_{\parallel}(s)$ at the distance s behind the bunch. The integral in this equation can be evaluated by analytic extension of the integrand into the complex plane κ, and changing the integration path to a closed one by a circle of infinitely large radius where the integrand tends to zero. For $s < 0$ (i.e. in front of the bunch) the contour must be closed in the lower half plane of κ. There $Z(\kappa)$ is an analytic function without poles, and thus the wake field is zero. This is a consequence of the fact that the bunch moves with the speed of light and the fields can not overtake the bunch.

For $s > 0$ (i.e. behind the bunch), the contour must be closed in the upper half plane of κ. There $Z(k)$ has poles $kb = \pm\sqrt{3} + j$ and a branch cut along the positive half of the imaginary axis, which is due to the fact that ν is a multi-valued function containing a square root of k.

6.4.3 The Long-Range Wake

First we consider distances s which are large compared to s_c. In the frequency domain this corresponds to low frequencies, and therefore the first term is dominant in Eq. (6.54). Then the result of the integration is

$$G_{\parallel}(s) = -\frac{1}{4\pi\varepsilon_0\sqrt{2\pi}b}(\frac{s_c}{s})^{3/2} \quad \text{for} \quad s > s_c . \tag{6.56}$$

The remarkable feature of this formula is that the wake accelerates a particle moving behind a bunch. If Eq. (6.56) were applicable in the limit $s \to 0$, the bunch would accelerate itself, which is clearly a non-physical conclusion. This contradiction is eliminated by the restriction $s > s_c$ which was used in deriving the result Eq. (6.56).

6.4.4 The Short Range Wake

To calculate the fields at smaller distances behind the bunch, we need to evaluate the integral Eq. (6.55) retaining all terms in its integrand. The results of such calculations can be found in refs.[Bane91, Henke90, Napoly93]. Our description here is based on ref. [Bane95], from which we reproduce here the main result. Integration of Eq. (6.55) gives the following result:

$$G_{\|}(s) = -\frac{4}{\pi\varepsilon_0 b}\left[\frac{1}{3}\mathrm{e}^{-\frac{s}{s_c}}\cos(\sqrt{3}\frac{s}{s_c}) - \frac{\sqrt{2}}{\pi}\int_0^\infty dx \frac{x^2}{x^6+8}\mathrm{e}^{-x^2\frac{s}{s_c}}\right] . \tag{6.57}$$

The first term is the sum of the residues at the poles $\kappa = \pm\sqrt{3} + j$. It resembles the field in a damped resonator with resonant frequency $\sqrt{3}c/2\pi s_c$ and quality factor $Q_r = \sqrt{3}/2$. The second term, which is positive definite, is the contribution of the cut along the positive half of the imaginary axis (the Riemann sheet $-\pi/2 < arg(\kappa) < 3\pi/2$ was chosen in this calculation). Physically it represents the field whose energy is dissipated in the lossy wall. At larger distances, $s/s_c \geq 4$, it dominates and approaches the long-range wake field Eq. (6.56).

The self-field $E_z(0) = -q/\pi\varepsilon_0 b^2$, which follows from Eq. (6.57), is exactly twice the value obtained from energy conservation considerations, Eq. (6.47). This is a manifestation of the general result referred to as *fundamental theorem of beam loading* [Wilson87]. It agrees with the mathematical convention that the value of a discontinuous function is obtained as half the sum of the limits from the two sides. When the function jumps from zero to a finite value, one thus has to take only half of it.

Both the longitudinal wake function, as well as its asymptotic approximation, are shown in Fig. 6.1 [Napoly93].

The derivation presented here is based on the assumption that the conductivity σ_c of the metal is independent of frequency. In general, this assumption breaks down at frequencies comparable with or larger than the inverse relaxation time τ of electrons in metal [Ashcroft76]. For good conductors there exists a range of frequencies which satisfies this condition, but where the anomalous skin effect does not yet play a role. For example, for copper $\tau \approx 2.7 \cdot 10^{-14}\mathrm{s}^{-1}$; therefore, at frequencies above 10 THz, which are important for bunch lengths smaller than $\approx 8\mu$m, the conductivity can no longer be assumed to be constant. However, such extremely short bunches are not (yet) used in accelerators. Even linear colliders which are presently designed have bunch lengths down to only about 0.1 mm.

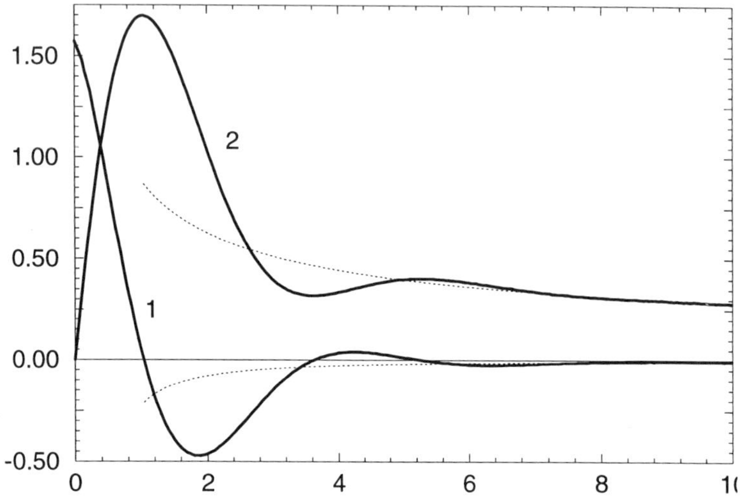

Figure 6.1: Longitudinal (1) and transverse (2) wake functions for circular cylindrical walls with finite resistivity, and their asymptotic approximations (dashed).

6.4.5 Impedance and Wake of a Ceramic Pipe

Here we consider the coupling impedance of a ceramic pipe with thick metal outside coating. To simplify the calculation and to separate effects of the ceramic from those of the metal coating, we will assume that it is perfectly conducting.

The radii of the vacuum channel will be called b, and that of the metal coating d. The solutions of the Maxwell equations for the vacuum region are the same as before, cf. Eqs. (6.38) and (6.39). To find the solutions for the region $b < r < d$, Eqs. (6.41) should be solved with the it permittivity of the ceramic given by:

$$\varepsilon = \varepsilon_0 n^2 \,, \tag{6.58}$$

where n is the refraction index of the ceramic material. Eqs. (6.42) are applicable in this case, with $E_z(d) = 0$ in order to satisfy the boundary condition at $r = d$. The function $G_z(r)$ should be chosen in the following way:

$$G_z(r) = \alpha \left[H_0^{(1)}(\nu r) H_0^{(2)}(\nu d) - H_0^{(2)}(\nu r) H_0^{(1)}(\nu d)\right] . \tag{6.59}$$

Here $H_0^{(1,2)}$ are the Hankel functions of the first and second kind and of order zero. The parameter ν is now given by

$$\nu = k\sqrt{n^2 - 1} \tag{6.60}$$

in accordance with Eq. (6.58). As before, the constants α and F_z are defined by the continuity condition at the ceramic surface $r = b$. For the constant F_z we get:

$$F_z = \frac{jq}{\pi b^2 \omega \varepsilon_0}[1 - \frac{2n^2}{kb\sqrt{n^2-1}} \frac{\mathcal{L}_1(b,d)}{\mathcal{L}_0(b,d)}]^{-1} , \tag{6.61}$$

where the following notations have been introduced:

$$\begin{aligned} \mathcal{L}_1(b,d) &\equiv H_1^{(1)}(\nu b)H_0^{(2)}(\nu d) - H_1^{(2)}(\nu b)H_0^{(1)}(\nu d) \\ \mathcal{L}_0(b,d) &\equiv H_0^{(1)}(\nu b)H_0^{(2)}(\nu d) - H_0^{(2)}(\nu b)H_0^{(1)}(\nu d) . \end{aligned} \tag{6.62}$$

The constant F_z is again proportional to the longitudinal coupling impedance. For the range of frequencies which satisfies the condition $\nu b \gg 1$ (which means that the wave length of the field in ceramic is much greater then the radius of the vacuum channel) we can use the asymptotic expressions for the Hankel functions as before. With $n' = \sqrt{n^2 - 1}$, we get the following result for the longitudinal impedance per unit length:

$$\frac{Z(k_0)}{L} = \frac{jZ_0}{\pi k_0 b^2}\left\{k_0 b n' - 2n^2 \cot[k_0 n'(d-b)]\right\}^{-1} . \tag{6.63}$$

The longitudinal wake function can be obtained from the expression for the impedance by applying an inverse Fourier transformation. The analytic properties of $Z(k)$ in the complex plane k then determine the result of the integration. The expression Eq. (6.63) has an infinite number of simple poles on the real axis. When the integration contour is closed in the lower half of the complex plane k, valid in front of the bunch, the result is therefore zero. That means again that there is no field ahead of a bunch moving with the speed of light.

On the other hand, for $s > 0$, i.e., behind the bunch, the integration contour must be closed in the upper half plane. A cut along the positive part of the imaginary axis and the poles then contribute to the integral. Assuming that the thickness of the dielectric satisfies the condition $d - b \gg \pi b/2n^2$, the integration produces the following expression for the wake function in a dielectric channel [Burov90]:

$$G_{\|}(s) = \frac{1}{\pi\varepsilon_0 b}\left[\exp\left(-\frac{s}{s_d}\right) - \frac{C_1}{2n^2}\right] . \tag{6.64}$$

This formula gives the wake potential behind a bunch at distance s. The effective length of the wake behind the bunch is:

$$s_d = \frac{b\sqrt{n^2-1}}{2n^2} . \tag{6.65}$$

The constant $C_1 = 0.37$ has been obtained by numerical integration.

6.5 Finite Wall Conductivity: Transverse

To obtain the transverse impedance and wake functions one has to perform calculations similar to those for the longitudinal ones but for azimuthal modes with the numbers $m \neq 0$.

6.5.1 Transverse *EM* Fields in Vacuum

We again distinguish two regions: $0 < r < a$ for the beam, and $a < r < b$ for the vacuum. In both regions the *EM* fields must be determined in such a way that the tangential electric, and normal magnetic, field components E_z, E_θ and H_r are continuous at $r = a$. Since in vacuum $\omega^2\varepsilon_0\mu_0 = k_0^2$, from the Maxwell equations follows:

$$\begin{aligned}
\frac{\partial \widetilde{E}_{zm}}{\partial r} &= +\frac{jmZ_0}{r}\widetilde{H}_{zm} , \\
\frac{\partial Z_0\widetilde{H}_{zm}}{\partial r} &= -\frac{jm}{r}\widetilde{E}_{zm} , \\
\frac{1}{r}\frac{\partial}{\partial r}\left(r\widetilde{E}_{rm}\right) + \frac{jmZ_0}{r}\widetilde{H}_{rm} &= \frac{q}{2\pi\varepsilon_0} + \left(jk_0 + \frac{m^2}{k_0 r^2}\right)\widetilde{E}_{zm} , \\
\frac{1}{r}\frac{\partial}{\partial r}\left(rZ_0\widetilde{H}_{rm}\right) - \frac{jm}{r}\widetilde{E}_{rm} &= +\left(jk_0 + \frac{m^2}{k_0 r^2}\right)Z_0\widetilde{H}_{zm} , \\
\widetilde{E}_{\theta m} &= -Z_0\widetilde{H}_r + \frac{m}{k_0 r}\widetilde{E}_{zm} ,
\end{aligned}$$

$$Z_0\widetilde{H}_{\theta m} = \widetilde{E}_r + \frac{m}{k_0 r} Z_0 \widetilde{H}_{zm} \ . \tag{6.66}$$

The first two equations have the following solutions

$$\widetilde{E}_{zm} = A r^m \ , \qquad Z_0\widetilde{H}_{zm} = -j\widetilde{E}_{zm} \ , \tag{6.67}$$

where A is a constant to be determined. Using these expressions the other field components can be determined.

Eqs. (6.66) allow to find the radial field components. Since $\widetilde{E}_{zm}$ and $\widetilde{H}_{zm}$ are proportional to r^m, the radial components $\widetilde{E}_{rm}$ and $\widetilde{H}_{rm}$ can be expressed as polynomials with terms proportional to r^{m+1}, r^{m-1} and r^{-m-1}. Keeping in mind the conditions at $r = 0$, we restrict the terms proportional to ar^{-m} to the region $r > a$ by using a step-function $\Theta(x)$, which equals unity for $x > 0$ and is zero for $x < 0$. Substituting the polynomials defined above into Eqs. (6.66) and comparing coefficients of terms of the same order, we find:

$$\begin{aligned}
\widetilde{E}_{rm} &= -jA\frac{k_0}{2(m+1)} r^{m+1} + \left[B + jA\frac{m}{2k_0}\right] r^{m-1} \\
&+ \frac{q}{4\pi\varepsilon_0}\left[\left(\frac{r}{a}\right)^{-m}\frac{\Theta(r-a)}{r} - \left(\frac{r}{a}\right)^{m}\frac{\Theta(a-r)}{r}\right] , \\
Z_0\widetilde{H}_{rm} &= A\frac{k_0}{2(m+1)} r^{m+1} + \left[A\frac{m}{2k_0} + jB\right] r^{m-1} \\
&- \frac{jq}{4\pi\varepsilon_0}\left[\left(\frac{r}{a}\right)^{-m}\frac{\Theta(r-a)}{r} + \left(\frac{r}{a}\right)^{m}\frac{\Theta(a-r)}{r}\right] ,
\end{aligned} \tag{6.68}$$

where B is a second constant to be determined. At the beam surface $r = a$, $\widetilde{H}_{rm}$ is continuous while $\widetilde{E}_{rm}$ is not. Similarly, Eqs. (6.66) define the θ-components of the fields:

$$\begin{aligned}
\widetilde{E}_{\theta} &= -A\frac{k_0}{2(m+1)} r^{m+1} - j\left[B + A\frac{m}{k_0}\right] r^{m-1} \\
&- \frac{jq}{4\pi\varepsilon_0}\left[\left(\frac{r}{a}\right)^{-m}\frac{\Theta(r-a)}{r} + \left(\frac{r}{a}\right)^{m}\frac{\Theta(a-r)}{r}\right] , \\
Z_0\widetilde{H}_{\theta m} &= jA\frac{k_0}{2(m+1)} r^{m+1} + \left[B - jA\frac{m}{2k_0}\right] r^{m-1} \\
&+ \frac{q}{4\pi\varepsilon_0}\left[\left(\frac{r}{a}\right)^{-m}\frac{\Theta(r-a)}{r} - \left(\frac{r}{a}\right)^{m}\frac{\Theta(a-r)}{r}\right] ,
\end{aligned} \tag{6.69}$$

The unknown coefficients A and B can be found from the continuity conditions at the metal surface $r = b$.

6.5.2 Transverse *EM* Field inside Metal

Now we return to Eqs. (6.2)-(6.5), in which we put $\tilde{\rho}_m = \tilde{J}_{zm} = 0$ since there are no free charges or currents inside the metal. To obtain equations for $\widetilde{E}_z$ and $\widetilde{H}_z$, one has to eliminate all other field components:

$$\begin{aligned} \frac{1}{r}\frac{\partial}{\partial r}\left(r\frac{\partial \widetilde{E}_{zm}}{\partial r}\right) + \left(\nu^2 - \frac{m^2}{r^2}\right)\widetilde{E}_{zm} &= 0 \,, \\ \frac{1}{r}\frac{\partial}{\partial r}\left(r\frac{\partial \widetilde{H}_{zm}}{\partial r}\right) + \left(\nu^2 - \frac{m^2}{r^2}\right)\widetilde{H}_{zm} &= 0 \,. \end{aligned} \tag{6.70}$$

The radial wave number ν is defined as follows:

$$\nu^2 = \omega^2(\varepsilon\mu - \varepsilon_0\mu_0) - j\omega\mu\sigma_c \,. \tag{6.71}$$

Assuming a dependence of all field components $\propto \exp(-jkz)$, the transverse ones can be expressed in terms of the longitudinal ones, using the Maxwell equations for the m-th multipolar Fourier transforms given in Eqs. (6.2)-(6.5):

$$\begin{aligned} \widetilde{E}_{rm} &= -\frac{jk_0}{\nu^2}\frac{\partial \widetilde{E}_{zm}}{\partial r} - \frac{m\omega\mu}{\nu^2 r}\widetilde{H}_{zm} \,, \\ \widetilde{H}_{rm} &= -\frac{jk_0}{\nu^2}\frac{\partial \widetilde{H}_{zm}}{\partial r} + \frac{m(\omega\varepsilon - j\sigma_c)}{\nu^2 r}\widetilde{E}_{zm} \,, \\ \widetilde{E}_{\theta m} &= \frac{j\omega\mu}{\nu^2}\frac{\partial \widetilde{H}_{zm}}{\partial r} - \frac{mk_0}{\nu^2 r}\widetilde{E}_{zm} \,, \\ \widetilde{H}_{\theta m} &= -\frac{j(\omega\varepsilon - j\sigma_c)}{\nu^2}\frac{\partial \widetilde{E}_{zm}}{\partial r} - \frac{mk_0}{\nu^2 r}\widetilde{H}_{zm} \,. \end{aligned} \tag{6.72}$$

The radial wave number ν for metals is very well approximated by Eq. (6.16). In general, the solutions of Eqs. (6.70) can be expressed in terms of the two kinds of Hankel functions $H_m^{(1)}(\nu r)$ and $H_m^{(2)}(\nu r)$ of order m. If we are mainly interested in the usual case that the skin depth is much smaller than the pipe radius b, the solutions which are continuous at $r = b$ with the vacuum fields Eqs. (6.67), can be approximated for $\nu b \gg 1$ by:

$$\widetilde{E}_{zm} = Ab^m \mathrm{e}^{-j\nu(r-b)} \,, \qquad Z_0\widetilde{H}_{zm} = -jAb^m \mathrm{e}^{-j\nu(r-b)} \,. \tag{6.73}$$

Retaining only the main terms of the order νb, neglecting $\omega\varepsilon$ compared to σ_c, the transverse field components become:

$$
\begin{aligned}
\widetilde{E}_{rm} &= -\frac{k_0}{\nu}\widetilde{E}_{zm} \, , \\
\widetilde{H}_{rm} &= -\frac{k_0}{\nu}\widetilde{H}_{zm} \, , \\
\widetilde{E}_{\theta m} &= \frac{\omega\mu}{\nu}\widetilde{H}_{zm} \, , \\
\widetilde{H}_{\theta m} &= \frac{j\sigma_c}{\nu}\widetilde{E}_{zm} \, .
\end{aligned}
\tag{6.74}
$$

Now we can find the unknown constants A and B by equating the azimuthal field components $\widetilde{E}_{\theta m}$ and $\widetilde{H}_{\theta m}$ on both sides of the boundary $r = b$.

Retaining only terms of lowest order in $1/\sigma_c$, assuming $\sigma_c/\omega\varepsilon_0 \gg 1$, the result can be written

$$
\begin{aligned}
A &= \frac{qa^m}{2\pi c b^{2m+1}} \left[\frac{Z_0\sigma_c}{\nu} - \frac{jk_0 b}{m+1} + \frac{jm}{k_0 b} \right]^{-1} , \\
B &= -A\frac{b^m}{2} \left[\frac{k_0 b}{m+1} + \frac{jZ_0\sigma_c}{\nu} \right] ,
\end{aligned}
\tag{6.75}
$$

The quantity qa^m is the m-th moment of the charge distribution. For a thick wall with $\nu(d-b) \gg 1$, the same result can be obtained simpler by using the Leontovich boundary condition Eq. (6.28).

Unlike the case of the monopole field, for $m > 0$ the transverse components of the Lorentz force $\widetilde{\mathbf{E}} - Z_0\widetilde{\mathbf{H}}$ in a resistive beam pipe does not vanish even when $\gamma \to \infty$. As one can see from Eqs. (6.68) and (6.69), it is proportional to constant A and remains finite even for $\gamma \to \infty$.

6.5.3 Transverse Impedance and Wake Function

The transverse impedance of a pipe with finite wall resistivity can be calculated using its definition Eq. (4.8). For a particle moving with the speed of light, this formula gives for the two components of the transverse impedance:

$$
\begin{aligned}
Z_{\perp rm}(\omega) &= -j\frac{2\pi R}{qa} \left[\widetilde{E}_{rm} - Z_0\widetilde{H}_{\theta m} \right] , \\
Z_{\perp \theta m}(\omega) &= \frac{j}{qa} \left[\widetilde{E}_{\theta m} + Z_0\widetilde{H}_{rm} \right] .
\end{aligned}
\tag{6.76}
$$

In this case, all field components - and hence also the radial coupling impedance $Z_{rm}(\omega)$ - are proportional to the constant A. To lowest order in $\omega\varepsilon_0/\sigma_c$, the transverse (radial) impedance is then given by the following formula:

$$Z_{\perp m}(\omega) = \frac{m r_b^{m-1} a^{m-1}}{2\pi b^{2m+1}} \frac{\nu}{k_0 \sigma_c} . \tag{6.77}$$

Including also the space-charge term, the transverse dipole impedance becomes:

$$Z_{\perp 1}(\omega) = \frac{R Z_0}{b^2} \left[\frac{-j}{\beta^2 \gamma^2} + (1+j) \frac{\mu_r \delta}{b} \right] . \tag{6.78}$$

The radial component of the wake function, often simply called *transverse wake function*, can be obtained by a Fourier transformation of the radial impedance. Using the approximate equation derived above, this yields the asymptotic expression for large distances $s \gg s_c$ behind the charge [Chao93]:

$$G_{rm}(s) = \frac{m a^m r_b^{m-1}}{\pi b^{2m+1}} \sqrt{\frac{c}{\pi \varepsilon_0 \sigma_c s}} \cos m\theta . \tag{6.79}$$

The azimuthal wake function is given by the same formula, with the cosine replaced by $-\sin m\theta$.

The transverse wake function for a resistive wall at any distance s behind the charge has also been calculated [Henry91] and is shown in Fig. 6.1, together with the longitudinal one, and their asymptotic approximations.

Appendix 6.A Multi-layer Walls

There exists a general algorithm to evaluate the coupling impedance of a straight, uniform beam in a concentric, circular-cylindrical vacuum chamber, whose walls may consist of a number of layers of different materials [Zotter72b]. Several geometries of practical interest can be treated in this manner.

The simplest case is that of a metallic vacuum chamber wall of finite thickness. The theory of the resistive wall impedance [Laslett65] is necessary to explain the experimentally observed *resistive wall instability*[5]. The original theory was made with the assumption that the wall thickness was much greater than the skin depth. However, in particular for large accelerators with low revolution frequencies, the skin depth of the lowest modes often exceeds the wall thickness. Furthermore, in some parts of the vacuum chamber very thin walls have to be used - e.g. in expansion bellows or experimental chambers - and even quite high frequency *EM* fields can penetrate them. Chamber walls consisting of a number of different metals have been proposed, e.g. a copper-clad, stainless steel *vacuum chamber liner* for the large hadron collider *LHC*.

Another important case is that of chamber walls made of ceramics or other insulating (*dielectric*) materials. Such walls are used e.g. for fast-cycling synchrotrons to suppress excessive *eddy currents* which would be induced in conducting walls by rapidly changing magnetic fields. Sometimes a thin metallisation is applied to the inside of such walls to avoid accumulation of static charges, and must be included in the calculations. Insulating materials are used in other accelerators in regions of the vacuum chamber where it is desirable that *EM* fields penetrate, such as windows for rf cavities, kickers, beam current monitors etc. Ceramic chambers have also been proposed as accelerating structures for linear colliders [Simpson89] where they are expected to produce smaller wakes due to higher-order modes than metallic, disk-loaded wave-guides. In general, many dielectric materials have low losses, which are usually specified by an *electric loss tangent*. Sometimes ferro-magnetic materials may be directly exposed to the fields of a particle beam, e.g. in certain accelerator designs where magnetic pole pieces either penetrate into - or even

[5]The resistive wall instability became known in the sixties through the publication of two seminal papers by a group at the University of California in Berkeley [Laslett65, Neil65]. It was discovered at almost the same time in the Institute for Nuclear Physics in Novosibirsk [Dikansky65]. However, the priority really belongs to a microwave engineer who had built and tested a *resistive wall amplifier* some 15 years earlier [Haeff49]. A complete theory of the effects of a resistive wall was published shortly thereafter [Birdsall53].

form part of - the vacuum chamber [Wilson72]. Furthermore, ferrites are often used as absorbers or for tuning rf cavities. The losses of such materials are usually bigger than in dielectrics and are specified by a *magnetic loss tangent.*

All these cases can be analysed with the algorithm described here.

When *EM* fields penetrate through the vacuum chamber wall, it is necessary to specify also the properties of the region outside. We will distinguish 3 cases:

1) air to infinity,

2) air to a perfect conductor,

3) air to a perfect magnet.

For simplicity, air is often replaced by vacuum in the calculations. However, it has a small, non-zero relative *electric susceptibility* $(\varepsilon_r - 1) = 6 \cdot 10^{-4}$ under standard atmospheric conditions. This is sufficient to reduce the light velocity in the medium to a value below the particle velocity for beams with energy for which $\gamma > 40$. This can give rise to *Cerenkov radiation* and to additional losses increasing the longitudinal impedance (cf. Fig. 6.2)

6.A.1 Derivation of the Algorithm

We consider a uniform, circular symmetric geometry, and determine the *EM* field components fulfilling the appropriate boundary conditions. We begin with the simplest case of the $m = 0$ mode, for which only the components E_z, E_r, and H_θ are different from zero. In the frequency domain, the two transverse components are related to the longitudinal one, cf. Eq. (6.42):

$$\begin{aligned} \widetilde{E}_r &= \frac{jk}{\chi^2}\frac{\partial \widetilde{E}_z}{\partial r} \\ \widetilde{H}_\theta &= \frac{j\omega\varepsilon}{\chi^2}\frac{\partial \widetilde{E}_z}{\partial r} , \end{aligned} \tag{6.80}$$

where the *radial propagation constant* χ is given by the square root with a positive real part of

$$\chi^2 = k^2 - \mu_c \varepsilon_c \omega^2 (= -\nu^2) , \tag{6.81}$$

cf. Eq. (6.10) and $k = \omega/\beta c$.

The *complex permittivity* $\varepsilon_c = \varepsilon_0 \varepsilon'$ of a material with relative dielectric constant ε_r, electric loss tangent $\tan \delta_E$, and conductivity σ_c is given by

$$\varepsilon' = \varepsilon_r (1 - j \tan \delta_E) - \frac{j\sigma_c}{\omega \varepsilon_0} , \tag{6.82}$$

which is strongly frequency dependent when $\sigma_c \neq 0$. Most metals used for vacuum chambers have such a high conductivity that only the last term in this expression is important over the frequency range of interest in accelerators. Conversely, most dielectrics have such a low conductivity that the last term is completely negligible. Similarly, the *complex permeability* $\mu_c = \mu_0 \mu'$ can be expressed by

$$\mu' = \mu_r (1 - j \tan \delta_M) , \tag{6.83}$$

where μ_r is the relative permeability, and $\tan \mu_M$ the magnetic loss tangent.

6.A.2 Circular Cylindrical Geometries

In the *cylindrical coordinate system*, appropriate for a geometry with axial symmetry, the radial and azimuthal components of the *vector wave equation* for the *EM* fields, cf. Eq. (2.10), are coupled and hence their solution is difficult to obtain. However, the longitudinal field components satisfy the *scalar wave equation* and are independent of the transverse components, which can be found subsequently by differentiation, cf. Eq. (6.80). Omitting the factor $\exp j(\omega t - kz)$, the wave equation for $\widetilde{E}_z$ can be written

$$\frac{1}{r}\frac{d}{dr}\left(r \frac{d\widetilde{E}_z}{dr}\right) - \chi^2 \widetilde{E}_z = -\frac{j\chi^2}{k\varepsilon_c}\tilde{\rho} . \tag{6.84}$$

The r.h.s. vanishes everywhere except in the beam region with non-zero charge density $\tilde{\rho}$. For a uniform charge density, a *particular solution* is simply $\widetilde{E}_z = j\rho/k\varepsilon_0$, independent of r. The general solution of the wave equation can be expressed as a superposition of this particular solution of the inhomogeneous equation, and solutions of the homogeneous one with free constants which should be chosen to satisfy the boundary conditions. The solutions of the homogeneous wave equation are *modified Bessel functions* of order zero and argument (χr).

We divide the geometry into q concentric regions with indices $1 \leq p \leq q$. The innermost, cylindrical region with $0 \leq r \leq a_1$ contains the beam. The intermediate, annular regions extend between the radii $a_{p-1} \leq r \leq a_p$, with

$2 \le p < q$. The outermost region ($r \ge a_{q-1}$) may extends to infinity, or to a perfect conductor (or magnet) at $r = a_q$. For simplicity we use the abbreviations $a_1 = a$ (beam radius), $a_2 = b$ (inner chamber radius), $a_3 = d$ (outer chamber radius) and $t = d - b$ (wall thickness).

The longitudinal electric field in the beam region can be written:

$$\widetilde{E}_z^{(1)}(r) = F_1 I_0(\tau r) + G_1 \ , \tag{6.85}$$

where $\tau = k/\gamma$ is the radial propagation constant in vacuum, and $G_1 = j\tilde{\rho}/k\varepsilon_0$ the particular solution of the wave equation for $\widetilde{E}_z$ in the beam region. Since there are no other free sources, the *EM* fields in all regions are proportional to it. In the beam region, which includes the axis $r = 0$, only modified Bessel functions of the first kind are permitted, since those of the second kind diverge for argument zero.

In all intermediate regions, the *EM* fields are given by superpositions of modified Bessel functions of both kinds. It is expedient to combine them into two new functions:

$$\begin{aligned} R_p(\chi_p r) &= K_0(\chi_p a_p) I_0(\chi_p r) - I_0(\chi_p a_p) K_0(\chi_p r) \ , \\ S_p(\chi_p r) &= -K_1(\chi_p a_p) I_0(\chi_p r) - I_1(\chi_p a_p) K_0(\chi_p r) \ , \end{aligned} \tag{6.86}$$

and their derivatives with respect to the argument R_p' and S_p'. At the *outer radius* of each region they have the convenient properties:

$$\begin{aligned} R_p(\chi_p a_p) &= S_p'(\chi_p a_p) = 0 \ , \\ R_p'(\chi_p a_p) &= -S_p(\chi_p a_p) = \frac{1}{\chi_p a_p} \ . \end{aligned} \tag{6.87}$$

With these functions, the longitudinal electric field in the intermediate regions $2 \le p < q$ can be written

$$\widetilde{E}_z^{(p)} = F_p R_p(\chi_p r) + G_p S_p(\chi_p r) \ , \tag{6.88}$$

where F_p and G_p are expansion coefficients to be determined by matching the fields at the boundaries. The azimuthal component of the magnetic field can be found from Eq. (6.80) and is given by:

$$\widetilde{H}_\theta^{(p)} = j\omega\varepsilon_0 \frac{\varepsilon_p'}{\chi_p} \left[F_p R_p'(\chi_p r) + G_p S_p'(\chi_p r) \right] \ . \tag{6.89}$$

If the outermost region ($p = q$) extends to infinity, only modified Bessel functions of the second kind are admissible, as those of the first kind diverge at infinity. The electric field then becomes:

$$\widetilde{E}_z^{(q)} = G_q K_0(\chi_q r) \ . \tag{6.90}$$

If this region extends to a perfect conductor, K_0 has to be replaced by the function $R_q(\chi_q r)$, and by $S_q(\chi_q r)$ if it extends to a perfect magnet.

6.A.3 Field Matching

The unknown coefficients F_p and G_p can be determined by matching the tangential field components $\widetilde{E}_z$ and $\widetilde{H}_\theta$ at all interfaces $r = a_p$, $1 \leq p < q$. However, since the coefficient $G_1 = j\tilde{\rho}/k\varepsilon_0$ is known, it is sufficient to match the ratios $Z_p = \widetilde{E}_z/\widetilde{H}_\theta$, called the *wave impedances*, in order to determine the ratio of the field coefficients F_p/G_p. For short we write without argument:

$$R_p^{(')} = R_p^{(')}(\chi_p a_{p-1}) \ , \quad S_p^{(')} = S_p^{(')}(\chi_p a_{p-1}) \ , \tag{6.91}$$

and $I_{0a} = I_0(\tau a)$, $I_{1b} = I_1(\tau b)$ etc. Since $\varepsilon'_{1,2} = 1$ and $\chi_{1,2} = \tau$, matching of the wave impedance at the beam edge $r = a$ yields:

$$\frac{j\omega\varepsilon_0}{\tau} Z_1 = \frac{F_1 I_{0a} + G_1}{F_1 I_{1a}} = \frac{F_2 R_2 + G_2 S_2}{F_2 R'_2 + G_2 S'_2} \ . \tag{6.92}$$

This equation can be solved for the coefficient F_1. Combining terms on the r.h.s. and using the Wronskian for modified Bessel functions one obtains:

$$F_1 = -\tau a \frac{F_2 R'_2 + G_2 S'_2}{F_2 I_{0b} + G_2 I_{1b}} G_1 \ . \tag{6.93}$$

Using Eqs. (6.87) to simplify the expressions on the l.h.s., of these expressions, the matching conditions at consecutive interfaces $r = a_{p-1}$ for $2 < p < q$ can be written:

$$j\omega\varepsilon_0 Z_{p-1} = -\frac{\chi_{p-1}}{\varepsilon'_{p-1}} \frac{G_{p-1}}{F_{p-1}} = \frac{\chi_p}{\varepsilon'_p} \frac{F_p R_p + G_p S_p}{F_p R'_p + G_p S'_p} \ . \tag{6.94}$$

These expressions can be simplified by introducing the abbreviations:

$$X_p = \frac{\chi_p \varepsilon'_{p-1}}{\varepsilon'_p} \chi_{p-1} \ , \qquad Y_p = \frac{G_p}{F_p} \ , \tag{6.95}$$

where the ratios X_p are completely determined by the material properties of the p-th region, and the ratios Y_p are proportional to the *wave admittances* at the p-th interface. They are determined by the recursive relations:

$$Y_{p-1} = -X_p \frac{R_p + Y_p S_p}{R'_p + Y_p S'_p} , \tag{6.96}$$

which expresses the ratios Y_{p-1} at the p-th interface in terms of Y_p at the next one to the outside. For an outermost region $p = q$ extending to infinity, the matching condition for the wave impedance at the inner edge $r = a_{q-1}$ becomes:

$$j\omega\varepsilon_0 Z_{q-1} = -\frac{\chi_{q-1}}{\varepsilon'_{q-1}} \frac{G_{q-1}}{F_{q-1}} = -\frac{\chi_q}{\varepsilon'_q} \frac{K_0(\chi_q a_{q-1})}{K_1(\chi_q a_{q-1})} , \tag{6.97}$$

which can also be written:

$$Y_{q-1} = X_q \frac{K_0(\chi_q a_{q-1})}{K_1(\chi_q a_{q-1})} . \tag{6.98}$$

If the outermost region extends to a perfect conductor (or perfect magnet) at $r = a_q$, the ratio of modified Bessel functions K_0/K_1 on the r.h.s. has to be replaced by R_q/R'_q (or S_q/S'_q).

The longitudinal impedance of a circular accelerator with circumference $2\pi R$ is defined as the ratio of the induced voltage, averaged over the beam cross section, $\tilde{V} = -2\pi R \langle \tilde{E}_z \rangle$ to the beam current $\tilde{I} = \pi a^2 \tilde{\rho} \beta c$. Integration over the beam cross section $r < a$ yields:

$$\langle \tilde{E}_z \rangle = G_1 + F_1 \frac{2I_{1a}}{\tau a} . \tag{6.99}$$

Substituting for F_1 from above, one gets the *longitudinal impedance* $Z_{\|}$. Usually one is more interested in the *reduced impedance* Z/n, where $n = \omega/\omega_0$ is the *harmonic mode number*. Then one obtains:

$$\frac{Z_{\|}}{n} = -\frac{2jZ_0}{\beta k^2 a^2} \left[1 - 2I_{1a} \frac{R'_2 + Y_2 S'_2}{I_{0b} + Y_2 I_{1b}} \right] = -\frac{2jZ_0}{\beta k^2 a^2} \frac{N_0 + Y_2 N_1}{I_{0b} + Y_2 I_{1b}} , \tag{6.100}$$

where we introduced the coefficients:

$$N_{0,1} = \left[1 - 2I_{1a} K_{1a} \right] I_{0,1b} \mp 2I_{1a}^2 K_{0,1b} . \tag{6.101}$$

These expressions can be used to evaluate the impedance of vacuum chambers with *laminated walls* for any number of layers. The wave admittances Y_2 can be obtained by solving Eqs. (6.96) and (6.98) recursively from outside inwards. In order to evaluate these expressions numerically, one needs to calculate modified Bessel functions of both kinds and in general of complex argument. These are available in the package *MATHEMATICA* [Wolfram89], which was used in the program *LAWAL* [Zotter72b] to compute and plot the reduced impedances Z/n shown in the figures of this Appendix. A similar package for computing the transverse impedance of a multi-layer, laminated wall is called *LAWAT* [Zotter69a].

6.A.4 Relativistic Limit

In most particle accelerators, the energy of the beam is high enough that for all mode numbers n

$$\tau a < \tau b = \frac{\omega b}{\beta\gamma c} \ll 1 , \tag{6.102}$$

even for the highest frequencies ω of interest. Then one can replace the (modified) Bessel functions of argument τa and τb by the first terms in their power series [Abramowitz65]

$$\begin{aligned} I_0(z) &= 1 + \frac{z^2}{4} + \dots , \\ I_1(z) &= \frac{z}{2}\left[1 + \frac{z^2}{8} + \dots\right. , \\ K_0(z) &= -\left[\left(\ln\frac{z}{2} + \Gamma\right)\left(1 + \frac{z^2}{4}\right) - \frac{z^2}{4} + \dots\right. , \\ K_1(z) &= \frac{1}{z}\left[1 + \frac{z^2}{2}\left(2\ln\frac{z}{2} + 2\Gamma - \tfrac{1}{2}\right) + \dots\right. , \end{aligned} \tag{6.103}$$

where $\Gamma \approx 0.5772$ is the *Euler constant.* With $\eta = b/a$ (> 1) we get the following approximations from the lowest order terms:

$$\begin{aligned} R_3 &= -\ln\eta , \quad & R_3' &= \frac{1}{\tau a} , \\ S_3 &= -\frac{1}{\tau b} , \quad & S_3' &= \frac{\eta^2 - 1}{2\eta} , \end{aligned} \tag{6.104}$$

and $N_1 = (\tau a)^2/2\tau b$, while one needs the second order terms to get

$$N_0 = \frac{(\tau a)^2}{4}\left(\frac{1}{2} + 2\ln\eta\right) = \frac{(\tau a)^2}{4} g_f , \tag{6.105}$$

where $g_f = 1/2 + 2\ln(b/a)$ is the *g-factor*, commonly used to describe the space-charge effect[6]. The impedance can then be written

$$\frac{Z_{\parallel}}{n} = -\frac{jZ_0}{2\beta\gamma^2}\,\frac{g_f + 2Y_2/\tau b}{1 + Y_2\tau b/2}\,. \tag{6.106}$$

In most cases the term $Y_2\tau b/2 \ll 1$ can be neglected and the *reduced impedance* becomes approximately:

$$\frac{Z_{\parallel}}{n} = -\frac{jZ_0}{2\beta\gamma^2}\left[g_f + \frac{2Y_2}{\tau b}\right]\,. \tag{6.107}$$

However, for some combinations of parameters, the real part of $Y_2\tau b/2$ may approach -1, e.g. for dielectric walls at certain frequencies. Then the denominator can become quite small, and large resonant impedances may occur.

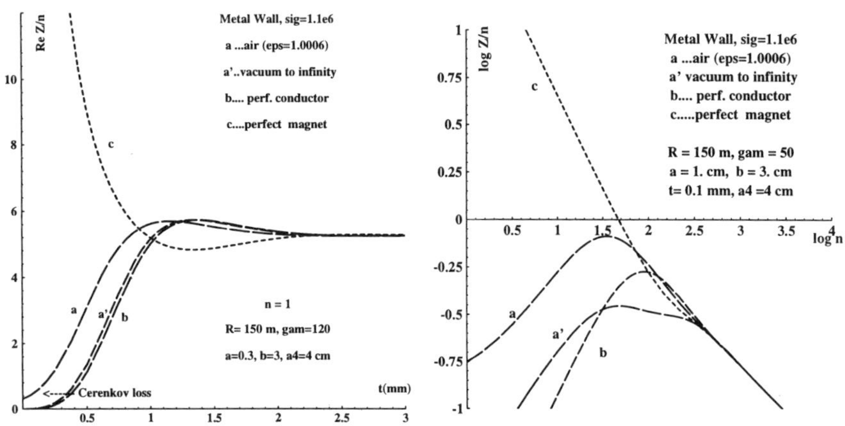

Figure 6.2: Real part of reduced impedance Z/n of a thin metal wall a) versus wall thickness t for mode number $n = 1$, and b) versus mode number n for wall thickness $t = 0.1$ mm.

6.A.5 Metal Walls of Finite Thickness

For most metal walls one may approximate the complex permittivity by $\varepsilon' = \sigma_c/j\omega\varepsilon_0$. Then the radial propagation constant in the metal can be written

[6]The constant term is 1/2, instead of the usual 1, since the induced voltage here is calculated by averaging over the beam cross section, rather than with the field on the axis.

$\chi = \chi_3 = (1+j)/\delta$, where $\delta = (2/\mu\omega\sigma_c)^{1/2}$ is the *skin depth*. When the pipe radius is large compared to the skin depth, the argument $|\chi b| \gg 1$, and one can use the *asymptotic expressions* for the modified Bessel functions to get:

$$R_3 = -S_3' = -\frac{\mathrm{Sinh}(\chi t)}{\chi\sqrt{db}} , \qquad S_3 = -R_3' = -\frac{\mathrm{Cosh}(\chi t)}{\chi\sqrt{db}} . \tag{6.108}$$

where $t = a_3 - a_2 = d - b$ is the *wall thickness*. The wave admittance parameter Y_2 is then given to a very good approximation by:

$$Y_2 = X_3 \frac{Y_3 + T}{1 + Y_3 T} , \tag{6.109}$$

where $T = \mathrm{Tanh}(\chi t)$. For a highly conducting metal wall with $|\varepsilon'| = \sigma_c/(\omega\varepsilon_0) \gg 1$ and hence $\chi \approx \beta k \sqrt{\varepsilon' \mu'}$:

$$\begin{aligned} X_3 &= \frac{\chi}{\tau\varepsilon'} \approx -\mu'\beta^2\gamma\frac{k}{\chi} , \\ X_3 Y_3 &= \tau d Y , \end{aligned} \tag{6.110}$$

where $Y \approx -\ln(\tau d)$ for air to infinity ($Y > 0$ for $\tau d \ll 1$) or $Y \approx \ln(a_4/a_3)$ for a perfect conductor at $r = a_4$. The real part of the reduced impedance is shown in Figs. 6.2, while the locus of the complex impedance of such walls, evaluated with the computer code *LAWAL* [Zotter72b], is shown in Figs. 6.3 for a typical proton (*SPS*) and an electron (*LEP*) accelerator.

For analytical estimates we can distinguish two limiting cases:

a) a wall thickness $t = d - b$ large compared to the skin depth, i.e. $|\chi t| \gg 1$, and hence $T \approx 1$. Then simply $Y_2 = X_3$, and one obtains the familiar expression for the impedance of an (infinitely) thick resistive wall

$$\frac{Z_{\parallel}}{n} = \frac{\beta Z_0}{2}\left[-\frac{jg}{\beta^2\gamma^2} + (1+j)\mu'\frac{\delta}{b}\right] . \tag{6.111}$$

The (relative) permeability μ' has been retained to emphasise that a magnetic material with $\mu' > 1$ will increase the resistive wall impedance, but actually scales as $\sqrt{\mu}$, since the skin depth δ is proportional to $\mu^{-1/2}$.

b) For a very thin wall with $t \ll \delta$, $\mathrm{Tanh}(\chi t) \approx \chi t$. Assuming a nonmagnetic metal with $\mu' = 1$, we define $\alpha = Y/\beta^2\gamma^2$ to get approximately

$$Y_2 = \beta^2\gamma k d \frac{\alpha + t/d}{1 - 2j\alpha t d/\delta^2} \tag{6.112}$$

Then the reduced impedance is:

$$\frac{Z_{\parallel}}{n} = -j\beta Z_0 \left[\frac{g}{2\beta^2\gamma^2} + \frac{\alpha + t/d}{1 - 2j\alpha td/\delta^2} \right] , \tag{6.113}$$

The impedance can also be written as:

$$Z(\omega) = j\omega L_0 + \frac{1}{1/R_1 + 1/(j\omega L_1)} , \tag{6.114}$$

which expression represents the impedance of an *equivalent circuit* consisting of an inductance L_0 in series with a resistance R_1 which is in parallel with a second inductance L_1. Equating the impedance obtained by multiplying Eq. (6.113) with $n = kR$, where R is the average machine radius, we get:

$$\begin{aligned} L_0 &= -\frac{\mu_0 R g}{2\beta^2\gamma^2} , \\ R_1 &= \frac{R\rho}{td} \left[1 + \frac{t}{\alpha d} \right] , \\ L_1 &= -\frac{\mu_0 R}{\beta^2\gamma^2} \left[\alpha + \frac{t}{d} \right] \end{aligned} \tag{6.115}$$

The inductance L_0 is due to the direct space charge. It is always negative, and therefore sometimes referred to as "capacitive" - although it does not have the frequency dependence of a capacitor ($Z \propto 1/j\omega C$). For negligible wall thickness, the resistance R_1 tends towards that of a conductor of length $2\pi R$ and cross section $2\pi bt$.

However, a second term in R_1 as well as the inductance L_1 are caused by the displacement current flowing outside the wall proper. This current shunts the wall resistance, and therefore - at least for a non-vanishing beam radius a and a finite energy factor γ - the real part of the impedance does not become infinite as could be expected by simply applying Ohm's law [Sacherer76] [7]. For vanishing wall thickness, the impedance tends to the finite value $Z/n = -[jZ_0 \ln(ka/\gamma) + C_0]/(\beta\gamma^2)$, where $C_0 = \ln 2 - \Gamma + 1/4 = 0.366$. R_1 is always positive since $\alpha = -\ln(\tau t)/\beta^2\gamma 2 > 0$ when $\tau t = kt/\gamma \ll 1$ which is fulfilled for very small wall thicknesses and high beam energies with a large but finite

[7] In reference [Sacherer76] it is clearly stated that the formula given there is valid "provided the impedance of alternate current paths outside the vacuum chamber is sufficiently high that all currents flow through the walls" - well, for thin walls they do not!

energy factor γ. With $\omega_c = R_1/L_1 = 1/(\mu_0\alpha\sigma_c tb)$, the real part of the reduced impedance can be written

$$\mathrm{Re}\frac{Z_{\|}}{n} = R_1\frac{\omega_0\omega}{\omega^2+\omega_c^2}\,. \tag{6.116}$$

It tends toward zero for vanishing frequency, and has a maximum at $\omega = \omega_c$, with the value max $(ReZ/n) = \omega_0 L_1/2$, independently of the resistance R_1. For somewhat higher frequencies $\omega > \omega_c$, it tends to $\omega_0 R_1/\omega$, before reaching the well-known inverse square-root dependence for frequencies so high that the skin depth becomes small compared to the wall thickness.

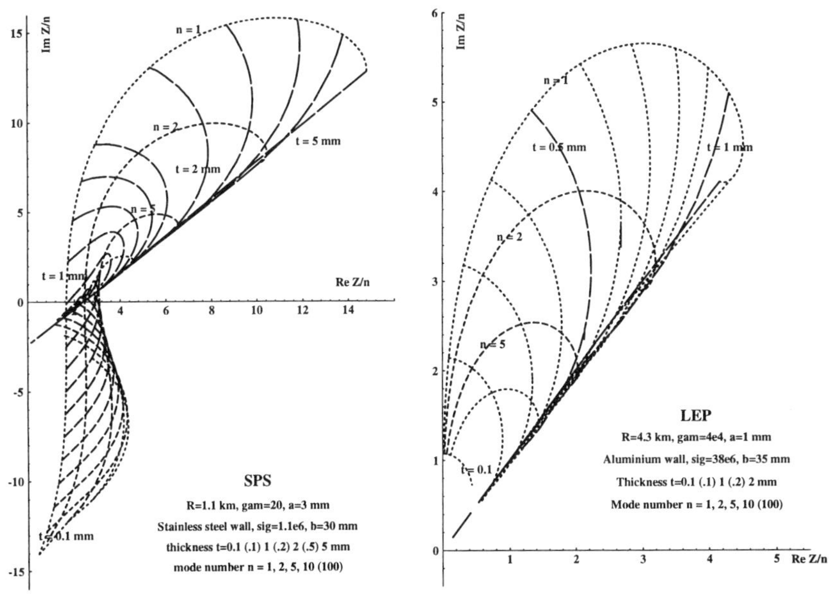

Figure 6.3: Im Z/n versus Re Z/n for a thin metal wall, for an intermediate ($\gamma = 40$) and a very high energy factor ($\gamma = 4\cdot 10^4$).

6.A.6 Ceramic Walls

When the dielectric constant is bigger than one, the radial propagation constant approaches $\chi = jk\sqrt{\beta^2\varepsilon_r - 1}$ (for $\mu' = 1, \delta_E = 0$). However, since ε_r

is never very large compared to unity for common materials, the condition $|\chi b| \ll 1$ is usually fulfilled up to quite high mode numbers. Then one gets approximately $Y_2 = \gamma kb \ln(1+\eta)\sqrt{\beta^2 - 1/\varepsilon_r}$, and hence the reduced impedance becomes

$$\frac{Z_{\|}}{n} = -j\beta Z_0 \left[\frac{g}{2\beta^2\gamma^2} + \ln \eta \sqrt{1 - \frac{1}{\varepsilon' \beta^2}} \right] , \tag{6.117}$$

which is purely imaginary for a lossless ceramic. For the usually small electric loss tangent, the real part of the impedance can be approximated by the expression $\mathrm{Re}Z/n = Z_0 \ln \eta \tan \delta_E / \varepsilon_r$.

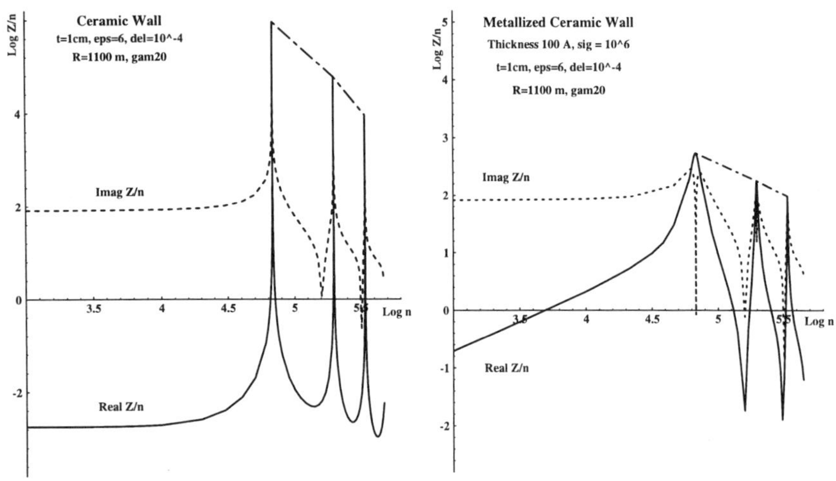

Figure 6.4: Z/n versus n for a ceramic wall, without and with metallisation.

However, for even higher frequencies, the real part of Y_2 becomes larger and can reach $2/\tau b$. Then the denominator becomes small which leads to a large resonant impedance. These resonances occur approximately when an odd multiple of a quarter wavelength in the dielectric equals the wall thickness t:

$$(2p+1)\frac{\nu}{4\sqrt{\beta^2\varepsilon' - 1}} = t \ . \tag{6.118}$$

The resonant impedances can reach very high values, as shown in Fig. 6.4a for a ceramic wall with $\varepsilon_r = 6$, $\tan \delta_E = 10^{-4}$ for a beam with $\gamma = 20$. However, the resonances are strongly suppressed by even a very thin metallisation [Piwinski77], e.g. with a layer of 100 Å as shown in Fig. 6.4b.

6.A.7 Transverse Impedance

A similar algorithm can be derived for the transverse impedance [Zotter69b] and has been used for the computer code *LAWAT* [Zotter69a]. The derivation of the algorithm is more complicated, since all six *EM* field components are in general excited by a transversely oscillating beam. Retaining the circular cylindrical geometry, the beam oscillations are replaced by surface charges and currents.

The transverse field components can again be expressed by the longitudinal ones using the Maxwell equations. The longitudinal field components E_z and H_z have to fulfil only scalar wave equations, and can be written as sums over products of modified Bessel functions for the radial, and exponentials for the longitudinal directions. There are in general four sets of expansion coefficients in each intermediate region, two for each kind of modified Bessel functions describing E_z, and similarly two for H_z. However, the beam region contains the axis and hence only modified Bessel functions of the first kind are permitted. When the outermost region extends to infinity, only modified Bessel functions of the second kind are allowed. If the outermost region extends to a perfect conductor or magnet, only particular combinations of Bessel functions of the two kinds are possible. Therefore in these regions there are only two sets of expansion coefficients.

For longitudinal oscillations, the beam and wave velocities are approximately the same[8], but for transverse (betatron) oscillations, the particle velocity factor is given by $\beta = v/c$, while the wave velocity factor of the n-th betatron mode is $\beta_w = \omega/kc = 1 - Q_\beta/n$, and is usually different from β. The combinations $\beta - \beta_w \equiv \beta_1$ and $1 - \beta\beta_w \equiv \beta_2$ appear in many expressions for the field components and the impedance.

The source terms for a vertically oscillating beam can be described by (the Fourier transforms of) a *surface charge* $\tilde{\eta}$ and a *surface current* $\widetilde{\mathbf{K}}$ at $r = b$, of which only the z-component is non-zero:

$$\tilde{\eta} = \tilde{\rho}\xi \sin\theta \ , \quad \widetilde{K}_z = \tilde{\eta}\beta c \ , \tag{6.119}$$

where ξ is the amplitude of the oscillation. These terms have to be included in the matching at the beam surface $r = a$, where $\tilde{\eta}/\varepsilon_0$ should be added to $\widetilde{E}_r$,

[8]Actually the longitudinal wave velocity differs from the beam velocity by $\pm v_p$, but the *plasma velocity* $v_p = \beta c/(1 \pm \omega_p/\omega)$ is usually very small compared to the beam velocity for relativistic particle beams in the frequency range of interest.

and $\widetilde{K}_z$ to that of $\widetilde{H}_\theta$. For vertical betatron oscillations there is also a vertical component of the volume current density:

$$\tilde{J}_y = -jk\tilde{\rho}\xi(\beta - \beta_w) \ , \quad \tilde{J}_r = \tilde{J}_y \sin\theta, \ \tilde{J}_\theta = \tilde{J}_y \cos\theta \ , \tag{6.120}$$

which need to be taken into account in deriving the transverse field components. The field components can all be normalised by a constant $E_0 = \tilde{\rho}\xi/\varepsilon_0$ which has the dimensions of the Fourier transform of an electric field strength.

It is generally sufficient to match four field components on the interfaces, since two of the six matching conditions are redundant, i.e. they are automatically fulfilled when the others are satisfied. At the outermost interface, these conditions determine four of the six sets of expansion coefficients. The remaining two are determined at the next interface, where again two remain undetermined until one arrives at the surface of the beam region. Since there are only two sets of free coefficients, the four matching conditions determine them uniquely. The coupling impedance can then be obtained from the integral over the transverse force averaged over the beam region which can be written

$$\left\langle \widetilde{F}_y \right\rangle = \gamma_w^2 \left[\beta_2^2 + j\frac{I_1(\tau a)}{ka}\left(\beta_1 E_1 + \beta_2 H_1\right) \right] \ , \tag{6.121}$$

where $\gamma_w = (1 - \beta_w^2)^{-1/2}$, and E_1, H_1 are the coefficients of the expansions of E_z and H_z in the beam region.

Thin Metal Walls

Here we present some results for a thin metal wall, whose transverse (dipole) impedance can be written[9]:

$$Z_\perp = \frac{\beta R Z_0}{b^2}\left[-\frac{j}{\beta^2\gamma^2}\left(\frac{b^2}{a^2} - 1\right) + \frac{\mu'}{tb/\delta^2 - j} \right] . \tag{6.122}$$

The equivalent longitudinal impedance $Z_{\parallel} = nb^2 Z_\perp / R$ (which has the correct dimensions) can again be expressed by that of the same equivalent circuit, but with different values for its elements:

$$L_0 = -\frac{\mu_0 R}{2\beta^2\gamma^2}\left(\frac{b^2}{a^2} - 1\right) ,$$

[9]The average machine radius R should not be confused with the resistance R_1 introduced in Eq. (6.115)- or the functions R_p introduced in Eq. (6.86).

$$R_1 = \frac{R\rho}{bt},$$

$$L_1 = \frac{\mu_0 R}{2}. \tag{6.123}$$

The transverse space-charge term, still negative, agrees with that known from the literature [Teng63]. The main difference to the longitudinal case is the fact that the parallel inductance L_1 is no longer reduced by a relativistic factor γ^2. At high energies, its influence therefore remains much stronger and can become quite important even in electron machines.

For ceramic walls, the transverse impedance shows a similar behaviour as the longitudinal one, with resonances at high frequencies when an odd multiple of a quarter of the reduced (transverse) wavelength equals the wall thickness. This effect seems to have been overlooked when it was claimed [Simpson89] that a lossless ceramic has a vanishing transverse loss factor. Nevertheless, it may well be smaller than that of an equivalent metallic, iris-loaded waveguide used commonly for particle acceleration in linacs.

Chapter 7

Perturbation Methods

The smooth, uniform, axially symmetric pipe considered in the previous Chapter is an exceptional case for which an exact analytical solution of the Maxwell equations exists. For most realistic structures only approximate methods have been developed so far.

When the deformation of the beam pipe wall can be assumed to be small in some sense or another, various *perturbation methods* can be applied. In the present Chapter several such perturbation methods are discussed, and approximate expressions for impedances of several important accelerator structures (bellows, holes, etc.) are derived.

The impedance of a smooth, straight, and perfectly conducting beam pipe tends to zero for an ultra-relativistic beam. This situation changes when the uniformity is broken by any kind of structure. In the present Chapter we consider the effects of small obstacles, such as holes, indentations, enlargements etc. A method applicable to cavities will be described in Chapter 9.

7.1 Impedance of a Pipe with a Small Hole

The approach used to evaluate the contribution of a hole to the impedance of a pipe is based on a theory developed for the diffraction of a plane *EM* wave on a small hole in a perfectly conducting, infinitely thin, planar screen [Bethe44]. A number of papers have used this approach to evaluate the impedance of vacuum chambers with holes [Kurennoy92, Kurennoy95a].

According to this theory, the Fourier transform of the diffracted *EM* field beyond a screen with a circular hole of radius h is expressed in terms of the amplitudes of the incident field $\mathbf{E}_0$ and $\mathbf{H}_0$ in the following way:

$$\begin{aligned} \tilde{\mathbf{E}} &= \frac{k_0^2 h^3}{3\pi} \frac{\exp(-jkR)}{R} \hat{\mathbf{R}} \times (2\tilde{\mathbf{H}}_0 + \tilde{\mathbf{E}}_0 \times \hat{\mathbf{R}}) , \\ \tilde{\mathbf{H}} &= \frac{k_0^2 h^3}{3\pi} \frac{\exp(-jkR)}{R} \hat{\mathbf{R}} \times (\tilde{\mathbf{E}}_0 - 2\tilde{\mathbf{H}}_0 \times \hat{\mathbf{R}}) . \end{aligned} \tag{7.1}$$

Here $\mathbf{R}$ is the radius-vector pointing to the observation point and $\hat{\mathbf{R}}$ is the unit vector in this direction. In particular, these expressions are valid in the plane of the hole where $\tilde{\mathbf{E}}_0$ is perpendicular and $\tilde{\mathbf{H}}_0$ is parallel to it. Their absolute values, calculated without a hole, are assumed to be constant over its area: $H_t = H_0, E_n = E_0$. This assumption is valid when the wave length is large in comparison to the hole size $kh = \omega h/c \ll 1$.

The reflected field, i.e. the field on the side of the incident wave, is given by expressions Eq. (7.1) in which the sign of the magnetic field should be reversed.

The diffracted field can also be expressed in terms of effective *'magnetic'* surface charges and currents induced by a plane *EM* wave in a screen. Their densities $\widetilde{\eta}_M$ and $\widetilde{\mathbf{K}}_M$ depend on both the incident field and the shape and size of the hole. In the plane of the hole, we define the radial vector $\mathbf{u}$ whose length u is measured from its center and the corresponding unit vector $\hat{\mathbf{u}}$. Then the following expressions have been found [Bethe44]:

$$\begin{aligned} \widetilde{\eta}_M(\mathbf{u}) &= -\frac{(\tilde{\mathbf{H}}_0 \cdot \mathbf{u})}{\pi^2 \sqrt{h^2 - u^2}} , \\ \widetilde{\mathbf{K}}_M(\mathbf{u}) &= -\frac{jZ_0 k \tilde{\mathbf{H}}_0}{\pi^2} \sqrt{h^2 - u^2} + \frac{\tilde{\mathbf{E}}_0 \times \mathbf{u}}{2\pi^2 \sqrt{h^2 - u^2}} . \end{aligned} \tag{7.2}$$

For a small hole of arbitrary shape, the *magnetic* current and charge densities can be replaced by effective electric $\widetilde{\mathbf{P}}$ and magnetic $\widetilde{\mathbf{M}}$ dipoles induced in the hole by the *EM* fields [Collin50]:

$$\widetilde{\mathbf{P}} = \frac{\varepsilon}{2} \int\int_S dS \; \mathbf{u} \times \widetilde{\mathbf{K}}_M , \qquad \widetilde{\mathbf{M}} = \frac{1}{2} \int\int_S dS \; \widetilde{\eta}_M \mathbf{u} , \tag{7.3}$$

where the integration is performed over the hole area S.

7.1.1 Magnetic and Electric Polarisabilities

We apply these formulae to the Fourier transforms of the *EM* fields $\widetilde{\mathbf{E}}_0$ and $\widetilde{\mathbf{H}}_0$ of an ultra-relativistic beam in a beam pipe with a hole. For this one needs to calculate the dipole moments $\tilde{\mathbf{P}}$ and $\widetilde{\mathbf{M}}$ induced by these fields. For an elliptic hole with semi-axes a_x and a_y, one gets at the surface of the beam pipe:

$$\begin{aligned} \tilde{\mathbf{P}} &= \hat{\mathbf{n}}\varepsilon_0\alpha_E\widetilde{E}_{0n} , \\ \widetilde{\mathbf{M}} &= \hat{\mathbf{u}}\alpha_{M||}\widetilde{H}_{0u} + \hat{\mathbf{v}}\alpha_{M\perp}\widetilde{H}_{0v} . \end{aligned} \tag{7.4}$$

Here $\hat{\mathbf{n}}, \hat{\mathbf{u}}, \hat{\mathbf{v}}$ are unit vectors in the directions normal to the plane of the hole, and of the magnetic and electric polarisations of the *EM* field which lie along the large and small ellipse axes, respectively. The purely geometric factors α_E and α_M are called *electric and magnetic polarisabilities* of the hole. For an elliptic hole with an eccentricity $\epsilon = (1 - a_x^2/a_y^2)^{1/2}$ they are [Collin50]:

$$\begin{aligned} \alpha_{M\perp} &= \frac{\pi a_x^3 \epsilon^2 (1-\epsilon^2)}{3\left[E(\epsilon) - (1-\epsilon^2)K(\epsilon)\right]} , \\ \alpha_{M||} &= -\frac{\pi a_x^3 \epsilon^2}{3\left[E(\epsilon) - K(\epsilon)\right]} , \\ \alpha_E &= -\frac{\pi a_x^3 (1-\epsilon^2)}{3E(\epsilon)} . \end{aligned} \tag{7.5}$$

Here $K(\epsilon)$ and $E(\epsilon)$ are the complete elliptic integrals of the first and second kind [Abramowitz65].

Suppose now that there is an angle ϕ between the large axis of an elliptic hole and the longitudinal axis of the beam pipe. Then

$$\begin{aligned} \tilde{\mathbf{P}} &= \hat{\mathbf{r}}\varepsilon_0\alpha_E\widetilde{E}_{0r} , \\ \tilde{\mathbf{M}} &= (\hat{\mathbf{z}}\alpha_{M||}\sin^2\phi + \hat{\theta}\alpha_{M\perp}\cos^2\phi)\widetilde{H}_{0\theta} , \end{aligned} \tag{7.6}$$

where $\hat{\mathbf{r}}$, $\hat{\theta}$ and $\hat{\mathbf{z}}$ are unit vectors in the directions of the coordinate axes r, θ and z.

In particular for a narrow, *longitudinal slot* ($\phi = 0$), the width w of which is much smaller than its length L, the limit $\epsilon \to 1$ in Eqs. (7.5) yields:

$$\alpha_{M\perp} = \frac{\pi L w^2}{24} , \quad \alpha_{M||} = 0 , \quad \alpha_E = -\frac{\pi L w^2}{24} . \tag{7.7}$$

For a long *transverse* slot ($\phi = \pi/2$), the dimensions of which satisfy the condition $w \ll L \ll b$, Eqs. (7.5) give:

$$\alpha_{M\perp} = \alpha_{M||} = \frac{\pi L^3}{24}\left[\ln\frac{4L}{w} - 1\right]^{-1} , \quad \alpha_E = -\frac{L w^2}{12} . \tag{7.8}$$

Finally, for a circular hole of radius h, for which $\epsilon = 0$, we get:

$$\alpha_{M||} = \alpha_{M\perp} = \alpha_M = \frac{4h^3}{3}\,, \qquad \alpha_E = -\frac{2h^3}{3}\,. \tag{7.9}$$

7.1.2 *EM* Fields of a Moving Charge

We consider a perfectly conducting beam pipe of circular cross section of radius b but without holes. Let a point charge q move with velocity of light parallel to the pipe axis $\mathbf{r} = 0$ at a transverse offset $\mathbf{r}_b$. The non-zero Fourier harmonics of the electric and magnetic fields at the pipe surface are:

$$\begin{aligned}\widetilde{E}_{0r}(b, r_b, \Delta\theta, z) &= \frac{Z_0 q}{2\pi b}\left[1 + \sum_{m=1}^{\infty}\left(\frac{r_b}{b}\right)^m \cos(m\Delta\theta)\right] e^{-jkz}\,, \\ \widetilde{H}_{0\theta}(b, r_b, \Delta\theta, z) &= \frac{q}{2\pi b}\left[1 + \sum_{m=1}^{\infty}\left(\frac{r_b}{b}\right)^m \cos(m\Delta\theta)\right] e^{-jkz}\,,\end{aligned} \tag{7.10}$$

where $\Delta\theta = \theta_b - \theta$ is the azimuth of a point on the pipe with respect to the charge. To find the *EM* fields induced by the presence of a hole, we apply the expressions Eq. (7.1) of the Bethe theory, in which H_t and E_n are replaced by the Fourier transforms $\widetilde{H}_\theta$ and $\widetilde{E}_r$:

For a small hole, the monopole and dipole terms are dominant and are usually sufficient for impedance calculations. The $m = 0$, *monopole fields* of a point charge travelling with the speed of light along the axis $r = 0$ are given by:

$$\widetilde{E}_{0z}(b, \theta, z) = 0\,, \quad \widetilde{E}_{0r}(b, \theta, z) = \frac{Z_0 q}{2\pi b} e^{-jkz}\,, \quad \widetilde{H}_{0\theta} = \frac{q}{2\pi b} e^{-jkz}\,. \tag{7.11}$$

However, this approximation is applicable only if the characteristic hole dimension h, besides $kh \ll 1$, satisfy two additional conditions: 1) $h \ll b$, since the wall curvature has not been taken into account in the calculations for a planar screen; 2) the wall thickness $t \ll h$, since in a wall with finite thickness also radial currents are excited.

7.1.3 Longitudinal Impedance of a Hole

A simple way to calculate the longitudinal impedance of the pipe with a hole is by using the Lorentz reciprocity theorem for two *EM* fields:

1) the space charge fields $\tilde{\mathbf{E}}_b, \tilde{\mathbf{H}}_b$ produced by a charge $\tilde{\rho}_b$ and current $\widetilde{\mathbf{K}}_b$ densities of the beam in a pipe without hole, and

2) the fieldS $\widetilde{\mathbf{E}}_M, \widetilde{\mathbf{H}}_M$ induced in the pipe by the 'magnetic' surface charge density $\widetilde{\eta}_M$ and the surface current density $\widetilde{\mathbf{K}}_M$ in a pipe without beam.

The Maxwell equations for the Fourier transforms of the *EM* fields in this case are:

$$\begin{aligned} \nabla \times \widetilde{\mathbf{E}}_b + j\mu\omega\widetilde{\mathbf{H}}_b &= 0 , \qquad \nabla \times \widetilde{\mathbf{H}}_b - j\varepsilon\omega\widetilde{\mathbf{E}}_b = \tilde{\mathbf{J}}_b , \\ \nabla \times \widetilde{\mathbf{E}}_M + j\mu\omega\widetilde{\mathbf{H}}_M &= -\tilde{\mathbf{J}}_M , \quad \nabla \times \widetilde{\mathbf{H}}_M - j\varepsilon\omega\widetilde{\mathbf{E}}_M = 0 . \end{aligned} \tag{7.12}$$

The magnetic current density $\tilde{\mathbf{J}}_M$ vanishes everywhere except at the hole surface where it is $\widetilde{\mathbf{K}}_M\delta(r-b)$. Repeating the method employed in Chapter 2 we obtain the following modified Lorentz reciprocity theorem:

$$\int_S d\mathbf{S} \cdot (\widetilde{\mathbf{E}}_b \times \widetilde{\mathbf{H}}_M - \widetilde{\mathbf{E}}_M \times \widetilde{\mathbf{H}}_b) = \int_V dV(\widetilde{\mathbf{H}}_b \cdot \tilde{\mathbf{J}}_M + \widetilde{\mathbf{E}}_M \cdot \tilde{\mathbf{J}}_b) . \tag{7.13}$$

The element of the surface $d\mathbf{S}$ is a vector directed inside and normal to the surface S enveloping the volume V.

The l.h.s. of this equation vanishes since each of the vectors represented by the cross product are perpendicular to the surface S. In the second term on the r.h.s. of this equation, the scalar product $\widetilde{\mathbf{E}}_M \cdot \tilde{\mathbf{J}}_b$ becomes simply $\widetilde{E}_{Mz}\tilde{J}_{bz}$ if the beam current is entirely longitudinal. Assuming that the electric field does not change significantly over the beam cross section, the integration over the volume can be split into two integrals. The first one, over the transverse coordinates, yields the total beam charge. The second one then becomes an integral of the synchronous component of the electric field over the longitudinal coordinate, which is proportional to the impedance. Hence:

$$Z_{\parallel} = \frac{1}{q^2} \int dV(\tilde{\mathbf{J}}_M \cdot \widetilde{\mathbf{H}}_b) . \tag{7.14}$$

For $\tilde{\mathbf{H}}_b$ we use the expression Eq. (7.11). The exponent in it can be expanded, retaining only terms of the order kz since their contribution to the impedance is of the same order as the main one:

$$\widetilde{H}_{b\theta} = \frac{q}{2\pi b}(1 - jkz) . \tag{7.15}$$

Then the longitudinal impedance becomes:

$$Z_{\parallel} = \frac{1}{2\pi bq} \int_0^{2\pi} d\phi \int_0^h udu(1 - jku\cos\phi)\widetilde{K}_{M\theta}(u) , \tag{7.16}$$

where ϕ is the polar angle in the plane of the hole and $\widetilde{K}_{M\theta}(u)$ is the θ-component of the 'magnetic' surface current density. According to Eq. (7.2):

$$\widetilde{K}_{\theta M} = jk\frac{Z_0 q}{\pi^2 b}\left[\sqrt{h^2-u^2} - \frac{u^2\cos^2\phi}{2\sqrt{h^2-u^2}}\right] . \tag{7.17}$$

Substituting this expression into Eq. (7.16) and performing the integration, we obtain an expression for the longitudinal impedance:

$$Z_{\parallel} = \frac{jkZ_0}{4\pi^2 b^2}[\int_0^h 4udu\sqrt{h^2-u^2} - \int_0^h \frac{u^3du}{\sqrt{h^2-u^2}}] . \tag{7.18}$$

The first integral in this equation is the magnetic, and the second the electric polarisability of the hole, cf. Eq. (7.9).

7.1.4 Impedance of a Hole as a Surface Integral

The longitudinal impedance of a hole in a pipe of radius b can also be expressed as a surface integral over the hole area S [Gluckstern92c]. To find this expression we start from the electric field component E_z. Its Fourier harmonic $\widetilde{E}_z(r,\theta,z)$ which is regular on the axis $r=0$ is:

$$\widetilde{E}_z(r,\theta,z) = \sum_m \int_{-\infty}^{\infty} dp\ A_m(p)\frac{J_m(\nu r)}{J_m(\nu b)} e^{-j(pz+m\theta)} . \tag{7.19}$$

Here $\nu = \sqrt{k_0^2 - p^2}$ is the radial propagation constant and $A(p)$ the unknown amplitude to be determined by the boundary conditions.

The inverse transform gives at $r = b$:

$$A_m(k) = \frac{1}{4\pi^2}\int_0^{2\pi} d\theta \int_{-\infty}^{\infty} dz\ \widetilde{E}_z(b,\theta,z) e^{j(kz+m\theta)} . \tag{7.20}$$

Evaluating Eq. (7.19) at $r=0$, the only non-zero term in the sum is that with $m=0$. Integration over z gives:

$$Z_{\parallel}(k) = -\frac{2\pi}{q}A_0(k) = -\frac{1}{2\pi q}\int_0^{2\pi} d\theta \int_{-\infty}^{\infty} dz\ \widetilde{E}_z(b,\theta,z)e^{jkz} , \tag{7.21}$$

For a perfectly conducting wall the tangential component of the electric field $\widetilde{E}_z(b,\theta,z)$ vanishes everywhere except inside the hole. Hence, the impedance can be written:

$$Z_{\parallel}^{hole}(k) = -\frac{1}{2\pi bq}\int_S dS\widetilde{E}_z(b,\theta,z)e^{jkz} . \tag{7.22}$$

Substituting now the reflected field $\widetilde{E}_z$ from Eq. (7.1), we obtain:

$$Z_{\parallel}^{hole}(k) = jkZ_0 \frac{\alpha_E + \alpha_M}{4\pi^2 b^2} \ . \tag{7.23}$$

For a circular hole this expression agrees with Eq. (7.18). Note, that the impedance in this approximation is purely imaginary and is proportional to frequency, i.e. it is inductive. It is proportional to the sum of the electric and magnetic polarisabilities.

In particular, for a circular hole of radius h, the longitudinal impedance becomes:

$$Z_{\parallel}^{circ}(k) = jk \frac{Z_0}{6\pi^2} \frac{h^3}{b^2} \ . \tag{7.24}$$

For a longitudinal slot, the impedance vanishes in this approximation. This is due to the fact that the induced current is almost not perturbed by such a slot. Taking terms of the next order into account in the expansion of Eqs. (7.5) gives:

$$Z_{\parallel}^{long\ slot}(k) = jk \frac{Z_0}{96\pi} \frac{w^4}{b^2 L} \left[\ln \frac{4L}{w} - 1 \right] \ . \tag{7.25}$$

For a transverse slot, which cuts the lines of induced current, the impedance is much larger:

$$Z_{\parallel}^{trans\ slot}(k) = jk \frac{Z_0}{96\pi} \frac{L^3}{b^2} \left[\ln \frac{4L}{w} - 1 \right]^{-1} \quad \text{for} \quad w \ll L \ll b \ . \tag{7.26}$$

7.1.5 Transverse Impedance of a Hole

Similar calculations can be performed also for the transverse impedance. The $m = 1$ dipole fields of a point charge travelling with the speed of light on a path parallel to the axis are:

$$\begin{aligned}
\widetilde{E}_{1z}(r,\theta,z) &= 0 \ , \\
\widetilde{E}_{1r}(r,\theta,z) &= Z_0 \frac{qr_b}{2\pi} \left[\frac{1}{r^2} + \frac{1}{b^2} \right] \cos(\Delta\theta) e^{-jkz} \ , \\
\widetilde{H}_{1\theta}(r,\theta,z) &= \frac{qr_b}{2\pi} \left[\frac{1}{r^2} + \frac{1}{b^2} \right] \cos(\Delta\theta) e^{-jkz} \ .
\end{aligned} \tag{7.27}$$

Applying the Panofsky-Wenzel theorem to Eq. (7.22), and integrating over θ we obtain:

$$Z_r = -\frac{1}{kqr_b} \int_{-\infty}^{\infty} dz \nabla_r \widetilde{E}_z |_{r=0} e^{jkz} \ . \tag{7.28}$$

Here r_b is the transverse displacement of the beam from the axis. Evaluating $\nabla_\perp \widetilde{E}_z|_{r=b}$ from Eq. (7.19), the transverse impedance of a hole can be written:

$$Z_\perp(k) = -\frac{2\pi}{q r_b k b}\left[A_1(k) + A_{-1}(k)\right] . \tag{7.29}$$

Substituting Eqs. (7.27) into the expressions for the reflected field, similar to the case for the longitudinal impedance described above, the transverse impedance of a hole is found to be constant:

$$Z_\perp(k) = jZ_0 \frac{(\alpha_M + \alpha_E)}{\pi^2 b^4} \cos(\Delta\theta) , \tag{7.30}$$

where $\Delta\theta = \theta_b - \theta$ was defined above. In particular, for a circular hole:

$$Z_\perp^{circ}(k) = jZ_0 \frac{2h^3}{3\pi^2 b^4} \cos(\Delta\theta) , \tag{7.31}$$

for a narrow longitudinal slot:

$$Z_\perp^{slot\ long}(k) = jZ_0 \frac{w^4}{24\pi b^4 L}\left[\ln\frac{4L}{w} - 1\right] \cos(\Delta\theta) , \tag{7.32}$$

and for a narrow transverse slot:

$$Z_\perp^{slot\ trans}(k) = jZ_0 \frac{L^3}{24\pi b^4}\left[\ln\frac{4L}{w} - 1\right]^{-1} \cos(\Delta\theta) . \tag{7.33}$$

7.1.6 Impedance of Holes in Thick Walls

The impedance of a hole was calculated so far assuming that the wall has a negligible thickness. When the wall has a non-zero thickness t the induced currents may flow also in the side walls of the hole, i.e. *in radial direction.* That produces additional electric and magnetic dipole moments, and consequently changes the polarisability and susceptibility of a hole with respect to a hole in the wall with negligible thickness. When the wave length of the field is large compared to the hole size, an evaluation of these quantities as well as of the impedance can be obtained using variational methods [Gluckstern91].

The results of numerical calculations for circular holes with radius h in a wall with thickness t [Gluckstern92b] are presented in Table 7.1. The electric and magnetic polarisabilities α_E^t and α_M^t decrease first rapidly with increasing wall thickness, but then approach a constant value for $t/h > 2$. For very thick

Table 7.1: Relative electric and magnetic polarisibilities and the impedance for a circular hole of radius h in a wall of thickness t.

t/h	$\alpha_E^t\ /\alpha_M$	$\alpha_M^t\ /\alpha_M$	$Z_\parallel^t\ /\ Z_\parallel$
0.0	0.500	1.000	1.000
0.1	0.459	0.871	0.824
0.3	0.438	0.778	0.680
0.6	0.431	0.732	0.602
1.0	0.430	0.715	0.570
2.0	0.429	0.710	0.562

walls, the coupling impedance ${Z_\parallel}^t$ approaches a value which is close to 56% of $Z_\parallel$ for a wall of zero thickness.

A similar table for elliptic holes can be found in ref.[Gluckstern93a]. These expressions have been used e.g. to estimate the impedances of the elongated holes in the vacuum chamber liner for the *LHC* and in the wall between the beam chamber and the pump chamber in *LEP* [Gluckstern96].

7.2 Effective Surface Impedance of Obstacles

In this Section we closely follow ref.[Balbekov93]. This paper is published as a preprint in Russian, and hence hardly known in the western world. The *EM* fields produced by a small obstacle are expressed in terms of an *effective surface impedance* which is determined by applying approximate boundary conditions.

7.2.1 Effective Surface Impedance of Single Obstacle

We consider a circular cylindrical vacuum chamber with radius b and an axially symmetric gap (obstacle) of length g. The whole structure is repeating periodically with the machine circumference $2\pi R$. The axial length is expressed by the machine azimuthal angle $\phi = z/R$.

To obtain the longitudinal impedance one needs to find the longitudinal component of the electric field $\widetilde{E}_z(r,\phi)$, which is excited by n-th Fourier harmonics of the beam current $\tilde{I}_n \exp(-jn\phi)$ flowing at the chamber axis $r = 0$. The time dependence $\propto \exp(j\omega t)$ of all quantities is omitted. The azimuthal mode number is proportional to the frequency: $n = \omega R/c$.

The electric field due to the presence of the gap, $\widetilde{E}_z(r, \phi)$, can be expanded into a series over all azimuthal modes:

$$\widetilde{E}_z(r, \phi) = \sum_p E_p(r) \exp(-jp\phi) \ . \tag{7.34}$$

As shown in Chapter 4, the longitudinal impedance is independent of the radial position r in the ultra-relativistic case and can be written:

$$Z_{\|}(\omega) = -\frac{2\pi R}{\tilde{I}_n} \widetilde{E}_n(b) \ . \tag{7.35}$$

From here on we shall omit to write the radial coordinate $r = b$ explicitly. In an axially symmetric structure, the only non-zero component of the magnetic field at the beam pipe radius is $\widetilde{H}_\theta = \tilde{I}/2\pi b$. It can also be expanded into azimuthal modes with the coefficients H_p. If the gap is small enough, i.e. if $g \ll b$ and $g \ll c/\omega$, the magnetic field inside the gap does not depend on the coordinate ϕ. Then $\widetilde{E}_z(\phi)$ is proportional to $\widetilde{H}_\theta$

$$\widetilde{E}_z(\phi) = -\hat{\zeta}(\phi)\widetilde{H}_\theta \ , \quad \text{for} \quad \phi \quad \text{inside the gap} \ . \tag{7.36}$$

This equation is similar to Eq. (6.28), where the coefficient of proportionality was the *surface impedance* ζ, expressed by the skin depth δ_s of the wall material:

$$\zeta = Z_0 \frac{|\omega|\delta_s}{2c} \left[1 + j\text{sgn}(\omega)\right] \ . \tag{7.37}$$

Therefore the coefficient of proportionality in Eq. (7.36) $\hat{\zeta}(\phi)$ will be called *effective surface impedance.* Contrary to the usual surface impedance ζ, which is constant, the effective surface impedance depends in general on the longitudinal coordinate ϕ. Under some reasonable assumptions, it can be approximated without obtaining the exact solution of the Maxwell equations.

In terms of the effective surface impedance, the *impedance of the obstacle* is defined as:

$$Z_{ob}(\omega) = \frac{1}{2\pi R} \int_{-g/2}^{g/2} \hat{\zeta}(z) dz \ . \tag{7.38}$$

The longitudinal electric field component, excited by the beam current in a beam pipe of finite conductivity, can be represented as the sum of a term describing the field in the smooth chamber $\widetilde{E}_z^{smth}$, and a term which arises due to the presence of the obstacle:

$$\widetilde{E}_z(\phi, r) = \widetilde{E}_z^{smth}(r)\mathrm{e}^{-jn\phi} + \sum_{p=-\infty}^{\infty} E_p \frac{J_0(\kappa_p r/b)}{J_1(\kappa_p)} \mathrm{e}^{-jp\phi} , \tag{7.39}$$

where

$$\kappa_p^2 = (kb)^2 - \left(\frac{pb}{R}\right)^2 . \tag{7.40}$$

Similarly, the azimuthal magnetic field at the chamber surface is written:

$$\widetilde{H}_\theta(\phi, b) = \widetilde{H}_\theta^{smth}\mathrm{e}^{-jn\phi} + \frac{j\omega b}{cZ_0} \sum_{p=-\infty}^{\infty} \frac{E_p}{G(\kappa_p)} \mathrm{e}^{-jp\phi} , \tag{7.41}$$

where we introduced the abbreviation

$$G(x) = xJ_0(x)/J_1(x) , \tag{7.42}$$

and $\widetilde{H}_\theta^{smth} = \widetilde{E}_z^{smth}(\phi, b)/\zeta$.

Substituting Eqs. (7.39) and (7.41) into Eq. (7.36) one obtains:

$$\sum_{p=-\infty}^{\infty} \left[1 + \frac{j\omega\varepsilon_0 b\tilde{\zeta}}{G(\kappa_p)}\right] E_p \mathrm{e}^{-jp\phi} = -[\zeta + \tilde{\zeta}(\phi)]\widetilde{H}_\theta^{smth}\mathrm{e}^{-jn\phi} . \tag{7.43}$$

We introduce the auxiliary function

$$F(\phi) = \frac{1}{\widetilde{H}_\theta^{smth}} \sum_{p=-\infty}^{\infty} \left(1 + \frac{j\omega\varepsilon_0 b\zeta}{G(\kappa_p)}\right) E_p \mathrm{e}^{j(n-p)\phi} . \tag{7.44}$$

Using the orthogonality of the exponential functions, one can solve the equation for the coefficients E_p to obtain the following integral equation:

$$F(\phi) = -(\hat{\zeta}(\phi) + \zeta) - \frac{j\omega\varepsilon_0 b}{2\pi}(\hat{\zeta}(\phi) - \zeta) \int_{-\pi}^{\pi} d\phi' F(\phi') \sum_{p=-\infty}^{\infty} \frac{\mathrm{e}^{j(n-p)(\phi-\phi')}}{G(\kappa_p) + j\omega\varepsilon_0 b\zeta} . \tag{7.45}$$

Neglecting the small surface impedance ζ compared to the larger effective one, one obtains the integral equation of ref.[Balbekov93]. The function $\hat{\zeta}(\phi)$ differs from zero only inside the gap, which was assumed to be short. Hence a first approximation can be obtained by ignoring the second, quadratic term, which yields simply $F(\phi) \approx -\tilde{\zeta}(\phi)$.

In the limit $g \to 0$ the series over p in Eq. (7.45) diverges. However, for any finite value of the gap length, the terms with large $|p| \gg 2\pi R/g$ are rapidly oscillating and hence we can limit the summation. For $|p| \ll 2\pi R/g$, the exponents in the sum can be replaced by unity, and thus the solution of Eq. (7.45) in the next approximation becomes:

$$F(\phi) = -\frac{\hat{\zeta}(\phi)}{1 + Z_{ob}(\omega)Y_{ch}(\omega)} , \tag{7.46}$$

where we introduced the (smooth) *chamber admittance*:

$$Y_{ch} = \frac{j\omega b^2}{4\pi R} \sum_{p=-\infty}^{\infty} \frac{1}{G(\kappa_p) + j\omega\varepsilon_0 b\zeta} . \tag{7.47}$$

We can now evaluate the average electric field component:

$$\langle \widetilde{E}_z \rangle = -\frac{b}{R} \frac{Z_{ob}(\omega) H_\theta^{(0)}}{[1 + Z_{ob}(\omega)Y_{ch}(\omega)]} \left[1 + \frac{j\omega\varepsilon_0 b\zeta}{G(\kappa_n)} \right]^{-1} . \tag{7.48}$$

Since ζ is small for a metallic wall, and $G(\kappa_n)$ is of order unity, the last factor can be omitted. The field $\widetilde{H}_\theta^{(0)} \approx \tilde{I}/2\pi b$. From Eq. (7.35) we then obtain the coupling impedance of a pipe with an obstacle:

$$Z_{\parallel}(\omega) = \frac{Z_{ob}(\omega)}{1 + Z_{ob}(\omega)Y_{ch}(\omega)} . \tag{7.49}$$

7.2.2 Equivalent Circuit

Eq. (7.49) corresponds to an equivalent circuit consisting of two impedances Z_{ob} and $Z_{ch} = 1/Y_{ch}$ connected in parallel. That justifies the interpretation of Z_{ch} as the impedance of the vacuum chamber. According to the definition Eq. (7.47), the chamber admittance is the sum of an infinite number of admittances $Y_n = 1/Z_n$ in parallel: $Y_{ch} = \sum_n Y_n$, where

$$\mathrm{Re}Z_n = Z_0 R \frac{|\omega|\delta_s}{2cb} , \qquad \mathrm{Im}Z_n = -jZ_0 R \left[\frac{G(k_n)c}{\omega b^2} - \frac{\omega\delta_s}{4\pi cb} \right] . \tag{7.50}$$

Using the summation formula [Watson66]

$$J_0(x) = \sum_{s=1}^{\infty} \left(1 - \frac{x^2}{j_{0s}^2} \right) , \tag{7.51}$$

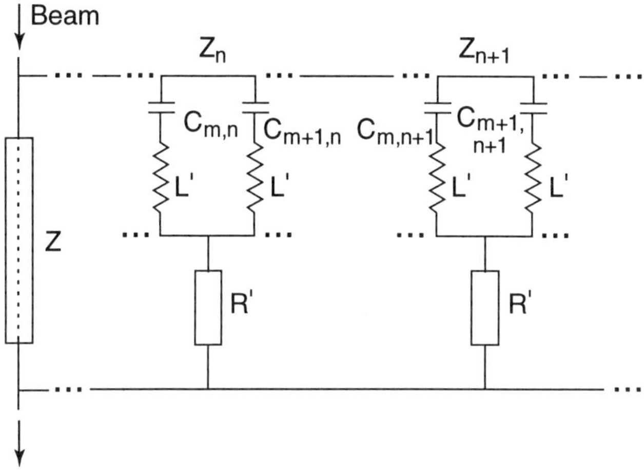

Figure 7.1: Equivalent circuit of chamber with obstacle.

where j_{0s} is the s-th zero of the Bessel function $J_0(x)$, the imaginary part of the admittance of the m-th branch can be represented in the following way:

$$Y_n = \sum_{s=1}^{\infty} Y_{sn} \quad ; \quad \mathrm{Im} Y_{sn} = -\frac{2\omega c}{Z_0 R(\omega_{sn}^2 - \omega^2)} , \tag{7.52}$$

where ω_{sn} are the eigenfrequencies of waves which can propagate in the axially symmetric chamber with radius b and period $2\pi R$:

$$\frac{\omega_{sn}^2}{c^2} = \frac{j_{0s}^2}{b^2} + \frac{n^2}{R^2} . \tag{7.53}$$

Eq. (7.52) corresponds to an LC series circuit with all inductances equal, $L_{sn} = Z_0 R/2c$,and the capacitances $C_{sn} = 2c/Z_0 R\omega_{sn}^2$. The complete equivalent circuit of a chamber with an obstacle is shown in Fig.7.1.

7.2.3 Infinite Straight Pipe

The results obtained in the previous subsection simplify for a straight, infinitely long pipe. The corresponding formulae are obtained in the limit $R \to \infty$. Then the sum over p in Eq. (7.47) is replaced by an integral over dp:

$$Y_{ob}(\omega) = \frac{2j\omega}{Z_0} \int_0^{2\pi b/g} \frac{dp}{G(\sqrt{(kb)^2 - p^2}) + jkb\zeta/Z_0} . \tag{7.54}$$

The upper limit of the integral is chosen in accordance with the truncation procedure discussed above.

For a perfectly conducting wall $\zeta \to 0$, and the equivalent circuit for the chamber becomes a simple, parallel RC circuit:

$$Y_{ob}(\omega) = G_{ob} + j\omega C_{ob} \,, \tag{7.55}$$

with the conductance

$$G_{ob} = \frac{2\pi}{Z_0} \sum_{n=1}^{N} \frac{1}{\sqrt{1 - j_{0n}^2/(kb)^2}} \,, \tag{7.56}$$

and the capacitance

$$C_{ob}(\omega) = \frac{4b}{Z_0 c} \sum_{n=N+1}^{\infty} \frac{1}{\sqrt{1 - j_{0n}^2/(kb)^2}} \arctan \frac{2\pi b/g}{\sqrt{1 - j_{0n}^2/(kb)^2}} \,. \tag{7.57}$$

Here N is the smallest number for which $j_{0N} > kb$.

The remarkable feature of this result is the appearance of a resistive term for frequencies above cutoff. Mathematically the real part of the admittance appears as the result of poles inside the integration contour in the integral in Eq. (7.54). From the physical point of view, this resistance has resonant character and arises due to overlap of resonances above cutoff.

7.2.4 Impedance below Cutoff

At low frequencies, when $kb \ll 1$, the impedance of the obstacle has inductive character. However, it is not very convenient to use the formula in Eq. (7.57). An approximation of the sum, which is valid in the limit $\zeta \to 0$, yields for this expression:

$$C_{ob}(\omega) = \frac{2b}{Z_0 c} \ln \frac{6b}{g} + \frac{2\pi b}{Z_0 c} \left[\frac{1}{\sqrt{j_{01}^2 - k^2 b^2}} - \frac{1}{j_{01}} \right] , \tag{7.58}$$

which agrees with exact numerical results to a few percent. It should be modified when the wall has finite conductivity. The admittance of the chamber with an obstacle then can be written:

$$Y_{ob}(\omega) = \frac{2j(kb)}{Z_0} \ln \frac{6b}{g} + \frac{2\pi j b(kb)}{Z_0} \left[\frac{1}{\sqrt{j_{01}^2 - k^2 b^2 + jkb|kb|\delta_s/b}} - \frac{1}{j_{01}} \right] . \tag{7.59}$$

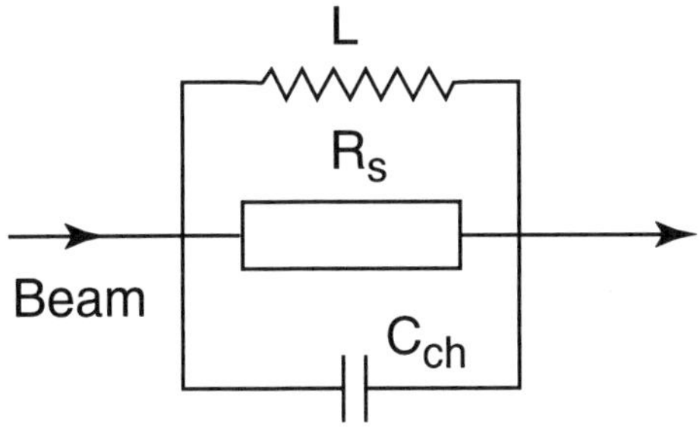

Figure 7.2: Equivalent circuit of chamber with obstacle below cutoff.

For low frequencies, the impedance of the obstacle can be represented as a resistance and an inductance in series:

$$Z_{ob} \approx j\omega L_{ob} + R_{ob} \ . \tag{7.60}$$

The inductance L_{ob} here is defined by the flux of the magnetic field of the beam, passing through the cross section S of the obstacle :

$$L_{ob} \approx \frac{Z_0 c}{2\pi} \frac{S}{cb} \ . \tag{7.61}$$

The resistance R_{ob} arises due to the fact that the path length of the current in the presence of an obstacle is longer than in a smooth pipe. For example, for a rectangular obstacle with area $S = g\,h$, the lengthening Δs is approximately $2h$. Taking into account that the current flows only through a thin layer with thickness δ_s, the *skin depth*, the resistance becomes:

$$R_{ob} \approx \frac{Z_0}{4\pi} \frac{\Delta s |\omega| \delta_s}{bc} \ . \tag{7.62}$$

The equivalent circuit of a chamber with an obstacle, valid at low frequencies, is shown in Fig.7.2. Eq. (7.58) shows that the capacitance increases sharply in the vicinity of the cutoff frequency. This produces a resonance slightly below it. For a small obstacle, this resonant frequency is given by:

$$\omega_{res} \approx \frac{j_{01} c}{b} \left[1 - \frac{j_{01}^2 S^2}{2b^4} \right] \ . \tag{7.63}$$

The half-width of the resonance is then given by

$$\delta\omega \approx \frac{j_{01} c \delta_s}{2b^2} \left[1 + \frac{j_{01}^2 S \Delta s}{b^3} \right] . \tag{7.64}$$

The shunt impedance at this frequency can be found from the expression

$$\frac{1}{R_s} \approx \frac{R_{ob}}{\omega^2 L_{ob}^2} + \mathrm{Re} Y_{ob} = \frac{\pi b^2 \delta_s \Delta s}{Z_0 j_{01} S^2} + \frac{\pi b^5 \delta_s}{Z_0 j_{01}^3 S^3} . \tag{7.65}$$

The first term describes the energy loss at the obstacle, the second one that for its vicinity, which usually dominates. The shunt impedance in such a case is proportional to the cube of the area of the obstacle:

$$R_s = \frac{Z_0 j_{01}^3 S^3}{\pi b^5 \delta_s} . \tag{7.66}$$

The impedance near the resonance can be written:

$$Z_{\parallel}(\omega) \approx -j Z_0 \frac{c j_{01}^4 S^3}{\pi b^7} \frac{\omega}{\omega^2 - \omega_{res}^2 - 2j\omega\delta_s \omega} . \tag{7.67}$$

The same results can be obtained in a more rigorous way using integral equations which we describe in Chapter 9.

7.3 Impedance of an Array of Holes

In many accelerators the beam is surrounded by a screen perforated with many small holes or slots. For example, often the vacuum chamber consists of a *beam chamber* separated by a screen from a *pump chamber*. These screens must have a sufficient number of openings to permit passage of residual gas molecules to be pumped. Screens with many holes have furthermore been proposed for cold vacuum chambers inside superconducting magnets, in order to avoid heating them by synchrotron radiation. Such screens, also called *liners*, again need to have a large number of openings to permit efficient cryo-pumping by the surrounding cold vacuum chamber.

When the distances D between holes are smaller than the relevant wave lengths λ of the diffracted radiation, the impedances of the holes cannot be considered independent. In this Subsection we calculate the coupling between holes, and derive expressions for the impedances of an array of holes under the assumption $D \ll \lambda$. This excludes resonant effects which would occur when the frequency equals $2\pi c/D$, but this is often higher than all frequencies in the bunch spectrum. One can distinguish two different cases - regular and irregular spaced holes. The expression for the longitudinal impedance of a single hole, Eq. (7.23), needs to be modified in two respects for an array of holes [Gluckstern92a]:

First, the impedance per unit length $Z_{\|}/L$ becomes proportional to the number of holes per unit length $2\pi b/(D_t D_z)$. Here D_t and D_z are the spacings between holes in transverse and longitudinal directions.

Second, the *EM* fields at a given hole are influenced by the presence of neighbouring holes, which changes its polarisability. We introduce *effective polarisabilities* $\hat{\alpha}$ to write the impedance as:

$$\frac{Z_{\|}}{L} = j\frac{Z_0 k}{2\pi b D_t D_z}(\hat{\alpha}_E + \hat{\alpha}_M) \ . \tag{7.68}$$

7.3.1 Effective Polarisability of Regular Array

Here we consider the case of regular hole spacing in a plane, repeating with periods p_x in the x and p_y in the y directions. The positions of the hole centres then are given by $\mathbf{r}_{sn} = mp_x\hat{\mathbf{x}} + np_y\hat{\mathbf{y}}$, where the integers n and m run from $-\infty$ to $+\infty$.

Both electric and magnetic effective polarisabilities $\hat{\alpha}$, for a hole in an array have been determined many years ago [Bennett53, Brown55]. The induced electric dipole moment $\mathbf{P}$ at the hole must again be proportional to the corresponding electric field $\mathbf{E}$,which is the sum of the beam field $\mathbf{E}_b$ and the diffracted field $\mathbf{E}_d$ due to all other holes of the array:

$$\mathbf{P} = \alpha(\mathbf{E}_b + \mathbf{E}_d) \ . \tag{7.69}$$

The field $\mathbf{E}_d$ arises as result of the dipole moment $\mathbf{P}$ and - in first approximation - must be proportional to it: $\mathbf{E}_d = C_\alpha \mathbf{P}$. The coefficient of proportionality depends only on the array geometry and on the dipole orientation. Hence, the effective polarisability of a hole in an array is given by:

$$\hat{\alpha} = \frac{\alpha}{1 - C_\alpha \alpha} \ , \qquad C_\alpha \alpha < 1 \ . \tag{7.70}$$

The same considerations are applicable to the magnetic dipole moment $\mathbf{M}$.

The general expression for the coefficients C_α has been obtained by Collin [Collin50]. For our purpose, it is enough to restrict the problem to the quasi-static case $D_{x,y} \ll \lambda$. We consider first the fields excited by dipoles which are oriented along the y axis. The electrostatic field E_y at $r = 0$ can be obtained from a scalar potential:

$$E_y = -\frac{\partial \Phi_y}{\partial y}|_{r=0}, \qquad \Phi_y = \hat{\mathbf{y}} \sum_{m,n\neq 0} \frac{\mathbf{r}_{sn}}{4\pi r_{sn}^3} \ . \tag{7.71}$$

A similar calculation can be done for the magnetostatic field. The calculation of these fields and the coupling coefficients $C_{x,y}$ can be found in ref.[Petracca95a]. For example:

$$C_y = \frac{6}{5\pi p_y^3} - \frac{8\pi}{p_y^3} k\left(\frac{2\pi p_x}{p_y}\right) , \tag{7.72}$$

where k is the zero-th order modified Bessel function of the second kind. For dipoles oriented parallel to the x axis, the periods p_x and p_y should be interchanged. Finally, for dipoles oriented perpendicular to the plane of the hole $C_z = -(C_x + C_y)$.

7.3.2 Longitudinal Impedance of a Regular Array

For an ultra-relativistic beam travelling in an axially symmetric pipe, the induced electric dipoles are parallel to the electric field and hence normal to the plane of the hole, while the magnetic dipoles are parallel to the magnetic field and to that plane. For holes with the same period in both directions, $p_x = p_y = p$, one gets:

$$C_E = -\frac{12}{5\pi p^3} + \frac{8\pi}{p^3} k(2\pi), \qquad C_M = \frac{6}{5\pi p^3} - \frac{8\pi}{p^3} k(2\pi) \ . \tag{7.73}$$

Combining expressions Eqs. (7.68) and (7.70), one obtains the longitudinal impedance of a periodic array of equidistant holes $D_x = D_y = D$. We give here the expression for the case of circular holes of radius h:

$$\frac{Z_\parallel^{arr}}{L} = jk \frac{Z_0 h^3}{3\pi b D^2} \left[\frac{2}{1 + C_2(h/D)^3} - \frac{1}{1 + C_1(h/D)^3} \right] , \tag{7.74}$$

where $C_1 = (8/3)[(2\pi k(2\pi) - 3/(5\pi)]$ and $C_2 = (8/3)[4\pi k(2\pi) - 3/(5\pi)]$.

In order to suppress strong resonant enhancement, irregular or "pseudo-random" arrays have been considered for vacuum chamber liners. The expressions for irregular arrays of holes can also be found in the literature [Kurennoy95b].

7.4 Periodic Wall Deformations

Various methods have been developed for the calculation of impedances of rotationally symmetric vacuum chambers with small, periodic variations of their radius, such as the convolutions of expansion bellows [Papiernik79, Krinsky80, Cooper82, Kheifets86b, Kurennoy90, Baarts94]. Most of these are limited to rather smooth wall deformations of small depth and not very steep slopes, e.g. sinusoidal variations of the radius. In general these methods have been derived for bunched beams of arbitrary but constant velocity, but the equations become much simpler in the ultra-relativistic case $v \to c$.

First we investigate the case of a rotationally symmetric wall with variable radius $b(z) = b_0[1+\epsilon s(z)]$, where b_0 is the mean radius of the pipe, ϵ is assumed to be small, and $s(z)$ is a periodic function with the period P. It is then most appropriate to use circular cylindrical coordinates r, θ and z. The function $s(z)$ can be expanded into the Fourier series:

$$s(z) = \sum_{p=-\infty}^{\infty} C_p \exp\left(jp\frac{2\pi z}{P}\right) , \quad C_0 = 0 . \tag{7.75}$$

For a beam moving with constant velocity $v = \beta c$ on the axis of such a structure, only the field components E_r, E_z, and H_θ are excited. For the monopole mode $m = 0$ they are independent of θ, and can be derived either from a single vector potential component, or from any one of the field components. Taking into account that the beam current density depends on the combination $\tau = t - z/v$, the wave equation in coordinates r, z, τ for an axially symmetric magnetic field H_θ can be written:

$$\frac{\partial}{\partial r}\frac{1}{r}\left[\frac{\partial(rH_\theta)}{\partial r}\right] + \frac{1}{v^2\gamma^2}\frac{\partial^2 H_\theta}{\partial \tau^2} + \frac{\partial^2 H_\theta}{\partial z^2} - \frac{2}{v}\frac{\partial^2 H_\theta}{\partial \tau \partial z} = \frac{\partial J_z(r,\tau)}{\partial r} . \tag{7.76}$$

The solutions of this equation can be expanded into Fourier-Bessel series in r and z with the same longitudinal period as the structure and into the Fourier integral in τ:

$$H_\theta = \int_{-\infty}^{\infty} d\nu \, \exp(2\pi j\nu\tau) \sum_{p=-\infty}^{\infty} \exp\left(\frac{2\pi jpz}{P}\right) \sum_{n=1}^{\infty} u_{np}(\nu) J_1\left(j_{0n}\frac{r}{b_0}\right) , \tag{7.77}$$

where j_{0n}, $n = 1, 2, 3...$ is the n-th zero of the Bessel function J_0.

The expansion coefficients $u_{np}(\nu)$ should be determined by applying the proper boundary conditions at the wall, e.g. for a perfectly conducting material the normal component of the magnetic field $H_n = 0$ must vanish on the wall:

$$\frac{\partial(rH_\theta)}{\partial r} = \frac{db(z)}{dz}\left[\frac{\partial(rH_\theta)}{\partial z} - \frac{1}{v}\frac{\partial(rH_\theta)}{\partial \tau}\right] . \tag{7.78}$$

Since the wall is not a coordinate surface, the resulting equation for $u_{np}(\nu)$ contains the unknown field strength, or rather its derivatives at $r = b_0$, on the right-hand side. By expanding the field into a Taylor series in the small parameter: $H_\theta = H_\theta^{(0)} + \epsilon H_\theta^{(1)} + \epsilon^2 H_\theta^{(2)} + ..$, the equations can be solved iteratively [Papiernik79].

The energy loss per period is obtained by keeping only the synchronous mode, but one needs to include terms in the expansion at least up to second order in ϵ. For $v = c$ one thus gets:

$$\Delta U = \frac{\pi\epsilon^2}{c\varepsilon_0}\sum_{p=-\infty}^{\infty} pC_pC_{-p}\sum_{n=1}^{\infty}\nu_{np}|\tilde{I}(\nu_{np})|^2 , \tag{7.79}$$

where C_p are the Fourier coefficients of the boundary shape,

$$\nu_{np} = \frac{cp}{2P} + \frac{P}{p}\frac{cj_{0n}^2}{8\pi^2 b_0^2} \tag{7.80}$$

are the mode frequencies, and $I(\nu)$ is the Fourier transform of the beam current.

By letting the period of the wall tend to infinity, one can also obtain expressions for a single obstacle in this way, as long as it is small enough that the perturbation expansion will converge. Furthermore it is possible to take into account the finite resistivity of metallic walls by applying the *Leontovich boundary conditions.*

In the long wave-length limit the coupling impedances are [Krinsky80]:

$$\begin{aligned} \frac{Z_\parallel}{n}(\omega) &= \epsilon^2\frac{j\omega_0 Z_0 P}{2\pi c}\sum_{p=-\infty}^{\infty}|C_p|^2\left(\frac{2\pi p b_0}{P}\right)^2\frac{I_1(u)}{uI_0(u)} , \\ Z_\perp(\omega) &= \epsilon^2\frac{j\omega_0 Z_0 P}{\pi b_0^2}\sum_{p=-\infty}^{\infty}|C_p|^2\left(\frac{2\pi p b_0}{P}\right)^2\left[\frac{I_1'(u)}{uI_1(u)} - \frac{I_1(u)}{u^3 I_1'(u)}\right] , \end{aligned} \tag{7.81}$$

where $u = \Lambda b_0$ and

$$\Lambda^2 = \left(\frac{2\pi p}{P}\right)^2 + \left(\frac{2\pi p}{P}\right)\left(\frac{2\omega}{c}\right) . \tag{7.82}$$

Resonances of the longitudinal impedance occur at frequencies $\omega_{np} = 2\pi\nu_{np}$ where $u^2 = -j_{0n}^2$. For a non-zero wall conductivity σ_c, the impedance peaks remain finite. In terms of the resistive wall impedance $Z_{rw} = Z_0(L/2\pi a)\zeta$, with the *surface resistivity* $\zeta = \sqrt{(\omega\varepsilon_0)/(2\sigma_c)}$, the peak and the quality factor of a resonance are:

$$R_{np} = \epsilon^2 \frac{|pC_p|^2 Z_0^2}{Z_{rw}(\omega_{np})} , \qquad Q_{np} = \frac{|p|Z_0}{2Z_{rw}(\omega_{np})} . \tag{7.83}$$

However, this method fails when the slope of the deformed walls is too large, e.g. for a square-wave shape of the wall which is often assumed for field matching or numerical computations with mesh programs. Then divergent terms may appear in the perturbation expansion, which can be evaluated by a modified approach [Krinsky81] which concentrates on the resonant terms.

A similar method is used in Chapter 12 to calculate the low-frequency impedance of bellows.

7.5 Method of Conformal Transformation

Another way to determine the impedance of structures with slowly varying boundaries $b(z)$ is to apply conformal transformations [Yokoya90], similar to a technique proposed in ref.[Kheifets81]. The region of applicability of this method is rather limited. For example, for a taper its slope b' should be smaller than the diffraction angle $(kb)^{-1}$. In order to avoid resonances, the method is also restricted to the high frequency region $kb \gg 1$. However, this inequality, as well as the author's claim that the total change of the boundary Δb can be of the order of b, are not correct, as will be discussed below, see Chapter 12.

For the axially symmetric structure with a variable radius $b(z)$ (see below) a small value of ϵ means a slowly varying boundary. A *conformal transformation* to new variables is chosen in such a way that with the accuracy up to ϵ^2 the pipe wall becomes a coordinate surface.

The set of curvilinear coordinates ζ, ρ which replace the cylindrical coordinates z, r are introduced according to the following relations:

$$\begin{aligned} z(\zeta,\rho) &= \frac{\xi}{\epsilon} - \epsilon\Big(\frac{bb'}{2}\rho^2 + u\Big) + O(\epsilon^3) , \\ r(\zeta,\rho) &= b\rho\Big[1 - \epsilon^2 (bb')^2 \frac{\rho^2}{6} + u'\Big] + O(\epsilon^4) , \end{aligned} \tag{7.84}$$

where $\xi(\zeta)$ and $u(\zeta)$ are defined by the implicit formulae

$$\begin{aligned}\zeta &= \int_0^{\xi} \frac{dz}{b(z)}\,, \qquad d\xi = b\, d\zeta\,, \\ u(\zeta) &= -\frac{bb'}{6} + \frac{b}{3}\int_0^{\xi} \frac{dz}{b(z)}(b')^2\,. \end{aligned} \tag{7.85}$$

Here b is assumed to be a function of ξ, which is a function of ζ but not of ρ. The prime denotes derivation with respect to ξ.

The Fourier components of the field produced by the m-th azimuthal harmonic of the source current are:

$$\widetilde{E}_z^S = c\widetilde{B}_z^S = 0\,, \quad \widetilde{E}_r^S = E_\theta^S = -c\widetilde{B}_r^S = c\widetilde{B}_\theta^S = \frac{\mathrm{e}^{-jkz}}{2\pi r^{m+1}}\,, \tag{7.86}$$

where the factors $\cos m\theta$ and $\sin m\theta$ have been omitted for brevity.

The projections of these fields on the boundary $\rho = 1$ are:

$$\begin{aligned}\widetilde{E}_\zeta^S &= -c\widetilde{B}_\zeta^S = \frac{\epsilon b' V}{\sqrt{1+\epsilon^2(b')^2}}\,, \\ \widetilde{E}_\rho^S &= -c\widetilde{B}_\rho^S = \frac{V}{\sqrt{1+\epsilon^2(b')^2}}\,, \\ \widetilde{E}_\theta^S &= c\widetilde{B}_\theta^S = V\,, \end{aligned} \tag{7.87}$$

where $V = \exp(-jkz)/2\pi b(z)$ and z is evaluated at the position $\zeta, \rho = 1$.

The total *EM* field is the sum of the source field Eq. (7.87) and the radiation field $\widetilde{\mathbf{E}}$ and $\widetilde{\mathbf{B}}$ (which obey the homogeneous Maxwell equations) with the boundary conditions:

$$\begin{aligned}\Big[\widetilde{E}_\zeta^S + \widetilde{E}_\zeta\Big]_{\rho=1} &= 0\,, \\ c\Big[\frac{\partial \widetilde{B}_\zeta}{\partial \rho}\Big]_{\rho=1} &= -\Big[m\widetilde{E}_\zeta^S + (\nabla_\zeta - jkb)(\widetilde{E}_\theta^S + c\widetilde{B}_\rho^S)\Big]_{\rho=1}\,. \end{aligned} \tag{7.88}$$

After the solution of the Maxwell equations up to the order of ϵ^2 is found, the longitudinal and transverse impedances are calculated in the usual way. The derivation and formulae in the general case can be found in ref.[Yokoya90].

Here we give the results for the longitudinal impedance of the monopole mode $m = 0$:

$$\begin{aligned} Z_{\|}^{m=0} &= Z_0\Big\{\frac{1}{2\pi}\ln\frac{b(\infty)}{b(-\infty)} + j\epsilon^2\frac{k}{4\pi}\int_{-\infty}^{\infty} dz(\frac{db}{dz})^2 \\ &- j\epsilon^2\frac{2k}{\pi}\Big[bb'\Big|_{\infty} - bb'\Big|_{-\infty}\Big]\sum_n \frac{1}{j_{0n}^3 J_1(j_{0n})}\Big\} \end{aligned} \tag{7.89}$$

and the transverse impedance of the dipole mode $m = 1$:

$$Z_{\perp}^{m=1} = j\frac{Z_0}{\pi}\epsilon^2\int_{-\infty}^{\infty} dz\left(\frac{db}{bdz}\right)^2 . \tag{7.90}$$

The first term in Eq. (7.89) describes the effect of pipes of different radii.

7.6 Beam Pipes with Complex Geometries

In this Section we derive some general formulae based on the Lorentz reciprocity theorem, in order to compute the coupling impedance in pipes with complex geometries in the transverse directions and/or non-trivial boundary conditions [Petracca93].

We consider a sheet of charge with surface charge density $\rho_\perp$ travelling in the positive z-direction with constant velocity $v = \beta c$. Its current density is then given by:

$$\mathbf{J}(\mathbf{r},t) = v\rho_\perp(x,y)\delta(z-vt)\hat{\mathbf{z}} , \tag{7.91}$$

and we expect all *EM* fields to have the same dependence on $z - vt$. We define the Fourier transform for any one of the *EM* field components

$$f(\mathbf{r},t) = \frac{1}{2\pi}\int_{-\infty}^{\infty} dk\ \tilde{f}(x,y,k)\exp[-jk(z-vt)] , \tag{7.92}$$

for which $\partial_z \sim -jk$, $\partial_t \sim jkv$.

7.6.1 Debye Potentials

The *EM* fields can be computed from the *Debye potentials* Φ and Ψ whose Fourier transforms are indicated as $\widetilde{\Phi}$ and $\widetilde{\Psi}$. They are solutions of the following *Helmholtz* equations, cf. Chapter 2:

$$\begin{aligned}\nabla_{\perp}^{2}\,\widetilde{\Phi}-k^{2}(1-v^{2}\varepsilon\mu)\,\widetilde{\Phi} &= -\frac{\rho_{\perp}}{\varepsilon}+C_{1}\,,\\ \nabla_{\perp}^{2}\,\widetilde{\Psi}-k^{2}(1-v^{2}\varepsilon\mu)\,\widetilde{\Psi} &= C_{2}\,,\end{aligned}\tag{7.93}$$

where $\nabla_{\perp}$ denotes the transverse gradient. The permittivity ε and the permeability μ are in general frequency dependent quantities. Boundary conditions on $\widetilde{\Phi},\widetilde{\Psi}$ can be obtained from those on the fields [Petracca94]. For an ultra-relativistic beam with $\gamma\to\infty$, the constants C_1, C_2 *cannot* be set to zero via a trivial transformation of the potentials.

The irrotational and solenoidal parts [Marcuvitz51b] of the transverse field are obtained from the Debye potentials by the relations:

$$\begin{aligned}\widetilde{\mathbf{E}}_{\perp}^{(irr)} &= -\nabla_{\perp}\widetilde{\Phi}\,, & \widetilde{\mathbf{E}}_{\perp}^{(sol)} &= -v\mu\,\hat{\mathbf{z}}\times\widetilde{\mathbf{H}}_{\perp}^{(irr)}\,,\\ \widetilde{\mathbf{H}}_{\perp}^{(irr)} &= -\nabla_{\perp}\widetilde{\Psi}\,, & \widetilde{\mathbf{H}}_{\perp}^{(sol)} &= v\varepsilon\,\hat{\mathbf{z}}\times\widetilde{\mathbf{E}}_{\perp}^{(irr)}\,,\\ \widetilde{E}_{z} &= jk(1-v^{2}\varepsilon\mu)\widetilde{\Phi}-\frac{C_{1}}{jk}\,, & \widetilde{H}_{z} &= jk(1-v^{2}\varepsilon\mu)\widetilde{\Psi}-\frac{C_{2}}{jk}\,.\end{aligned}\tag{7.94}$$

In a pipe with longitudinally invariant cross section, the fields produced by a point charge q moving with constant velocity βc parallel to the axis at the transverse position $\mathbf{r}_b$, are functions of $\mathbf{r}_b, \mathbf{r}, (z-\beta ct)$, where $\mathbf{r}$ is the transverse coordinate of the field-point. The wake force exerted by this charge on another point charge e, travelling at the (axial) distance s behind it on a similar trajectory with transverse coordinates $\mathbf{r}$:

$$\widetilde{\mathbf{F}} = e\left[\widetilde{\mathbf{E}}+\beta c\,\hat{\mathbf{z}}\times\widetilde{\mathbf{H}}\right]_{z-\beta ct\,=-s}\tag{7.95}$$

is a function of $\mathbf{r}_b$, $\mathbf{r}$, and s only. This force can be derived from the wake function per unit length $G_{||}$ as:

$$\widetilde{\mathbf{F}}(\mathbf{r}_b,\mathbf{r},-s) = -eq\,\nabla G_{||}(\mathbf{r}_b,\mathbf{r},-s)\,,\tag{7.96}$$

where $\nabla=\{\hat{\mathbf{z}}\partial_z,\nabla_{\mathbf{r}}\}$.

In pipes with longitudinally varying cross sections, the fields depend separately on z and t, and the Lorentz force thus is *not* only a function of s.

However, one can divide the pipe into subsections $L_i \leq z \leq L_{i+1}$, $i = 1, 2, ...$ with constant cross section, and define:

$$\widetilde{\mathbf{F}}_i = \int_{L_i}^{L_{i+1}} \widetilde{\mathbf{F}} dz \ = \ -\nabla V_i, \quad i = 1, 2, ... \tag{7.97}$$

where $V_i = eqG_i$ is a function of $\mathbf{r}_b, \mathbf{r}$, and s only, and G_i are parts of the wake function [Chao93].

For pipes with a longitudinally invariant cross section, one can chose equal intervals $\Delta L = L_{i+1} - L_i$, and then the wake function per unit length $G_1 = G_{||}/\Delta L$ is not dependent on the axial position. It can be derived from the electric *Debye potential* produced by the leading charge q, which is most readily established by noting that $\partial_s = -\partial_z$. Switching from $(z - \beta ct)$ to the k domain in Eq. (7.92), using Eqs. (7.94), and then reverting back to the s domain yields:

$$G_1(\mathbf{r}_b, \mathbf{r}, s) \ = \ \frac{1}{q} \, (1 - \beta^2) \, \widetilde{\Phi}(\mathbf{r}_b, \mathbf{r}, -s). \tag{7.98}$$

It is convenient to expand the wake function into a double McLaurin series with respect to $\mathbf{r}_b$ and $\mathbf{r}$:

$$\begin{aligned} G_1(\mathbf{r}_b, \mathbf{r}, s) \ &= \ G_1(0,0,s) + \nabla_{\mathbf{r}_b} G_1(\mathbf{r}_b, \mathbf{r}, s)\big|_{\mathbf{r}_b=\mathbf{r}=0} \cdot \mathbf{r}_b \\ &+ \ \nabla_{\mathbf{r}} G_1(\mathbf{r}_b, \mathbf{r}, s)\big|_{\mathbf{r}_b=\mathbf{r}=0} \cdot \mathbf{r} + \text{higher order terms} \, . \end{aligned} \tag{7.99}$$

Note that $G_1(\mathbf{r}_b, 0, s)$ has a minimum at $\mathbf{r}_b = 0$, hence $\nabla_{\mathbf{r}_b} G_1(\mathbf{r}_b, \mathbf{r}, s)\big|_{\mathbf{r}_b=0} = 0$, provided $\mathbf{r} = 0$ is a symmetry axis. Therefore the dominant contributions to the longitudinal and transverse force in the limit $\mathbf{r}_b \to 0$, $\mathbf{r} \to 0$ are given by:

$$\begin{aligned} \widetilde{F}_s \ &= \ -eq \, \partial_s \, G_1(0,0,s) \, , \\ \widetilde{\mathbf{F}}_\perp \ &= \ -eq \, \nabla_{\mathbf{r}_b} \nabla_{\mathbf{r}} \, G_1(\mathbf{r}_b, \mathbf{r}, s)\big|_{\mathbf{r}_b=\mathbf{r}=0} \cdot \mathbf{r}_b \, , \end{aligned} \tag{7.100}$$

which are related to the monopole and dipole terms of the leading charge distribution, respectively.

The Fourier transforms of the wake forces are the impedances (per unit-length), and are defined as follows [Chao93, Wilson89]:

$$\begin{aligned} \frac{Z_{||}(\mathbf{r}_b, \mathbf{r}, \omega)}{L} \ &= \ \frac{1}{\beta c} \int_{-\infty}^{\infty} ds \frac{\widetilde{F}_s}{eq} \exp{(jks)} \, , \\ \frac{\mathbf{Z}_\perp(\mathbf{r}_b, \mathbf{r}, \omega)}{L} \ &= \ \frac{j}{\beta c |\mathbf{r}_b|} \int_{-\infty}^{\infty} ds \frac{\widetilde{\mathbf{F}}_\perp}{eq} \exp{(jks)} \, , \end{aligned} \tag{7.101}$$

where $k = \omega/\beta c$. From this and Eqs. (7.93) one obtains:

$$\frac{Z_{\|}(\mathbf{r}_b, \mathbf{r}, \omega)}{L} = -jk \, \frac{(1-\beta^2)}{\beta c q} \, \widetilde{\Phi}(\mathbf{r}_b, \mathbf{r}, k), \tag{7.102}$$

from which the standard definition of the longitudinal impedance is obtained by setting $\mathbf{r}_b = \mathbf{r} = 0$.

Similarly, in the limit $\mathbf{r} \to 0$ and $\mathbf{r}_b \to 0$, the dominant multipole can be found, i.e. the dipole term of the transverse impedance:

$$\frac{Z_{\perp}(0,0,\omega)}{L} = -j \, \frac{(1-\beta^2)}{\beta c q} \, \nabla_{\mathbf{r}_b} \nabla_{\mathbf{r}} \widetilde{\Phi}(\mathbf{r}_b, \mathbf{r}, k)\Big|_{\mathbf{r}_b=\mathbf{r}=0} \tag{7.103}$$

7.6.2 The Lorentz Reciprocity Theorem

Let $\widetilde{\Phi}_0(\mathbf{r}, \mathbf{r}_b)$ and $\widetilde{\Phi}(\mathbf{r}, \mathbf{r}_b)$ be the electrical Debye potentials produced by the bunch charge densities:

$$\rho_0 = q\delta(\mathbf{r} - \mathbf{r}_b)\delta(z - \beta c t) \; , \quad \rho = q\delta(\mathbf{r} - \mathbf{r}_b)\delta(z - \beta c t) \tag{7.104}$$

in two vacuum pipes differing in the geometry and/or electrical properties of the boundaries.

In the following we shall assume that $\widetilde{\Phi}_0$ is the Debye potential in a perfectly conducting pipe, with a geometry for which the Helmholtz equation is separable (e.g. circular). On the other hand $\widetilde{\Phi}$ refers to a pipe with a complex geometry for which the Helmholtz equation is not separable, and/or with non-trivial boundary conditions. We call the problem described by $\widetilde{\Phi}_0$ and $\widetilde{\Phi}$ as *unperturbed* and *perturbed*, respectively, and assume that the perturbations are *small* in a sense to be clarified below. Then:

$$\begin{aligned} \nabla_{\perp}^2 \widetilde{\Phi}_0^* - k^2(1-\beta^2)\widetilde{\Phi}_0^* &= -\frac{q}{\varepsilon_0}\delta(\mathbf{r} - \mathbf{r}_b) \; , \\ \nabla_{\perp}^2 \widetilde{\Phi} \; - k^2(1-\beta^2)\widetilde{\Phi} &= -\frac{q}{\varepsilon_0}\delta(\mathbf{r} - \mathbf{r}_b) \; , \end{aligned} \tag{7.105}$$

where an asterisk means complex conjugation.

Multiplying the first of Eqs. (7.105) by $\widetilde{\Phi}$, and the second one by $\widetilde{\Phi}_0^*$, subtracting and integrating over the perturbed pipe cross section S, one gets the following reciprocity formula [Petracca95a]:

$$\begin{aligned} \int_S dS \, \left(\widetilde{\Phi} \nabla_{\perp}^2 \widetilde{\Phi}_0^* - \widetilde{\Phi}_0^* \nabla_{\perp}^2 \widetilde{\Phi}\right) &= \oint_C d\ell \; \hat{\mathbf{n}} \cdot \left(\widetilde{\Phi} \nabla_{\perp} \widetilde{\Phi}_0^* - \widetilde{\Phi}_0^* \nabla_{\perp} \widetilde{\Phi}\right) \\ &= \frac{q}{\varepsilon_0}\Big[\widetilde{\Phi}_0^*(\mathbf{r}_b, \mathbf{r}_b) - \widetilde{\Phi}(\mathbf{r}_b, \mathbf{r}_b)\Big] \; , \end{aligned} \tag{7.106}$$

where $d\ell$ is the length element of is the boundary C of S, $\hat{\mathbf{n}}$ is the unit vector normal to the boundary C, and the second Green theorem has been used to convert the surface integral into a contour one. It is expedient to rewrite the l.h.s. of Eq. (7.106) in terms of fields instead of potentials. For this we use Eqs.(7.94):

$$\begin{aligned} \nabla_\perp \widetilde{\Phi}_0^* \cdot \hat{\mathbf{n}} &= -\widetilde{E}_{0n}^{*(irr)} , & \nabla_\perp \widetilde{\Phi} \cdot \hat{\mathbf{n}} = -\widetilde{E}_n^{(irr)} , \\ \widetilde{\Phi}_0^* &= \frac{-\widetilde{E}_{0z}^*}{jk(1-\beta^2)} , & \widetilde{\Phi}\Big|_C = \frac{\widetilde{E}_z\Big|_C}{jk(1-\beta^2)} = \frac{Z_{wall}\,\widetilde{H}_t\Big|_C}{jk(1-\beta^2)} . \end{aligned} \tag{7.107}$$

where the subscripts n and t denote the normal and the tangential field components with respect to C. The second equality in Eq. (7.107) follows from the *Leontovich boundary condition* [Senior81] at the pipe wall:

$$\hat{\mathbf{n}} \times \left[\hat{\mathbf{n}} \times \widetilde{\mathbf{E}} - Z_{wall} \widetilde{\mathbf{H}} \right]_C = 0 . \tag{7.108}$$

With the *free space admittance* $Y_0 = 1/Z_0$, the last expression can also be written:

$$\widetilde{\Phi}\Big|_C = -\frac{Y_0 Z_{wall}}{jk(1-\beta^2)} \left[\beta \widetilde{E}_n^{(irr)} + \beta^{-1} \widetilde{E}_n^{(sol)} \right]_C . \tag{7.109}$$

For a resistive wall with conductivity σ_c at frequency $\omega = kv$, the complex impedance is

$$Z_{wall} = Z_0 \left(1 - j \frac{\sigma_c}{kv\varepsilon_0} \right)^{-1/2} . \tag{7.110}$$

7.6.3 Pipes with Perturbed Boundaries

Now we derive expressions needed for computing the coupling impedance in pipes with complex transverse geometry and boundary conditions [Petracca94], based on the Lorentz reciprocity theorem.

In Eqs. (7.107) and (7.109), the fields $\widetilde{E}_{0n}$ and $\widetilde{E}_{0z}$ refer to the *unperturbed* situation, while $\widetilde{E}_n$, $\widetilde{E}_z$, $\widetilde{H}_t$ refer to the *perturbed* one.

Substituting Eqs. (7.107) into Eqs. (7.109) one gets:

$$\begin{aligned} \frac{1}{jk(1-\beta^2)} \oint_C d\ell \, \Big\{ \, & \widetilde{E}_{0z}^*(\mathbf{r}, \mathbf{r}_b) \widetilde{E}_n^{(irr)}(\mathbf{r}, \mathbf{r}_b) - \frac{Z_{wall}}{Z_0} \widetilde{E}_{0n}^{*(irr)}(\mathbf{r}, \mathbf{r}_b) \times \\ & \left[\beta \widetilde{E}_n^{(irr)}(\mathbf{r}, \mathbf{r}_b) + \beta^{-1} \widetilde{E}_n^{(sol)}(\mathbf{r}, \mathbf{r}_b) \right] \Big\} = -\frac{q}{\varepsilon_0} \left[\widetilde{\Phi}_0^*(\mathbf{r}_b, \mathbf{r}_b) - \widetilde{\Phi}(\mathbf{r}_b, \mathbf{r}_b) \right] . \end{aligned} \tag{7.111}$$

In the limit $\mathbf{r}$ and $\mathbf{r}_b \to 0$ we get from this and Eq. (7.102):

$$\begin{aligned}\frac{Z_{\|}(\omega)}{L} &= Z_{\|0}(\omega) - \frac{\varepsilon_0}{\beta c q^2} \oint_C d\ell \, \Big\{ \widetilde{E}^*_{0z}(\mathbf{r},0)\, \widetilde{E}^{(irr)}_n(\mathbf{r},0) \\ &+ \frac{Z_{wall}}{Z_0} \widetilde{E}^{(irr)*}_{0n}(\mathbf{r},0) \left[\beta \widetilde{E}^{(irr)}_n(\mathbf{r},0) + \beta^{-1} \widetilde{E}^{(sol)}_n(\mathbf{r},0)\right] \Big\} \,, \qquad (7.112)\end{aligned}$$

where we used the well known fact that the longitudinal impedance of a perfectly conducting straight pipe is zero.

Furthermore, using the Panofsky-Wentzel theorem by first applying the differential operator[1] $\nabla_{\mathbf{r}_b} \cdot \nabla_{\mathbf{r}_b}$ to both sides of the identity (7.111), then letting $\mathbf{r}$ and $\mathbf{r}_b \to 0$, and finally using Eq. (7.103) one gets:

$$\begin{aligned}\frac{Z_{\perp}(\omega)}{L} &= \frac{Z_{\perp 0}(\omega)}{L} + \frac{\varepsilon_0}{\beta c k q^2} \oint_C d\ell \, \Big\{ \nabla_{\mathbf{r}_b} \widetilde{E}^*_{0z}(\mathbf{r},0) \cdot \nabla_{\mathbf{r}_b} \widetilde{E}^{(irr)}_n(\mathbf{r},0) \\ &- \frac{Z_{wall}}{Z_0} \nabla_{\mathbf{r}_b} \widetilde{E}^{(irr)*}_{0n}(\mathbf{r},0) \cdot \nabla_{\mathbf{r}_b} \left[\beta \widetilde{E}^{(irr)}_n(\mathbf{r},0) + \beta^{-1} \widetilde{E}^{(sol)}_n(\mathbf{r},0)\right] \Big\} \,. \qquad (7.113)\end{aligned}$$

The impedances are independent of the charge, since the fields in Eqs. (7.112) and (7.113), which relate the *perturbed* and the *unperturbed* coupling impedances, are proportional to the charge q. They show the different roles played by the boundary and material perturbations in a particularly clear way. The r.h.s.'s of Eqs. (7.112) and (7.113) consist of two terms:

- the first term is non-zero when the *unperturbed* axial field component $\widetilde{E}_{0z}$ is not identically zero on the boundary C, and accounts for the effect of a geometrical perturbation of the boundary. Therefore the first integral in Eqs. (7.112) and (7.113) effectively spans only the *geometric perturbation* $C - C_0$;

- the second integral, on the other hand, is non-zero when Z_{wall} is not identically zero on the boundary, and accounts for the effect of the constitutive properties of the wall.

In general, the *local* boundary conditions could also vary from point to point along the boundary C, e.g. for walls consisting of different materials.

[1] The transverse impedance is in general a tensor [Palumbo87].

7.6.4 Approximate Perturbative Formulae

Eqs.(7.112) and (7.113) are exact, but contain unknown fields. If the fields in the *perturbed* geometry were known, one could calculate the coupling impedances directly from their definitions Eqs. (7.102) and (7.103). Nevertheless, these equations can be useful as approximate expressions for the impedances if one replaces the fields in the perturbed geometry by those produced in the unperturbed one by the same source term.

The error introduced by the substitution $\widetilde{\mathbf{E}} \to \widetilde{\mathbf{E}}_0$ is small, as it is at least of second order in the parameters $\delta \approx measure\ (C - C_0)$ and $\Delta \approx {Z_{wall}}^{pert}/{Z_{wall}}^{unpert}$ which were assumed to be small.

When the true coupling impedance depends *linearly* on Δ and/or δ, this procedure actually gives the exact solution. This happens, e.g., for a circular pipe of radius b with a resistive wall. In this case Eqs. (7.112) and (7.113) yield the results discussed in Chapter 6:

$$\frac{Z_{\parallel}}{L} = -\frac{Z_{wall}}{2\pi b}, \qquad \frac{Z_{\perp}}{L} = \frac{Z_{wall}}{2\pi k b^3}. \tag{7.114}$$

An approximate expressions for impedances of a beam pipe with a square cross section and rounded corners, which may be made of different metals, can be found in ref.[Petracca95b].

Appendix 7.A Trapped Modes near Cutoff

Trapped modes just below the cutoff frequency of the beam pipe may exist due to small obstacles such as widening or holes in the wall. Such trapped modes are narrow resonances, which have been predicted analytically [Stupakov94, Kurennoy95b] and observed in measurements [Caspers95]. Trapped modes may have adverse effects on the accelerator performance; e.g. they may cause (additional) heating of the beam pipe and, under certain conditions, they may influence the motion of a bunch in a (multi-bunch) accelerator. Here we present some results of calculations of the impedance associated with these modes.

We consider an axially symmetric, perfectly conducting beam pipe of radius b, with a small, axially symmetric *enlargement* of the area S of the cross section in the rz plane. Such an obstacle shifts the resonant frequencies of the modes downwards. In particular, there are usually one or more modes with resonances slightly below the cutoff frequency of the beam pipe. Such a mode cannot propagate in the pipe, and therefore remains localised around the obstacle. On the other hand, a small *indentation* shifts the frequency upwards, and "trapping" of modes does not occur.

For small obstacles the shift of the frequency of the trapped mode is also small. Under these circumstances, the *EM* field penetration into the beam pipe, characterized by the *localisation length* L, may become large in comparison to the pipe radius. The importance of the impedance contributions of such fields, with frequencies Ω_{ms} slightly below the cutoff frequencies of the beam pipe $\omega_{ms} = j_{ms}c/b$, in particular for vacuum chamber "liners" with many holes, were pointed out in ref. [Stupakov94].

At large distances $|z| \gg b$ from the obstacle, the *EM* field components of the lowest TM axially symmetric ($m = 0$) *trapped mode* are approximated by:

$$\begin{aligned}
\widetilde{E}_r^{trap} &= \operatorname{sign}(z)\frac{j_{01}k_{01}}{b}J_1(\frac{j_{01}r}{b})\exp(-k_{01}|z|) , \\
\widetilde{E}_z^{trap} &= (\frac{j_{01}}{b})^2 J_0(\frac{j_{01}r}{b})\exp(-k_{01}|z|) , \\
\widetilde{H}_\theta^{trap} &= -j\frac{j_{01}k}{Z_0 b}J_1(\frac{j_{01}r}{b})\exp(-k_{01}|z|) . \qquad (7.115)
\end{aligned}$$

The *localisation length* L is the inverse of the decrement $k_{01} = \sqrt{\omega_{01}^2 - \Omega_{01}^2}/c$. Using the reciprocity theorem for the *EM* fields excited by a beam in a pipe with and without enlargement [Stupakov94] one finds approximately:

$$\delta\omega_{01} \approx \omega_{01}\frac{j_{01}^2}{2}(\frac{S}{b^2})^2 \ . \tag{7.116}$$

and the decrement

$$k_{01} \approx \frac{j_{01}^2 S}{b^3} \ . \tag{7.117}$$

For metal walls with large but finite conductivity σ_c, the resonance acquires a small imaginary part Γ_{01} describing the energy dissipation in the walls. The damping rate of the oscillations then is:

$$\Gamma_{01} \approx \frac{\omega_{01}\delta_s}{2b} \ , \tag{7.118}$$

where $\delta_s = \sqrt{2/(\mu_0\sigma_c\omega_{01})}$ is the skin depth. The trapped mode disappears when Γ_{01} becomes larger than $\delta\omega_{01}$.

The longitudinal impedance produced by the trapped mode can be represented by a resonance [Kurennoy95b] with frequency $\Omega_{01} = \omega_{01} - \delta\omega_{01}$:

$$Z_{\parallel}^{trap}(\omega) = \frac{-2j\Omega_{01}\Gamma_{01}R_{s01}^{trap}}{\omega^2 - (\Omega_{01} + j\Gamma_{01})^2} \ , \tag{7.119}$$

where the shunt impedance

$$R_{\parallel 01}^{trap} = \frac{4Z_0 j_{01} S^3}{\pi\delta_s b^5 J_1^2(j_{01})} \ . \tag{7.120}$$

In the vicinity of the resonance, the last two results are in agreement with Eqs. (7.66) and (7.67) obtained in Section 7.2. In the limit of the perfect conductivity $\delta_s \to 0$, the resonant impedance is described by a delta function.

Higher order modes can be trapped in a similar way. The lowest transverse *TM* resonance occurs near the cutoff frequency $\omega_{11} = j_{11}c/b$, where j_{11} is the first root of the Bessel function J_1. The transverse impedance of the trapped mode becomes a vector in the direction of the bunch displacement $\mathbf{r}_b$, and is described again by a resonator with frequency Ω_{11}:

$$\mathbf{Z}_{\perp}^{trap}(\omega) = \frac{-2j\Omega_{11}\Gamma_{11}R_{\perp 11}^{trap}}{\omega^2 - (\Omega_{11} + j\Gamma_{11})^2}\,\hat{\mathbf{r}}_b \ , \tag{7.121}$$

The corresponding transverse shunt impedance (per unit displacement) is:

$$R_{\perp 11}^{trap} = \frac{2 Z_0 j_{11} S^3}{\pi \delta_s b^6 J_0^2(j_{11})} \; . \tag{7.122}$$

Similar considerations and results apply also to the case of a small hole in the beam pipe, but in this case the structure is no longer axially symmetric. In the longitudinal case, the equations for a hole remain the same as Eqs. (7.116) and (7.120) for an enlargement, if the area S is replaced by $\alpha_M/(4\pi b)$, where α_M is the magnetic polarisability of the hole. For a circular hole with radius h it is $4h^3/3$.

In the transverse case, S should be replaced by $\alpha_M/(2\pi b)$ and, in addition, the unit vector $\hat{\mathbf{r}}_b$ by $\hat{\mathbf{r}}_h \cos(\theta_b - \theta_h)$, where $\hat{\mathbf{r}}_h$ is the 2-D unit vector in the direction of the hole, and θ_b, θ_h are the azimuthal angles in the transverse plane of the beam and the hole, respectively. Hence, the deflecting force depends on the angle between the directions of the beam offset and the hole.

Chapter 8

Field Matching Techniques

The wave equation is a (vectorial) partial differential equation and thus cannot be solved directly. However, in a limited number of coordinate systems it is possible to reduce it to a system of ordinary differential equations by the method called *separation of variables.* If the structure under consideration is a coordinate surface in such a system, analytical solutions can be found. Only very few structures can be described in this way - e.g. uniform cylindrical pipes with circular or elliptic cross sections. However, one can also find analytic solutions for the *EM* fields excited by a particle beam passing a structure which can be divided into a number of *regions*, which are bounded by *piecewise constant coordinate surfaces.* They are generally given in terms of infinite series over transcendental functions, which are often difficult to evaluate numerically. But the integrals of their products are always orthogonal over the region considered, a property which can be used to simplify the resulting equations.

The *EM* fields have to fulfill boundary conditions at the walls of the structure, and also have to be continuous across the interfaces between neighbouring regions. The equations expressing these *matching conditions* contain infinite series with unknown expansion coefficients for each region. Using the orthogonality of the expansion functions these coefficients in any one region can be obtained explicitly in terms of infinite sums over coefficients in adjacent regions. In this manner, one can reduce the problem to a single system of linear, algebraic equations for one set of coefficients, from which one can determine all the others. This approach is generally known as the *field matching technique.*

8.1 Rotationally Symmetric Structures

Field matching is most often applied to axially symmetric structures with piecewise circular cylindrical boundaries. The cylindrical coordinate system r, θ, and z permits separation of the variables of the wave equations, and the solutions are expressed in terms of infinite series of simple and well known exponential and Bessel functions, although in general of complex argument. Additional constraints reduce the number of possible solutions, such as the *radiation condition*, which imposes that only waves traveling away from the source can be excited by a beam, or the requirement that the fields must be periodic with the circumference of a circular machine. The matching conditions then lead to infinite systems of linear equations for the expansion coefficients, which are reduced by substitutions to a single, but still infinite, system of equations for just one set of coefficients.

When this infinite system of equations is truncated to a finite size for numerical evaluation, the solutions do not necessarily converge to the correct values, unless the truncation is chosen such that the solutions fulfill the *edge condition* [Meixner49], which requires that the divergence of certain field components at sharp corners or edges follows a definite power law depending only on their shape. If these conditions are not fulfilled, incorrect results may be obtained even when truncating at a very large number of terms.

The *EM* fields of a bunch of charged particles, moving with constant velocity along the axis of a metallic pipe of constant cross-section, can be visualised as a wave packet accompanying the charge. Electric field lines originate at the bunch and end at the pipe surface, where *induced charges* - often incorrectly called *image charges* - appear to be traveling at the same speed as the particles in the beam. In reality, the charges in the wall are displaced only by small amounts from their equilibrium positions around the stationary ions of the metal lattice, just as in normal electric conduction. The charge maxima travel with the *phase velocity* equal to the beam velocity. When the wall is perfectly conducting, the electric field lines must be at right angle to it, while the magnetic field lines, which circulate around a moving charge, must remain tangential to the wall surface. For ultra-relativistic velocities, the *EM* fields are limited to a disk of the same thickness as the length of the bunch.

We now consider what happens when the diameter of an otherwise smooth beam pipe suddenly increases or decreases. The induced charges must change the direction of their (apparent) movement, which corresponds to a transverse acceleration, and they will therefore radiate *EM* fields. The energy required to create these fields has to be supplied by the beam, but the beam trajectory and energy will not be influenced measurably if its motion is ultra-relativistic.

The *EM* fields in non-uniform geometries can also be analysed from a different point of view, which produces an equivalent result, e.g. one can consider the diffraction of waves into which the *EM* fields can be decomposed. This approach will be discussed in more detail in one of the following Chapters.

Here we discuss the approach imposing appropriate boundary conditions, and matching the *EM* fields at all interfaces between different regions. These conditions have to be fulfilled by the *total EM fields*, consisting of the sum of those created directly by the charges in a bunch, often called *space charge fields*, and the *radiated fields* caused by the induced charges at the boundaries. The latter fields are solutions of the homogeneous Maxwell (or wave) equations, i.e. with no charge or current sources present.

After a bunch has passed a structure, some of the *EM* fields continue to oscillate in it, thereby creating what are appropriately called *wake fields*. After a sufficiently long time, most *EM* fields excited by the bunch passage cancel each other due to multiple reflections from the walls leading to addition at arbitrary phases. Only those with frequencies close to the *eigenfrequencies* of the structure, will interfere constructively when reflected back and forth. They represent the main bulk of *EM* fields left in the structure behind the charge. When these frequencies lie below the cutoff frequency of the beam pipe ω_{co}, the corresponding fields cannot leave the cavity. For small resistive losses, as is usually the case for well-conducting metal walls, they oscillate for many periods, corresponding to a high *quality factor* $Q \gg 1$. That creates sharp peaks in the coupling impedance, which is therefore called *narrow-band impedance*, in contrast to the *broad-band impedance*, corresponding to the short-range wake over distances too small to resolve single resonances.

8.1.1 Wave Equation in Cylindrical Coordinates

As discussed in Chapter 2, in a region with constant permittivity ε and permeability μ, the Maxwell equations can be transformed into the wave equations [Jackson75]. Assuming a time dependence $\exp(j\omega t)$, they become the *Helmholtz equations*:

$$\begin{aligned}
[\Delta + \varepsilon\mu\omega^2]\,\widetilde{\mathbf{E}} &= j\omega\mu\tilde{\mathbf{J}} + \frac{1}{\varepsilon}\,\nabla\,\tilde{\rho}\,, \\
[\Delta + \varepsilon\mu\omega^2]\,\widetilde{\mathbf{H}} &= -\,\nabla\times\,\tilde{\mathbf{J}}\,,
\end{aligned} \tag{8.1}$$

where Δ is the vector Laplacian operator. In all but Cartesian coordinates, the electric, as well as the magnetic, field components in different coordinate directions are in general coupled by these equations, or by the boundary conditions. In circular cylindrical coordinates for axially symmetric structures, however, the wave equation for the longitudinal component of the electric field does not contain any transverse field components and becomes decoupled from them:

$$\frac{1}{r}\frac{\partial}{\partial r}\left(r\frac{\partial \widetilde{E}_z}{\partial r}\right)+\frac{\partial^2 \widetilde{E}_z}{\partial z^2}+\frac{1}{r^2}\frac{\partial^2 \widetilde{E}_z}{\partial \theta^2}+\varepsilon\mu\omega^2\widetilde{E}_z=\frac{1}{\varepsilon}\frac{\partial\tilde{\rho}}{\partial z}+j\omega\mu\tilde{J}_z\,. \tag{8.2}$$

A similar equation holds for the longitudinal magnetic field $\widetilde{H}_z$, but the r.h.s. is the z-component of $\nabla\times\,\tilde{\mathbf{J}}$, which vanishes for a purely longitudinal particle motion. Hence the corresponding TE modes are *not* excited by such a current.

For rotationally symmetric ($m=0$) TM modes in regions without sources, the transverse field components can be obtained from the longitudinal electric field by the relations:

$$\begin{aligned}\left(\frac{\partial^2}{\partial z^2}+\varepsilon\mu\omega^2\right)\widetilde{E}_r &= \frac{\partial^2\widetilde{E}_z}{\partial r\partial z}\,,\\ \left(\frac{\partial^2}{\partial z^2}+\varepsilon\mu\omega^2\right)Z_0\widetilde{H}_\theta &= -j\sqrt{\varepsilon\mu}\omega\frac{\partial\widetilde{E}_z}{\partial r}\,.\end{aligned} \tag{8.3}$$

8.1.2 Separation of Variables

Here we concentrate on rotationally symmetric *EM* fields with azimuthal mode number $m=0$. Then all derivatives with respect to the azimuthal variable θ vanish. We consider a coasting beam of circular cross section with radius a, uniform charge density $\tilde{\rho}$, and purely axial velocity $v=\beta c$ where $\beta\to 1$. It is modulated by a (small) perturbation of the charge density $\rho(z)=\tilde{\rho}\exp(-jkz)$, where a factor $\exp(j\omega t)$ is understood for all source or field components. This perturbation is a wave, traveling in the positive z- direction with the speed of the beam $\omega/k=\beta c$. The only non-zero component of the corresponding current density is $\tilde{J}_z=\beta c\tilde{\rho}$.

In vacuum, $\varepsilon=\varepsilon_0$ and $\mu=\mu_0$ and, with $\varepsilon_0\mu_0=1/c^2$ the r.h.s. of the wave equation for $\widetilde{E}_z$ becomes $(-jk/\varepsilon_0+j\omega\mu_0\beta c)\tilde{\rho}=-jk\tilde{\rho}\exp(-jkz)/\gamma^2\varepsilon_0$. When $\tilde{\rho}$ is independent of the radial coordinate r, i.e. for a uniform transverse distribution, a particular solution of the inhomogeneous wave equation can have the same property. In this case the operator on the l.h.s. of the wave equation becomes $-k^2 + \omega^2/c^2 = -k^2/\gamma^2 \equiv -\tau^2$, and a particular solution is given by:

$$\widetilde{E}_z^{(p)} = \frac{j\tilde{\rho}}{k\varepsilon_0}\exp(-jkz) \tag{8.4}$$

In order to be able to fulfill the boundary conditions, one has to add a solution of the homogeneous wave equation. We seek solutions in product form $\widetilde{E}_z(r,z) = R(r)\,Z(z)$, which permits separation of a partial differential equation into a set of ordinary ones. They contain an arbitrary separation constant, and the general solution is found by summing over all its possible values. One thus obtains the two differential equations:

$$\begin{aligned} \frac{1}{r}\frac{d}{dr}\left(r\frac{dR}{dr}\right) - \left(\lambda^2 - \frac{\omega^2}{c^2}\right)R &= 0 \\ \frac{d^2Z}{dz^2} + \lambda^2\, Z &= 0\,, \end{aligned} \tag{8.5}$$

where λ^2 is the separation constant. The solutions of the axial equation can be written $Z(z) = \exp(\pm j\lambda z)$. Periodicity with the circumference $2\pi R$ of a circular machine requires that $\lambda = p/R = k_p$ should take only the discrete values, where p is an arbitrary integer. Thus all field components are superpositions of terms with equally spaced longitudinal wave numbers, which are usually called *space harmonics* and are typical for periodic structures. In particular, for the harmonic with the same mode number as the perturbation, which we call n, $k_n = k = \omega/\beta c$.

The radial functions are then given by the differential equation:

$$\frac{1}{r}\frac{d}{dr}\left(r\frac{dR}{dr}\right) - \tau_p^2 R = 0\,, \tag{8.6}$$

where

$$\tau_p = \sqrt{k_p^2 - \omega^2/c^2} \quad \text{with} \quad k_p = p/R \tag{8.7}$$

is the *radial propagation constant.* The solutions of this equation are modified Bessel functions of order zero, I_0 or K_0, with argument $\tau_p r$. They could also be written as ordinary Bessel functions of argument $j\tau_p r$. However, when the frequency is not too high, i.e. $\omega < k_p c$, the modified Bessel functions are preferable since their arguments are then real.

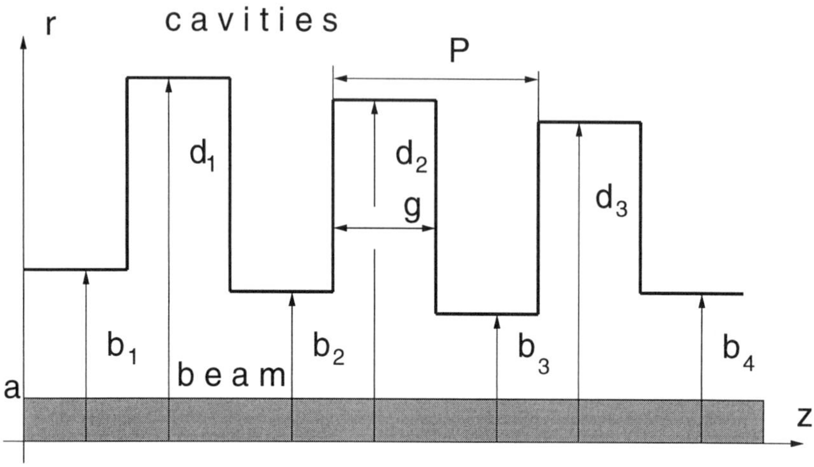

Figure 8.1: Periodic cavity chain excited by beam current.

8.2 Periodically Widened Beam Pipes

We first apply this technique to a coasting beam inside a so-called "corrugated" vacuum chamber with circular cross section [Keil72b]. This geometry consists of a straight, cylindrical beam of radius a, inside a concentric beam pipe of radius b, which is widened to radius d over a length g, repeating with period $P = 2\pi R$. The structure forms a chain of cylindrical cavities connected by coaxial *beam tubes*, as sketched in Fig. 8.1.

We divide its volume into 3 regions bounded by coordinate surfaces:

i) the "beam region" $0 \leq r \leq a$, identified by the superscript I;

ii) the "pipe region" $a < r \leq b$, identified by the superscript II;

iii) the "cavity region" $b < r \leq d$ and $0 < z_M < g$, which will be identified by the superscript III, where $z_M = z - MP$ is the *local longitudinal coordinate* in the M-th cavity.

In the beam region, only modified Bessel functions of the first kind I_0 are admissible, since those of the second kind K_0 diverge for $r \to 0$. The general solution is found by summing the products of axial and radial functions over all permitted propagation constants k_p or τ_p, i.e. over all integers p. Each term is multiplied by a yet unknown expansion coefficient A_p which will be determined by the boundary conditions. A particular solution of the inhomogeneous equation, given by Eq. (8.4), must be added in this region. The multiplicative factor $j\tilde{\rho}/k\varepsilon_0$ can be absorbed in the expansion coefficients. The transverse field component $\widetilde{H}_\theta$ can be found from Eq. (8.3), whereas $\widetilde{E}_r$ will not be needed here since it is sufficient to match any 2 of the 3 non-zero field components, the third one is then automatically continuous. With $\partial/\partial z = -jk_p$ and $\partial^2/\partial z^2 + \omega^2/c^2 = \tau_p^2$ one can form the derivatives term by term to get:

$$\begin{aligned} \widetilde{E}_z^I &= \sum_{p=-\infty}^{\infty} \left[A_p I_0(\tau_p r) + \delta_{pn}\right] \exp(-jk_p z) , \\ Z_0 \widetilde{H}_\theta^I &= \frac{j\omega}{c} \sum_{p=-\infty}^{\infty} A_p \frac{I_1(\tau_p r)}{\tau_p} \exp(-jk_p z) . \end{aligned} \tag{8.8}$$

Here δ_{pn} is the *Kronecker symbol*, which is zero for $p \neq n$ and unity for $p = n$.

In the pipe region surrounding the beam, the fields can be written as superpositions of modified Bessel functions of both kinds, with the coefficients B_p and C_p:

$$\begin{aligned} \widetilde{E}_z^{II} &= \sum_{p=-\infty}^{\infty} \left[B_p I_0(\tau_p r) + C_p K_0(\tau_p r)\right] \exp(-jk_p z) , \\ Z_0 \widetilde{H}_\theta^{II} &= \frac{j\omega}{c} \sum_{p=-\infty}^{\infty} \left[B_p \frac{I_1(\tau_p r)}{\tau_p} - C_p \frac{K_1(\tau_p r)}{\tau_p}\right] \exp(-jk_p z) . \end{aligned} \tag{8.9}$$

The above expressions, Eqs. (8.8) and (8.9), represent the decomposition of the *EM* fields into *traveling waves*. Terms with $p > 0$ correspond to diffracted, and those with $p < 0$ to reflected waves. Equating the field components $\widetilde{E}_z$ and $\widetilde{H}_\theta$ at the beam edge $r = a$ term by term yields directly the unknown coefficients B_p and C_p expressed by A_p. The result can be written:

$$\begin{aligned} \widetilde{E}_z^{II} &= \sum_{p=-\infty}^{\infty} \left[A_p I_0(\tau_p r) + \tau a T_0(\tau r)\delta_{pn}\right] \exp(-jk_p z) , \\ Z_0 \widetilde{H}_\theta^{II} &= \frac{j\omega}{c} \sum_{p=-\infty}^{\infty} \left[A_p \frac{I_1(\tau_p r)}{\tau_p} + aT_1(\tau r)\delta_{pn}\right] \exp(-jk_p z) , \end{aligned} \tag{8.10}$$

where $\tau \equiv \tau_n = (k^2 - k^2\beta^2)^{1/2} = k/\gamma$ and

$$T_{0,1}(\tau r) = K_1(\tau a) I_{0,1}(\tau r) \pm I_1(\tau a) K_{0,1}(\tau r) . \tag{8.11}$$

These function have the properties $T_0(\tau a) = 1/\tau a$, while $T_1(\tau a) = 0$.

In order to fulfill the boundary conditions at both sides of the cavity, at $z_M = 0$ and $z_M = g$, we replace the exponential function by a circular one to describe the axial dependence in region III. For perfectly conducting side walls, the radial electric field component must vanish there. Taking $\widetilde{E}_r \propto \sin \mu_s z$, this will happen if we chose

$$\mu_s = \pi s/g \ , \tag{8.12}$$

where s is an arbitrary integer. It may be taken as positive without loss of generality since terms with index $-s$ are indistinguishable from those with $+s$. However, although the radial electric field is zero for $s = 0$, one must include this term since neither of the corresponding components $\widetilde{E}_z$ and $\widetilde{H}_\theta$ vanish, though they are independent of z. The radial functions can be written as judiciously chosen combinations of modified Bessel functions of both kinds:

$$\begin{aligned} R_{0,1}(\Gamma r) &= K_0(\Gamma d) I_{0,1}(\Gamma r) \mp I_0(\Gamma d) K_{0,1}(\Gamma r) \ , \\ S_{0,1}(\Gamma r) &= -K_1(\Gamma d) I_{0,1}(\Gamma r) \pm I_1(\Gamma d) K_{0,1}(\Gamma r) \ , \end{aligned} \tag{8.13}$$

which have the properties $R_0(\Gamma d) = S_1(\Gamma d) = 0$ and $R_1(\Gamma d) = -S_0(\Gamma d) = 1/\Gamma d$, which follows from the Wronskian of the modified Bessel functions [Abramowitz65].

With the expansion coefficients D_s and E_s, we can then write the *EM* fields in the cavity region as expansions into *standing waves*:

$$\begin{aligned} \widetilde{E}_z^{III} &= \sum_{s=0}^{\infty} \left[D_s R_0(\Gamma_s r) + E_s S_0(\Gamma_s r) \right] \cos \mu_s z \ , \\ Z_0 \widetilde{H}_\theta^{III} &= \frac{j\omega}{c} \sum_{s=0}^{\infty} \left[D_s \frac{R_1(\Gamma_s r)}{\Gamma_s} - E_s \frac{S_1(\Gamma_s r)}{\Gamma_s} \right] \cos \mu_s z \ , \\ \widetilde{E}_r^{III} &= \sum_{s=0}^{\infty} \frac{\mu_s}{\Gamma_s} \left[D_s R_1(\Gamma_s r) - E_s S_1(\Gamma_s r) \right] \sin \mu_s z \ , \end{aligned} \tag{8.14}$$

where the *radial propagation constants* Γ_s are given by

$$\Gamma_s^2 = \mu_s^2 - \omega^2/c^2 \ . \tag{8.15}$$

If the outer cavity wall at $r = d$ is perfectly conducting, the tangential electric field must vanish there, i.e. $\widetilde{E}_z^{III} = 0$. Since $R_0(\Gamma_s d) = 0$, while $S_0(\Gamma_s d) = -1/\Gamma d$ is non-zero, it follows that $E_s = 0$, and only the coefficients D_s still need to be determined. If the pipe walls are perfectly conducting, the fields must fulfil the following conditions at $r = b$:

- $\widetilde{E}_z^{II} = \widetilde{E}_z^{III}$,
- $\widetilde{H}_\theta^{II} = \widetilde{H}_\theta^{III}$ for $0 < z < g$,
- $\widetilde{E}_z^{II} = 0$ for $g < z < 2\pi R$.

Since the *EM* fields in region II are periodic with $2\pi R$, the matching conditions are also fulfilled at all other cavity gaps if we replace z by $z_M = z \pm 2\pi MR$ in region III. Omitting the arguments of all functions and replacing them by corresponding subscripts, we obtain the relations:

$$\sum_{p=-\infty}^{\infty} \Big[A_p I_{0b} + \tau a T_{0b}\delta_{ps}\Big] e^{-jk_p z} = \begin{cases} \sum_{s=0}^{\infty} D_s R_{0b} \cos \mu_s z & \text{for } 0 < z < g \; , \\ 0 & \text{for } g < z < 2\pi R \; , \end{cases}$$

$$\sum_{p=-\infty}^{\infty} \Big[A_p \frac{I_{1b}}{\tau_p} + \; aT_{1b}\delta_{ps}\Big] e^{-jk_p z} = \; \sum_{s=0}^{\infty} D_s \frac{R_{1b}}{\Gamma_s} \cos \mu_s z \;\; \text{for } 0 < z < g \; . \quad (8.16)$$

Multiplying the first of these equations with $\exp{(jk_{p'} z)}$ and integrating over z from 0 to $2\pi R$ yields, due to the orthogonality of the exponential functions in that interval, an expression for the coefficients A_p in terms of an infinite sum, containing the unknown coefficients D_s:

$$A_p \; I_{0b} + \tau a \; T_{0b} \; \delta_{ps} = \alpha \sum_{s=0}^{\infty} \; N_{ps} \; D_s \; R_{0b} \; . \quad (8.17)$$

Here $\alpha = g/2\pi R$ is the *circumference factor*, and the *kernel* is:

$$N_{ps} = \int_0^g \frac{dz}{g} e^{jk_p z} \cos \mu_s z = \frac{(\pi \alpha p) \; e^{j\pi\alpha p}}{(\pi\alpha p)^2 - (\pi s/2)^2} \begin{cases} \sin(\pi\alpha p) & \text{for } s \text{ even} \; , \\ j\cos(\pi\alpha p) & \text{for } s \text{ odd} \; . \end{cases} \quad (8.18)$$

Similarly, the second Eq. (8.16) can be solved for the coefficients D_s by multiplying by $\cos(\nu_s z)$ and integrating from 0 to g. Due to the orthogonality of the circular functions in that interval one gets:

$$\epsilon_{s0} \frac{R_{1b}}{\Gamma_s} D_s = \sum_{p=-\infty}^{\infty} \left[\frac{I_{1b}}{\tau_p} A_p + aT_{1b}\delta_{ps}\right] N_{ps}^* \; , \quad (8.19)$$

where ϵ_{s0} is the *Neumann symbol* which equals 1 for $s = 0$ and 2 otherwise, and N_{ps}^* is the complex conjugate of N_{ps}.

8.2.1 Coupling Impedance

The coupling impedance $Z_{\|}(\omega)$ is defined as the integral over the longitudinal electric field component normalised by the beam current and averaged over the beam cross section. The integration should extend sufficiently far on both sides of a structure that the fields become negligible, or around the whole circumference of circular machines. For these, the impedance divided by the mode number $n = \omega/\omega_0$ is usually more interesting. By integrating E_z in Eq. (8.8) the latter can be expressed in terms of the expansion coefficient A_n:

$$\frac{Z_{\|}}{n}(\omega) = \frac{2jZ_0}{\beta(ka)^2}\left[\frac{2I_1(\tau a)}{\tau a}A_n + 1\right] \approx \frac{2jZ_0}{\beta(ka)^2}\left[A_n + 1\right] , \qquad (8.20)$$

where the last approximation is good for $ka \ll \gamma$, which is usually the case for not extremely high frequency. The impedance is thus entirely determined by the synchronous component of the electric field, traveling with the speed of the beam (and hence its modulation). This is the only field component which interacts on average with the beam.

The equations for the expansion coefficients can be written more compactly in matrix form, but one has to remember that the index p runs from $-\infty$ to $+\infty$, while s runs from 0 to ∞. We furthermore introduce the abbreviations:

$$\begin{aligned}
\bar{A}_p &= A_p I_0(\tau_p b) , & \bar{D}_s &= D_s R_0(\Gamma_s b) , \\
\bar{B}_p &= \tau a T_0(\tau b)\delta_{ps} , & \bar{C}_p &= \frac{a}{b}T_1(\tau b)\delta_{ps} , \\
I_{pp} &= \left[\frac{I_1(x)}{x\, I_0(x)}\right]_{\tau_p b} , & R_{ss} &= \epsilon_{s0}\left[\frac{R_1(x)}{xR_0(x)}\right]_{\Gamma_s b} ,
\end{aligned} \qquad (8.21)$$

where ϵ_{s0} was defined after Eq. (8.19). With this and Eq. (8.17) one then obtains:

$$\begin{aligned}
\bar{A}_p + \bar{B}_p &= \alpha \sum_{s=0}^{\infty} N_{ps}\bar{D}_s \\
R_{ss}\,\bar{D}_s &= \sum_{p=-\infty}^{\infty} (I_{pp}\bar{A}_p + \bar{C}_p)N^*_{ps} .
\end{aligned} \qquad (8.22)$$

In matrix notation, one can write these equations as:

$$\begin{aligned}
\bar{A} + \bar{B} &= \alpha\, N\bar{D} \\
R\,\bar{D} &= N^+(I\bar{A} + \bar{C}) ,
\end{aligned} \qquad (8.23)$$

where $N^+_{sp} = N^*_{ps}$ is the Hermitian conjugate of N_{ps}. Substituting the second equation into the first one, an infinite system of linear equations for the unknown coefficients A_p is obtained which can be written:

$$[U - \alpha W I]\bar{A} = -\bar{B} + \alpha W \bar{C} \ , \tag{8.24}$$

where U is the unit matrix and $W = NR^{-1}N^+$. The coefficients $\bar{A}$ can be found, at least in principle, by inverting this infinite system of linear equations[1].

For $\alpha = 0, \bar{A}_n = -\bar{B}_n$, and one obtains the expression for the impedance of a smooth, perfectly conducting beam pipe already derived before. In the approximation $ka \ll \gamma$, one finds $A_n = -\bar{B}_n \to -1$ and the impedance of a smooth pipe, given by Eq. (8.20), vanishes.

Here we will evaluate the additional contributions due to the corrugation, which is given by $A_n + B_n$. When $\alpha \ll 1$ one gets in first order:

$$\bar{A}_n + \bar{B}_n = \alpha W_{nn}[\bar{C}_n - I_{nn}\bar{B}_n] = \alpha W_{nn}\frac{a}{b}\frac{I_{1a}}{\tau b \ I_{0b}} \ , \tag{8.25}$$

where the last expression has been simplified with the help of the relation $I_{0b}T_{1b} - I_{1b}T_{0b} = 2I_{1a}/\tau b$ following from the Wronskian [Abramowitz65].

At low frequencies $\omega \ll c/g$, we may furthermore replace the sum for W_{nn} by its lowest term $|N_{n0}|^2 R_{00}^{-1}$. From Eq. (8.18) one finds for $s = 0$:

$$|N_{n0}|^2 = \left[\frac{\sin \pi\alpha n}{\pi\alpha n}\right]^2 \approx 1 \ . \tag{8.26}$$

For $\Gamma_0^2 = -(\omega b/c)^2 \ll 1$, one can use the small argument approximations for modified Bessel functions to get $R_{0b} \approx \ln(b/d)$ and $R_{1b} \approx 1/\Gamma_0 b$. This yields:

$$W_{nn} \approx R_{00}^{-1} = (\beta k b)^2 \ln\frac{d}{b} \ . \tag{8.27}$$

Under these conditions, also $\bar{A}_n \approx A_n$, and one gets the additional coupling impedance from Eq. (8.20):

$$\frac{Z_{\parallel}}{n}(\omega) = jZ_0\frac{\beta g}{\pi R}\ln\frac{d}{b} \qquad \text{for} \quad kb \ll 1 \ . \tag{8.28}$$

---

[1]It is equally valid to eliminate $\bar{A}$ and solve the Eqs. (8.23) for $\bar{D}$, but to calculate the impedance one has to obtain $\bar{A}$ subsequently from the first one of them. Nevertheless, the convergence properties are sometimes more favourable in this way.

For shallow corrugations with a depth $d-b \ll b$, one can expand the logarithm and replace it by $(d/b-1)$, in agreement with the impedance of a disk-loaded wave guide [Marcuvitz51a].

At higher frequencies resonant effects may appear. Since we have neglected the wall resistivity, they will cause delta-function peaks, which occur when the determinant of the system of linear equations vanishes. The search for such zeros is performed numerically e.g. by the computer codes *KN7C* [Keil72a] for longitudinal and *TRANSVRSE* [Zotter79b] for transverse modes. These programs step up in frequency until the determinant changes sign, and then find the exact position of the zero. Sign changes due to (first order) poles are eliminated by division with an appropriate function with poles at the same positions, which are determined analytically. There remains the possibility of missing some zeros when an even number of them lie inside the chosen interval. The result can be improved by reducing the step size, or by inspecting the remaining function after dividing out also the zeros already found.

It is also possible to solve the above equations including a finite conductivity of the cylindrical walls of the cavity [Keil72b], which then leads to finite heights and widths of the resonances. A finite conductivity of the beam tubes can also be included, but is of less interest since the resonant fields are mainly localised in the cavities. Finite conductivity in the cavity side walls leads to slightly more complicated expressions and does not change the character of the solutions. The computation of fields in axially symmetric cavities of more general cross sections in the rz plane can better be done with purely numerical computer codes such as *SUPERFISH* [Halbach76] or *URMEL* [Weiland82], which has now been incorporated into *MAFIA* [Weiland88].

An interesting result of these calculations is the fact that the real part of the impedance becomes proportional to $\omega^{-3/2}$ at high frequencies. This agrees with the frequency dependence of the *optical resonator model* which was derived originally for laser cavities [Vainshtein69], but can be applied to the *EM* fields of an infinite chain of cavities [Sessler68, Brandt82] as will be discussed in the Chapter 9 on integral methods. This behaviour should be compared to the $\omega^{-1/2}$ dependence of the impedance of a single cavity with infinite beam tubes, which will be discussed below.

In general, one must use approximate numerical methods to solve the infinite system of linear equations. For this one needs to truncate the matrices to a finite size. A simple estimate shows that this size should be large compared to the mode number n, which can be many thousands for big accelerators. Handling such large matrices would be impractical even with modern, high-speed computers, but it is possible to transform the variables in such a manner that equations with better convergence properties are obtained [Keil72b, Warnock82]. Care must also be taken to fulfill the *edge conditions* [Meixner49], otherwise the solutions may converge to incorrect values even for quite large matrix sizes [Mitra71, Hahn76].

8.3 Single Cavity or Collimator

We describe now a slightly different way of subdividing the volume of an axially symmetric structure into regions which is better suited for determining the impedance of a single cavity or a collimator with beam tubes, in particular if these have unequal diameters. The geometry will be divided by radial cuts into a left beam pipe with radius b_1 for $z < -g/2$ (region 1 with superscript (1)), the cavity (or the collimator) with radius d for $|z| < g/2$ (region 2), and the right beam pipe with radius b_2 for $z > g/2$ (region 3) as shown in Fig. 8.2. We further assume that a point charge q is traveling along the axis of

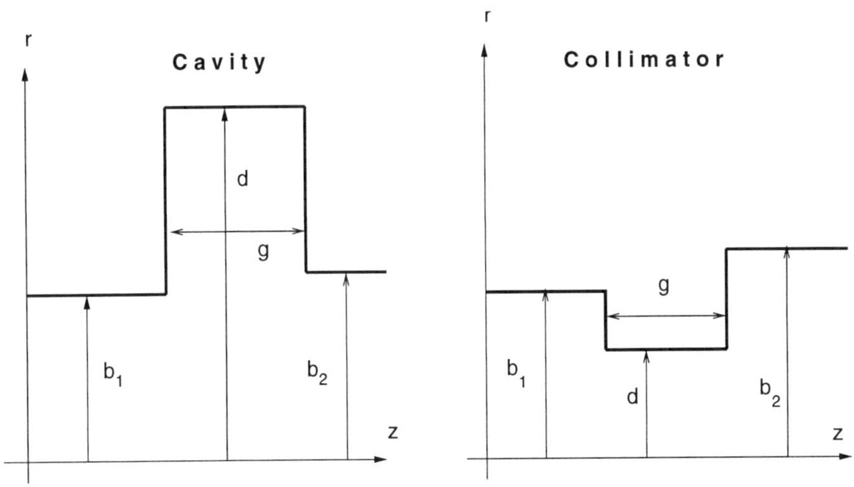

Figure 8.2: Single cavity and collimator with concentric beam tubes.

the structure with velocity $v = \beta c$, with $\beta \to 1$. The Fourier transform of its charge and current density are then given by $\beta c\tilde{\rho} = \tilde{J}_z = q\exp(-jkz)\delta(r)/2\pi r$. The solutions of the Maxwell equations, which satisfy the boundary conditions on a cylindrical wall with constant radius b, for forward and reflected waves (superscripts $\pm$) can be written:

$$\begin{aligned}
\frac{\widetilde{E}_z^{\pm}}{E_\gamma} &= -jG_0(b,r)\,\mathrm{e}^{-jkz} - j\sum_{s=1}^{\infty} C_s^{\pm}\, J_0(j_{0s}\frac{r}{b})\, \exp(\mp j\lambda_{bs}z)\ , \\
\frac{\widetilde{E}_r^{\pm}}{E_\gamma} &= \gamma G_1(b,r)\,\mathrm{e}^{-jkz} \pm \sum_{s=1}^{\infty} \frac{\lambda_{bs}b}{j_{0s}}\, C_s^{\pm} J_1(j_{0s}\frac{r}{b})\, \exp(\mp j\lambda_{bs}z)\ , \\
\frac{Z_0\widetilde{H}_\theta^{\pm}}{E_\gamma} &= \beta\gamma G_1(b,r)\,\mathrm{e}^{-jkz} + \frac{\omega b}{c}\sum_{s=1}^{\infty} \frac{C_s^{\pm}}{j_{0s}} J_1(j_{0s}\frac{r}{b})\, \exp(\mp j\lambda_{bs}z)\ , \quad (8.29)
\end{aligned}$$

where

$$E_\gamma = \frac{\omega\mu_0 q}{2\pi\beta^2\gamma^2}\ , \quad (8.30)$$

and $\tau = k/\gamma$, j_{0s} is the s-th zero of the zero-order Bessel function J_0, $\lambda_{bs} = \sqrt{k_0^2 - j_{0s}^2/b^2}$. The functions G_i are combination of modified Bessel functions, similar to the function R_i defined in Eq. (8.13):

$$G_{0,1}(b,r) = I_0(\tau b)K_{0,1}(\tau r) \mp K_0(\tau b)I_{0,1}(\tau r)\ , \quad (8.31)$$

which have the properties $G_0(b,b) = 0$ and $G_1(b,b) = 1/\tau b$.

The radiation condition requires that only forward waves are excited in region 3, with coefficients C_s^+, and only reflected waves in region 1, with coefficients C_s^-. In region 2 both forward and reflected waves are allowed, and we have to replace the tube radius b by the cavity radius d everywhere.

Each term in the sums for the fields in the beam tubes describes either

(a) an outgoing wave propagating in the $\pm z$ direction if $kb > j_{0s}$, or

(b) an evanescent wave if $kb < j_{0s}$.

For any given frequency, there exist only a finite number of propagating waves, but an infinite number of evanescent ones.

To obtain a compact system of equations, we will call all unknown expansion coefficients $B_s^{(L)}$, with indices $L = 1,2,3,4$, with the following meaning:

$B_s^{(1)} = C_s^-$ in region 1, $B_s^{(2)} = C_s^+$ for the outgoing, $B_s^{(3)} = C_s^-$ for the reflected wave in region 2, $B_s^{(4)} = C_s^+$ in region 3. The dimensionless coefficients $B_s^{(L)}$ can be determined from the boundary and continuity conditions in the planes $z = \pm g/2$ between adjacent regions:

(a) the radial component of the electric field at the (perfectly conducting) side walls of the cavity should be equal to zero for $b < r < d$

(b) all three components of the *EM* field, which are excited by an axially symmetric beam, should be continuous across the opening $r < b$. However, it is sufficient to match any two of them, since the Maxwell equation then guarantee the continuity of the third one.

At the boundary $z = -g/2$ one obtains the two equations:

$$\widetilde{E}_r^{(2)} = \begin{cases} \widetilde{E}_r^{(1)} & \text{for} \quad 0 < r < b_1 \ , \\ 0 & \text{for} \quad b_1 < r < d \ , \end{cases}$$

$$\widetilde{E}_z^{(2)} = \widetilde{E}_z^{(1)} \quad \text{for} \quad 0 < r < b_1 \ . \tag{8.32}$$

For a collimator, with $d < b_1$, one has to interchange superscripts 1 and 2. Analogous conditions at the other interface, $z = g/2$, are obtained by replacing the superscript (1) with (3) and b_1 with b_2. The unknown coefficients $B_s^{(L)}$ can be determined by substituting the expressions for the field components into these equations. The coefficients can then be obtained by multiplying each equation with $rJ_0(j_{0t}r/b_i)$ and integrating over r from 0 to b_i. Due to the orthogonality of the Bessel functions, only the term with $t = s$ is left.

The infinite system of linear equations can be written in matrix form:

$$\sum_{N=1}^{4}\sum_{s=1}^{\infty} A_{LN}^{(st)} \, B_s^{(N)} = P_L^{(t)} \ , \qquad L = 1, ..., 4, \quad t = 1, 2, \ldots, \infty \ . \tag{8.33}$$

These equations constitute an infinite system of linear equations for the unknown coefficients $B_s^{(N)}$, while the coefficients $A_{LN}^{(st)}$ and $P_L^{(t)}$ on the r.h.s. are abbreviations for expressions shown in Table 8.1 [Kheifets87], both for cavities $(d > b_i)$ and collimators $(d < b_i)$ with unequal beam pipe radii. With the abbreviation $\kappa_i \equiv b_i/d$ we get for $j_{0s} \neq \kappa_i j_{0t}$

$$\phi_{st}(\kappa_i) = \frac{j_{0s} J_0(j_{0t}\kappa_i) J_1(j_{0s})}{j_{0s}^2 - \kappa_i^2 j_{0t}^2} \tag{8.34}$$

while it is simply $\phi_{st}(\kappa_i) = J_1^2(j_{0s})/2$ for $j_{0s} = \kappa_i j_{0t}$.

Approximate solutions of this system can be obtained by truncating the equations at $s_{tr} = t_{tr}$. The choice of the actual value of the matrix size depends on the frequency. The higher the frequency, the larger the dimension needed to represent the details of the field behaviour accurately. For cavities and beam pipe radii of a few cm, $s_{tr} \approx 20$ is acceptable for frequencies up to 30

Table 8.1: Coefficients $A_{LN}^{(st)}$ and r.h.s. $P_L^{(t)}$ for a Cavity and a Collimator.

Cavity

L	$A_{L1}^{(st)}$	$A_{L2}^{(st)}$	$-A_{L3}^{(st)}$	$A_{L4}^{(st)}$	$-P_L^{(t)}$
1	$2\kappa_1^2\lambda_{b_1s}d\phi_{st}(\kappa_1)$	$\lambda_{ds}dJ_1^2(j_{0s})\delta_{st}$	$\lambda_{ds}dJ_1^2(j_{0s})\delta_{st}E_-$	0	$\frac{-J_0(j_{0t}\kappa_1)}{I_0(\tau b_1)[j_{0t}^2+(\tau d)^2]}$
2	$j_{0s}^2J_1^2(j_{0s})\delta_{st}$	$-2\kappa_1^2j_{0s}^2\phi_{st}(\kappa_1)$	$2\kappa_1^2j_{0s}^2\phi_{st}(\kappa_1)E_-$	0	$\frac{\kappa_1^2j_{0t}J_1(j_{0t})I_0(\tau b_1)F(b_1)\tau d}{\gamma[j_{0t}^2+(\tau b_1)^2]}$
3	0	$\lambda_{ds}dJ_1^2(j_{0s})\delta_{st}E_+$	$\lambda_{ds}dJ_1^2(j_{0s})\delta_{st}$	$2p_2^2\lambda_{b_2s}d\phi_{st}(\kappa_2)$	$\frac{-J_0(j_{0t}\kappa_2)}{I_0(\tau b_2)[j_{0t}^2+(\tau d)^2]}$
4	0	$-2\kappa_2^2j_{0s}^2\phi_{st}(\kappa_2)E_+$	$2\kappa_2^2j_{0s}^2\phi_{st}(\kappa_2)$	$-j_{0s}^2J_1^2(j_{0s})\delta_{st}$	$\frac{\kappa_2^2j_{0t}J_1(j_{0t})I_0(\tau b_2)F(b_2)\tau d}{\gamma[j_{0t}^2+(\tau b_2)^2]}$

Collimator

L	$A_{L1}^{(st)}$	$A_{L2}^{(st)}$	$-A_{L3}^{(st)}$	$A_{L4}^{(st)}$	$-P_L^{(t)}$
1	$\lambda_{b_1s}dJ_1^2(j_{0s})\delta_{st}$	$\frac{2\lambda_{ds}d}{\kappa_1^2}\phi_{st}(\kappa_1^{-1})$	$\frac{2\lambda_{ds}d}{\kappa_1^2}\phi_{st}(\kappa_1^{-1})E_-$	0	$\frac{J_0(j_{0t}/\kappa_1)}{I_0(\tau b_1)[j_{0t}^2+(\tau d)^2]}$
2	$\frac{2j_{0s}^2}{\kappa_1^2}\phi_{st}(\kappa_1^{-1})$	$-j_{0s}^2J_1^2(j_{0s})\delta_{st}$	$j_{0s}^2J_1^2(j_{0s})\delta_{st}E_-$	0	$\frac{j_{0t}J_1(j_{0t})I_0(\tau b_1)F(b_1)\tau d}{\gamma\kappa_1^2[j_{0t}^2+(\tau b_1)^2]}$
3	0	$\frac{2\lambda_{ds}d}{\kappa_2^2}\phi_{st}(\kappa_2^{-1})E_+$	$\frac{2\lambda_{ds}d}{\kappa_2^2}\phi_{st}(\kappa_2^{-1})$	$\lambda_{b_2s}dJ_1^2(j_{0s})\delta_{st}$	$\frac{J_0(j_{0t}/\kappa_2)}{I_0(\tau b_2)[j_{0t}^2+(\tau d)^2]}$
4	0	$-j_{0s}^2J_1^2(j_{0s})\delta_{st}E_+$	$j_{0s}^2J_1^2(j_{0s})\delta_{st}$	$\frac{-2j_{0s}^2}{\kappa_2^2}\phi_{st}(\kappa_2^{-1})$	$\frac{j_{0t}J_1(j_{0t})I_0(\tau b_2)F(b_2)\tau d}{\gamma\kappa_2^2[j_{0t}^2+(\tau b_2)^2]}$

$$E_+ = \exp[+jg(k-\lambda_{ds})], \quad \kappa_1 = b_1/d, \quad \kappa_2 = b_2/d$$
$$E_- = \exp[-jg(k+\lambda_{ds})], F(b) = K_0(\tau d)/I_0(\tau d) - K_0(\tau b)/I_0(\tau b)$$

GHz. However, precautions must be taken to ensure the convergence of the truncated series to the correct value.

The eigenfrequencies ω_t can be found from the *characteristic equation*

$$\det\left[A_{LN}^{(st)}(\omega_t) - \omega_t\delta_{st}\delta_{LN}\right] = 0\ . \tag{8.35}$$

This equation has no solutions below the *cavity cutoff frequency* $j_{01}c/d$, where $j_{01} \approx 2.405$ is the first zero of the Bessel function of order zero. This means that charges cannot excite fields in that frequency range. Above that frequency, but below the pipe cutoff ν_{01}/b, solutions exist at one or more eigenfrequencies. The real part of the longitudinal impedance has peaks at these frequencies, which become δ-functions for perfectly conducting walls. The exact values of the eigenfrequencies depend slightly on the truncation, but they converge to the correct value with increasing matrix size. In a smooth pipe all $P_L^{(t)} = 0$, and only the trivial solution $B_s^{(N)} = 0$ can exist since Det $A_{LN}^{(st)} \neq 0$, i.e. no radiated fields are excited in a smooth pipe.

An equivalent system of equations, in particular for equal beam pipe radii, can be obtained by matching the fields at the surface $r = b$ [Henke85]. For small beam holes, matching at $z = \pm g/2$ is preferable, because in this case the field in the structure is close to that of a pillbox cavity. This type of matching is also more convenient when the side tubes have different radii. With equal side tubes radii, the two types of matching converge to the same solution from different sides [Hahn76].

8.3.1 Longitudinal Impedance

Once the coefficients $B_s^{(N)}$ have been found as solutions of Eqs. (8.32), one can obtain the longitudinal component of the electric field $\widetilde{E}_z(z)$ and hence the longitudinal impedance:

$$\begin{aligned}
\frac{\pi Z_{\parallel}(k)}{Z_0} &= -\sum_{s=1}^{\infty}\Bigg\{B_s^{(1)} J_0\left(\frac{j_{0s}r}{b}\right)\frac{(k-\lambda_{bs})d}{1+(\tau b/j_{0s})^2}\exp\left[-j\frac{g}{2}(k+\lambda_{bs})\right] \\
&+ 2j\, B_s^{(2)} J_0\left(\frac{j_{0s}r}{d}\right)\frac{(k+\lambda_{ds})d}{1+(\tau d/j_{0s})^2}\ \sin\left[+\frac{g}{2}(k-\lambda_{ds})\right] \\
&+ 2j\, B_s^{(3)} J_0\left(\frac{j_{0s}r}{d}\right)\frac{(k-\lambda_{ds})d}{1+(\tau d/j_{0s})^2}\ \sin\left[+\frac{g}{2}(k+\lambda_{ds})\right] \\
&- B_s^{(4)} J_0\left(\frac{j_{0s}r}{b}\right)\frac{(k+\lambda_{bs})d}{1+(\tau b/j_{0s})^2}\ \exp\left[+j\frac{g}{2}(k-\lambda_{bs})\right]\Bigg\}\ .
\end{aligned} \tag{8.36}$$

This expression is valid for both cavities and collimators, if the corresponding matrix elements given in Table 8.1 are used for determining the coefficients $B_s^{(N)}$. In the ultra-relativistic limit $\gamma \to \infty$, the terms $\tau d > \tau b$ can be neglected compared to j_{0s}. The remarkable feature of this expression is that it then does not depend on r, in spite of its explicit appearance (cf. Chapter 4). In particular, for a cavity with equal side tube radii b, a convenient choice is $r = b$, where $J_0(j_{0s}r/b)$ vanishes. Then the pipe regions $z > g/2$ and $z < -g/2$ do not contribute at all to the value of the integral:

$$Z_{\parallel}^{cav}(k) = -2j\frac{Z_0}{\pi}\sum_{s=1}^{\infty} J_0\left(\frac{j_{0s}b}{d}\right)\Big\{B_s^{(2)}(k+\lambda_{ds})d\sin\left[\frac{g}{2}(k-\lambda_{ds})\right] + B_s^{(3)}(k-\lambda_{ds})d\sin\left[\frac{g}{2}(k+\lambda_{ds})\right]\Big\} . \quad (8.37)$$

Similarly, for a collimator a convenient choice is $r = d < b$. In this case, the region $-g/2 < z < g/2$ does not contribute to the integral. The impedance is then given by:

$$Z_{\parallel}^{coll}(k) = -\frac{Z_0}{\pi}\sum_{s=1}^{\infty} J_0\left(\frac{j_{0s}d}{b}\right)\Big\{B_s^{(1)}(k-\lambda_{bs})d\exp\left[-\frac{jg}{2}(k+\lambda_{bs})\right] + B_s^{(4)}(k+\lambda_{bs})d\exp\left[\frac{jg}{2}(k-\lambda_{bs})\right]\Big\} . \quad (8.38)$$

In general, solutions of Eq. (8.32) can be found only numerically. Two computer codes, for cavities and for collimators, exist for this purpose [Kheifets87]. In Fig. 8.3 we show real and imaginary parts of the impedance of a cavity with infinite side tubes, calculated with one of these codes. Approximate solutions are found by truncating the system to a finite size. One can expect that these approximations are valid for wave lengths larger than the diameter of the opening $2b$. A matrix size of 20x20 is usually sufficient to obtain good accuracy up to $kb \approx 3$. The results were found to be almost independent of the matrix size up to the maximum of 100x100 which was allowed by the computer. The impedance of a collimator (for the *SLAC* storage ring *PEP*) is shown in Fig. 8.4 to illustrate its different behaviour. Other calculations of the impedance of a cavity with beam pipes using field matching can be found in several papers [Warnock82, Vos86, van Rienen85].

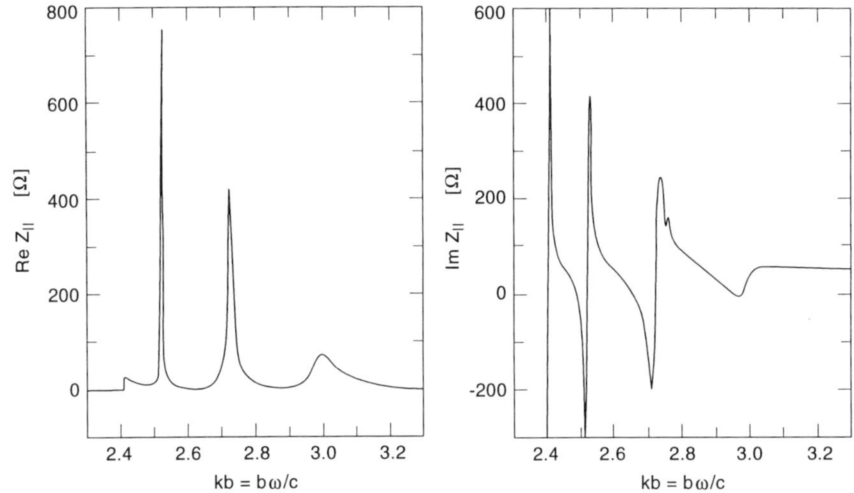

Figure 8.3: Real and imaginary parts of impedance of single cavity with side tubes as function of frequency parameter $kb = \omega b/c$ in region above cutoff.

8.3.2 Transverse Impedance

Transverse fields are excited in the structure when the beam has an offset from the axis ($r_b \neq 0$). The *EM* field expansions into space harmonics for a dipole mode ($m = 1$) are similar to the monopole case considered in the previous Section. The main difference is that both TM and TE modes are excited simultaneously by the charge and current densities.

The matching conditions are the same as for monopole modes, and a similar infinite system of linear equations is obtained, but with twice the number of unknown coefficients [Kheifets87]. They can again be approximately solved by truncation to a finite size. After the expansion coefficients have been found, the radial component of the transverse dipole impedance in the ultra-relativistic approximation can be calculated from its definition:

$$Z_r = \frac{j}{q r_b} \int_{-\infty}^{\infty} dz \, (\widetilde{E}_r - Z_0 \widetilde{H}_\theta) e^{jkz} \tag{8.39}$$

Fig. 8.5 gives an example of results obtained by this method. It shows the real part of the transverse impedance for a cavity with small beam holes.

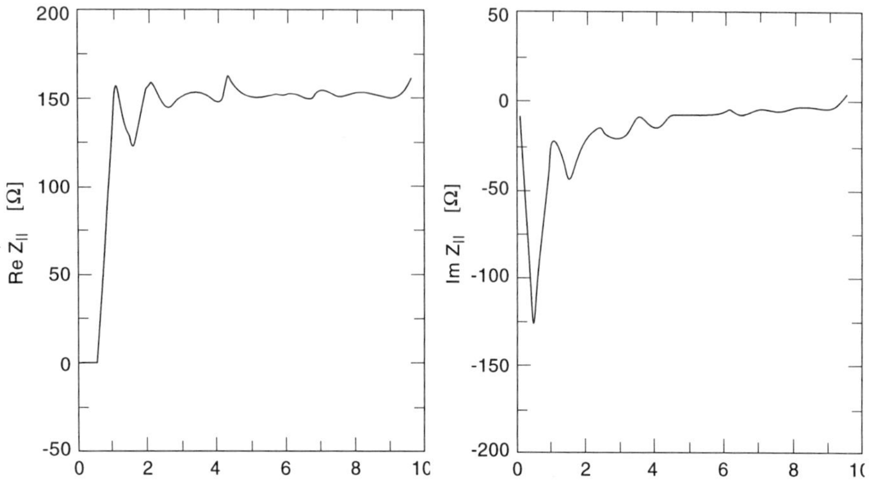

Figure 8.4: Real and imaginary parts of collimator impedance as function of frequency parameter $kb = \omega b/c$.

8.4 Method of Effective Surface Impedance

An approximate solution for the narrow-band impedance of a cavity can be obtained with a particular perturbation method using an effective surface impedance [Heifets88a, Heifets90a].

In a cavity with an opening to a waveguide, modes above the cutoff frequency of the beam pipe are coupled to propagating modes. This produces a non-zero width of the corresponding narrow-band resonances of the impedance, in addition to the width due to the finite wall resistivity. The existence of the narrow-band impedance is only possible if this coupling is small. A perturbation theory in this small parameter can therefore be developed.

In lowest approximation, the field pattern inside the cavity is the same as that of a closed cavity, and tangential components of the electric field are zero at the opening. In the next approximation, matching of the normal component of the electric field defines the amplitudes of the longitudinal components of the waves propagating in the waveguide. The transverse components there are then uniquely defined and also define the tangential components of the field at the cavity opening. As a result, the relation between normal and tangential

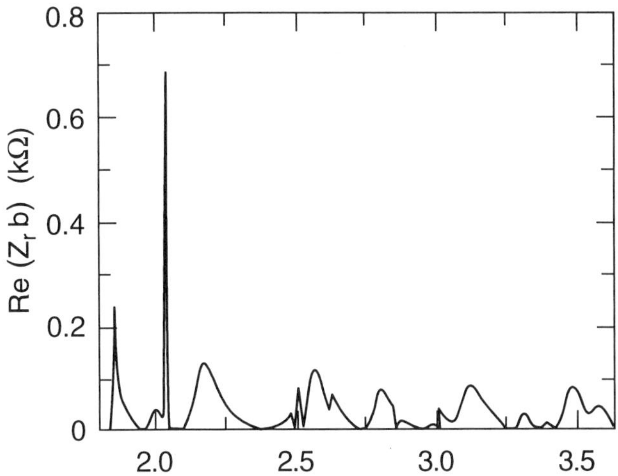

Figure 8.5: Real part of radial impedance of a cavity, multiplied by the beam pipe radius b, as function of the frequency parameter $kb = \omega b/c$ for $d/b = 6.6$, $g/b = 4$.

components of the field at the opening inside the cavity can be written for each radial mode with index s as:

$$E_{rs} = \zeta_s H_{\theta s} \ , \tag{8.40}$$

where the coefficient $\zeta_s = Z_0\sqrt{1-(j_{0s}/kb)^2}$ is the *effective surface impedance* of the opening. The frequency shift of a s-th mode caused by the opening can be calculated using the formula [Landau60] for the frequency shift due to a surface impedance ζ_s:

$$\Delta\omega_s = j\frac{\oint_O dS\zeta_s|\mathbf{H}_s|^2}{\oint_S d\mathbf{S}\cdot\mathbf{r}(\mu|\mathbf{H}_s|^2 - \varepsilon|\mathbf{E}_s|^2)} \ , \tag{8.41}$$

where the integration in the numerator and denominator are performed over the opening O and the surface S of the cavity. Narrow-band impedances exist only when the beam pipe radius is small in comparison to the cavity radius. In this case, we expect that the field pattern inside the cavity is only slightly perturbed, and is similar to that of the closed cavity [Dôme85].

This fact may be utilized to find an approximation suitable for solving the system of equations Eqs. (8.32). To make them more suitable for the perturbation solution, we exclude coefficients $B_s^{(1)}$ and $B_s^{(4)}$ from it, and make the following substitution:

$$\begin{aligned} B_s^{(2)} &= -j(D_s^+ + D_s^-)\exp\left[-j\left(\frac{kg}{2} - \chi_s\right)\right], \\ B_s^{(3)} &= -j(D_s^+ - D_s^-)\exp\left[+j\left(\frac{kg}{2} + \chi_s\right)\right], \end{aligned} \tag{8.42}$$

where $\chi_s = g\lambda_{bs}/2$, with $\lambda_{bs} = \sqrt{k_0^2 - j_{0s}^2/b^2}$ as before. This substitution corresponds to a decomposition of the field into standing waves. The coefficients $D_s^\pm$ are amplitudes of the longitudinal even (cosine-like) and odd (sine-like) modes, respectively. They satisfy two separate systems of equations:

$$\begin{aligned} D_s^+ &= -\frac{j}{\sin\chi_s}\left[P_s \sin\frac{kg}{2} + \sum_t T_{st} D_t^+ \cos\chi_t\right], \\ D_s^- &= -\frac{j}{\cos\chi_s}\left[P_s \cos\frac{kg}{2} - \sum_t T_{st} D_t^- \sin\chi_s\right]. \end{aligned} \tag{8.43}$$

With $\kappa_i = b_i/d$, the coefficients T_{st} and P_s are defined as follows:

$$T_{st} = \frac{2gj_{0t}^2}{b\chi_s}\kappa_i^4 \frac{J_0(j_{0s}\kappa_i)J_0(j_{0t}\kappa_i)}{J_1^2(j_{0s})} \sum_{s'} \frac{\sqrt{k^2b^2 - j_{0s'}^2}}{[j_{0s'}^2 - (j_{0s}\kappa_i)^2][j_{0s'}^2 - (j_{0t}\kappa_i)^2]}, \tag{8.44}$$

and

$$P_s = \frac{g}{2b\chi_s}\frac{J_0(j_{0s}\kappa_i)}{2j_{0s}^2 J_1^2(j_{0s})}. \tag{8.45}$$

In terms of the coefficients D_s^+ and D_s^- the impedance becomes:

$$\begin{aligned} Z(\omega) &= -\frac{Z_0}{\pi}\frac{g}{b}\sum_s j_{0s}^2 J_0(j_{0s}\kappa_i)\left[\left(D_s^+ + D_s^-\right)\frac{\sin(kg/2 - \chi_s)}{kg/2 - \chi_s}\right. \\ &+ \left.\left(D_s^+ - D_s^-\right)\frac{\sin(kg/2 + \chi_s)}{kg/2 + \chi_s}\right]. \end{aligned} \tag{8.46}$$

In the unperturbed cavity we define the dimensionless coefficients $\chi_{sm}^{(0)}$ by the expressions:

$$\begin{aligned} \sin\chi_{sm}^{(0)} &= 0 : \quad \chi_{sm}^{(0)} = m\pi \qquad \text{for } s \text{ even}, \\ \cos\chi_{sm}^{(0)} &= 0 : \quad \chi_{sm}^{(0)} = (m + 1/2)\pi \qquad \text{for } s \text{ odd}. \end{aligned} \tag{8.47}$$

In the vicinity of the eigenfrequencies of the unperturbed cavity only diagonal modes $D_s^{\pm}$ need to be retained in the sums in Eqs. (8.42), which yields:

$$D_s^+ = \frac{-jP_s \sin(kg/2)}{\sin\chi_s - jT_{ss}\cos\chi_s}\,, \quad D_s^- = \frac{-jP_s \cos(kg/2)}{\cos\chi_s + jT_{ss}\sin\chi_s}\,. \tag{8.48}$$

Other amplitudes $D_s^{\pm}$, where $s \neq t$, describe the mixing of the modes of a closed cavity and are zero in this approximation. It can be refined, however, by substituting these equations into the r.h.s. of Eqs. (8.42), which yields $D_t^{\pm} \neq 0$. Repeating the substitution further refines the approximation.

Eqs. (8.47) have resonances with frequencies ω_{sm} and widths Γ_{sm}, which are given by:

$$\omega_{sm} = c\sqrt{\frac{j_{0s}^2}{b^2} + \frac{4\pi^2 m^2}{g^2}}\,, \qquad \Gamma_{sm} = \frac{8c^2\chi_{sm}^{(0)}}{g^2\omega_{sm}}\, T_{ss}\,. \tag{8.49}$$

From this follows an estimate of the external quality factor $Q = 2\Gamma_{sm}/\omega_{sm}$. The expression at the r.h.s. of the second equation is the ratio of the energy flow $\dot{W}_{sm}$ of a mode labeled sm, given by the ratio of the integral of the *Poynting vector* over the cross section of the beam pipe to the *stored energy* W_{sm} in this mode:

$$\Gamma_{sm} = -\frac{\dot{W}_{sm}}{W_{sm}}\,. \tag{8.50}$$

The loss factors $\kappa_{\parallel sm}$ can be found from Eqs. (8.45) and (8.47):

$$\begin{aligned}\kappa_{\parallel sm} &= \frac{4}{\pi\varepsilon_0 g}\,\frac{J_0^2(j_{0s}\kappa_i)}{j_{0s}^2 J_1^2(j_{0s})}\sin^2\frac{\omega_{sm}g}{2c} \quad \text{for } \chi_{sm}^{(0)} = m\pi\,, \\ \kappa_{\parallel sm} &= \frac{4}{\pi\varepsilon_0 g}\,\frac{J_0^2(j_{0s}\kappa_i)}{j_{0s}^2 J_1^2(j_{0s})}\cos^2\frac{\omega_{sm}g}{2c} \quad \text{for } \chi_{sm}^{(0)} = (m+1/2)\,\pi\,. \end{aligned} \tag{8.51}$$

The longitudinal impedance, calculated by this approximate method, agrees quite well with numerical results obtained with the code *URMEL* [Weiland82] as shown in ref. [Heifets90a].

8.5 Impedance of a Step

An infinitely long, straight, cylindrical pipe with an abrupt change of its cross section (a 'step') can be considered as the limit $g \to \infty$ of a cavity or a collimator, or as the limit $g \to 0$ for such a structure with unequal beam tubes. Hence, the impedance for the rotationally symmetric geometries can be obtained directly from the results described in the previous Section [Kheifets87]. The coupling impedance of a single step in a planar geometry was evaluated already in an earlier paper [Hereward75]. For a single step, the impedance contains not only terms due to the induced charges in the wall, but also a term due to the potential energy stored in the field accompanying the beam, which have a different volume due to the change of cross section. Contrary to the coupling impedance proper, this term has a non-zero value at zero frequency, and will thus accelerate (or decelerate) the beam [Friedmann79].

In circular accelerators, a single step cannot appear by itself since the vacuum chamber is periodic, at least with the machine circumference. An increase of the cross section must always be followed by a decrease further downstream. Nevertheless, if the distance between the two steps is large enough, the fields produced by each of them do not interfere, and the two steps can be treated separately.

We distinguish two cases when considering a step:

Step-in: A charge coming out of a larger pipe with radius b_1, and entering a smaller one with radius $b_2 < b_1$. A subscript *'in'* is used in this case.

Step-out: A charge exiting from a smaller pipe with radius b_1 and entering a larger one with radius $b_2 > b_1$. A subscript *'out'* is used in this case.

Both these cases are included in the solutions derived in the previous section. The impedance of a charge passing through a step can be obtained from the expression for that of a cavity of zero length, but with unequal beam pipes $b_2 \neq b_1$. We thus find the coupling impedance for a charge entering a smaller pipe:

$$Z_{in}(k) = -Z_0 \frac{b_2}{\pi} \sum_{s=1}^{\infty} \left[B_s^{(1)}(k - \lambda_{b_1 s}) + B_s^{(4)}(k + \lambda_{b_2 s}) \right] . \tag{8.52}$$

For a charge exiting the narrow pipe one should interchange b_1 and b_2:

$$Z_{out}(k) = -Z_0 \frac{b_1}{\pi} \sum_{s=1}^{\infty} \left[B_s^{(1)}(k - \lambda_{b_2 s}) + B_s^{(4)}(k + \lambda_{b_1 s}) \right] . \tag{8.53}$$

The coefficients $B_s^{(1)}$ and $B_s^{(4)}$ are obtained by solving the infinite system of linear algebraic equations:

$$\sum_{s=1}^{\infty} \left[T_{ts} + \delta_{ts} b_1 \lambda_{b_1 l} J_1^2(j_{0t}) \right] B_s^{(1)} = P_t , \tag{8.54}$$

and similar for $B_s^{(4)}$ with a minus sign in front of the second term. Here $P_t = J_0(j_{0t}\kappa_i)/j_{0t}^2$ and $\kappa_{12} = b_1/b_2$.

$$T_{ts} = 4j_{0s}^2\kappa_{12}^4 J_0(j_{0s}\kappa_{12})J_0(j_{0t}\kappa_{12})\sum_{s'=0}^{\infty}\frac{b_2\lambda_{b_2 s'}}{(j_{0s'}^2 - j_{0s}^2\kappa_{12}^2)(j_{0s'}^2 - j_{0t}^2\kappa_{12}^2)}\,. \tag{8.55}$$

It is instructive to consider two limiting cases:

a) when the step vanishes, i.e. $\kappa_{12} = 1$, then $P_t = 0$ for all t and hence $B_t^{(1)} = B_t^{(4)} = 0$, i.e. no radiated fields are excited.

b) when the pipe is closed, i.e. $\kappa_{12} = 0$, one can find the exact solution which yields [Kheifets86a]

$$B_s^{(1)} = \frac{-2j}{j_{0s}^2 J_1^2(j_{0s})\sqrt{k^2b_2^2 - j_{0s}^2}}\,, \tag{8.56}$$

while $B_s^{(4)} = 0$. This agrees with the expression for the radiation produced in a Faraday cup.

Approximate solutions of Eq. (8.54) can be found numerically by truncating it to a finite size. For moderate steps, a 20x20 matrix is sufficient to obtain reasonable accuracy for not too high frequencies ($kb \leq 3$). Since the coefficients B_s fall off quite rapidly with increasing s, the results do not change much with a larger matrix size. The behaviour of the narrow-band impedance [Kheifets86a] is illustrated in Fig. 8.6, where the real and imaginary parts of the longitudinal impedance are plotted for $\kappa_{12} = 0.1$ as function of kb. The resonant character of the impedance is clearly exhibited.

The impedances of narrowing and widening steps are similar, except that the latter is shifted up by a constant amount. The shift is proportional to the difference of the *EM* field energy of a particle traveling in pipes with radii b_1 and b_2. For a step-in, i.e. de-creasing pipe diameter, the radiated energy is taken from the larger field energy in a bigger pipe. As a result, the loss factor is negative, corresponding to a gain of energy. This increase of particle energy can also be interpreted as the result of acceleration due to the attraction by the image charges in the flange of the step [Chan87]. For a step-out, i.e. a widening beam pipe, a restoration of the particle field takes place. The energy for this is taken from the particle energy, and the beam slows down. The loss is larger than the gain in the corresponding step-in, so the total loss factor of two steps is always positive if the beam pipe radius is the same on both sides.

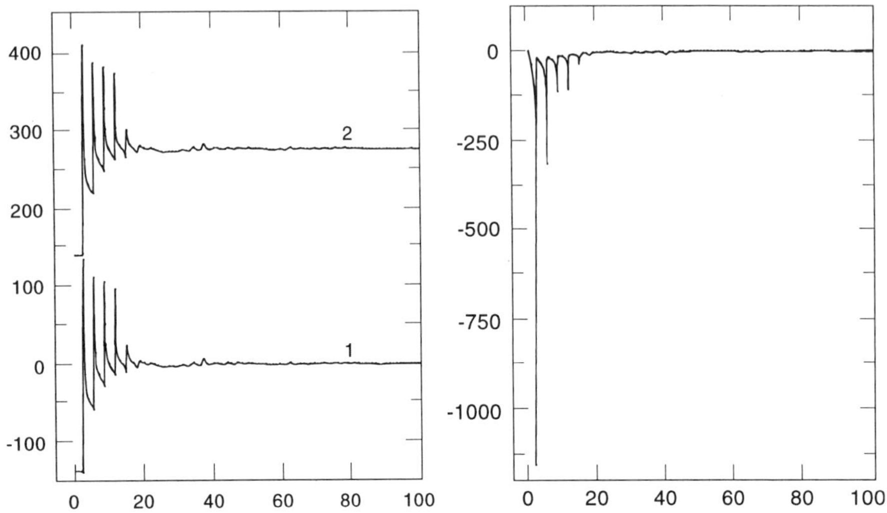

Figure 8.6: Real and imaginary parts of the impedances of step-in [1] and step-out [2] as function of the frequency parameter kb.

8.6 Multi-Element Structures

We consider here a *multi-element structure* consisting of an arbitrary, but finite number N of circular cylindrical cavities connected by $N-1$ beam tubes between them and with two additional ones extending to infinity on both sides. Short beam tubes between the cavities are often called *irises*. The dimensions of all these elements may be different.

We divide such a geometry into $2N+1$ piecewise constant regions with $2N$ interfaces. In each region, the solutions of the wave equation for axially symmetric modes $m=0$ have 2 unknown coefficients, in total $4N+2$. They can be determined from matching two field components at the $2N$ interfaces, and the two *radiation conditions* in the side pipes extending to infinity which is also $4N+2$. Matching of the third field component is *redundant*, i.e. it is automatically fulfilled by matching the other two. Here we choose the field components $\widetilde{E}_z$ and $\widetilde{E}_r$ to fulfill the matching conditions.

For structures with equal beam tube radii, one can alternatively divide the geometry radially at $r=b$ into N cavity regions, and a single beam pipe region, as has been done in the first section of this chapter for a periodic structure. The total number of coefficients then is $2N + 2$, and the number of matching conditions $2N$ at the cavities and two more radiation conditions, again sufficient to determine the coefficients uniquely.

For not axially symmetric modes with $m \geq 1$, both TM and TE modes can be excited by a beam, and thus the number of unknowns doubles. At the same time, it is necessary to match four field components at the interfaces, and fulfill the radiation conditions for two of them in each side pipe, which also doubles the number of equations.

In what follows we demonstrate this method for two structures: a multi-cavity, disk-loaded accelerator section with irises of different radii, and its particular case of an infinite periodic array of cavities with beam tubes.

8.6.1 Basic Equations for Field Amplitudes

Here we derive the equations for the axially symmetric EM fields $m = 0$. For a single cavity, they were already stated in Eqs. (8.29). The factor E_γ was used there for normalisation. To simplify the equations, we assume here that the particle velocity is ultra-relativistic $\gamma \to \infty$. Then the factor $E_\gamma \propto 1/\gamma^2$ vanishes, and we use instead $E_\infty = \gamma^2 E_\gamma = \mu_0 \omega q / 2\pi$ for normalisation. Furthermore $G_0(b, r) \to 0$ and $G_1(b, r) \to \gamma / kr$. For the fields in the n-th cavity of a multi-element structure, with the radius $d^{(n)}$, one obtains:

$$\begin{aligned}
\frac{\widetilde{E}_z^{(n)}}{E_\infty} &= -j \sum_{s=1}^{\infty} J_0\left(\frac{j_{0s} r}{d^{(n)}}\right) X_s^{(n)}(\bar{z}) \,, \\
\frac{\widetilde{E}_r^{(n)}}{E_\infty} &= \frac{\mathrm{e}^{-jkz}}{kr} + \sum_{s=1}^{\infty} \frac{\lambda_s^{(n)} d^{(n)}}{j_{0s}} J_1\left(\frac{j_{0s} r}{d^{(n)}}\right) Y_s^{(n)}(\bar{z}) \,, \\
\frac{Z_0 \widetilde{H}_\theta^{(n)}}{E_\infty} &= \frac{\mathrm{e}^{-jkz}}{kr} + k d^{(n)} \sum_{s=1}^{\infty} J_1\left(\frac{j_{0s} r}{d^{(n)}}\right) X_s^{(n)}(\bar{z}) \,,
\end{aligned} \tag{8.57}$$

for $n = 1, 2, \ldots N$. The dimensionless propagation functions $X_s^{(n)}$ and $Y_s^{(n)}$ are defined for $0 < \bar{z} < L^{(n)}$ as follows:

$$\begin{aligned}
X_s^{(n)}(\bar{z}) &= C_s^{(n)} \exp(j\lambda_s^{(n)} \bar{z}) + D_s^{(n)} \exp(-j\lambda_s^{(n)} \bar{z}) \,, \\
Y_s^{(n)}(\bar{z}) &= \frac{d\, X_s^{(n)}(\bar{z})}{\lambda_s^{(n)}\, d\, \bar{z}} \,.
\end{aligned} \tag{8.58}$$

The propagation constants are defined by $\lambda_s^{(n)} = [k^2 - (j_{0s}/d^{(n)})^2]^{1/2}$. For $k < j_{0s}/d^{(n)}$, its imaginary part should be chosen negative in order to satisfy the radiation condition if the fields in the beam regions, extending to $\pm\infty$, are described by similar equations. In the above expressions, the *global* longitudinal coordinate z is calculated from the beginning of the structure, and should be distinguished from the *local* longitudinal coordinate $0 < \bar{z} < L^{(n)}$, measured from the left side of each particular region of length $L^{(n)}$. This also defines the phases of the complex amplitudes $C_s^{(n)}$ and $D_s^{(n)}$. Similar equations apply to the $(N-1)$ irises between the cavities, with $d^{(n)}$ replaced by $b^{(n)}$, and with another set of coefficients and propagation functions. The matching equations can be solved by first eliminating the coefficients in the iris regions, and are expressed in terms of the coefficients of the cavity regions alone.

Each term of the field expansion of E_z into radial modes satisfies the boundary condition at the surface $r = d^{(n)}$ individually. The boundary and continuity conditions at the remaining surfaces $\bar{z} = 0$ and $\bar{z} = L^{(n)}$ determine the unknown coefficients. They are the same as Eqs. (8.31), but with local values of the parameters. The resulting infinite systems of linear algebraic equations can be solved approximately by truncating it to a finite number s_{tr}. In order to get good accuracy over the frequency range of interest, the truncated matrices should be large enough. For structures with many regions, one then obtains a very large system of linear equations.

We can rewrite the matching equations in matrix form after grouping the coefficients $C_s^{(n)}$ and $D_s^{(n)}$ into N column vectors $F^{(n)}$ of length $2s_{tr}$. For $n = 2, 3 \dots N$ one obtains $N-1$ equations of the form:

$$A_{(n)}^{(n)} F^{(n)} = A_{(n-1)}^{(n)} F^{(n-1)} + A_{(n+1)}^{(n)} F^{(n+1)} + G^{(n)} \ . \tag{8.59}$$

The matrices A, as well as the source vector G, are determined by the boundary conditions, and expressions for them are given in ref.[Heifets89b]. These matrices express the coupling of the fields at the entrance and exit holes of each cavity. Since the coefficients in the n-th region are determined by those in the neighbouring ones, with indices $n-1$ and $n+1$, the total matrix of the whole system of equations is block-diagonal. One method to solve such a system of equations with large but *sparse* matrices is called *Gaussian elimination with back substitution* [Samarskii83, Samarskii89].

This method was used to develop the numerical code *PROGON* [Heifets89b] which computes the *EM* fields in a coaxial chain of circular cylindrical cavities and irises of arbitrary dimensions. For the special case of an infinite periodic structure, it finds the *Brillouin* or *dispersion diagram*. For a finite, periodic or aperiodic structure, it solves for *EM* fields excited either by an external rf generator or by a charged particle. Both the longitudinal monopole ($m = 0$) and transverse dipole ($m = 1$) impedances [Kheifets95] can be found.

8.6.2 Infinite Periodic Array

Here we demonstrate the method developed above on a periodic chain of cavities and irises, consisting of an infinite number of identical cells, which we already analysed in a previous section. The matrices for each cell then are the same and their superscript n can be omitted. Furthermore the coupling coefficients at both beam holes are equal, and hence $A_{(n-1)} = A_{(n+1)}$. Finding the field in such a structure is tantamount to solving the eigenvalue problem, and the equations to be solved are homogeneous:

$$A_{(n)}F^{(n)} = A_{(n-1)}\left[F^{(n-1)} + F^{(n+1)}\right] . \qquad (8.60)$$

According to the *Floquet theorem*, the fields in adjacent cavities, with indices n and $n \pm 1$, are equal except for a constant *phase advance per cell* ψ:

$$F^{n\pm1} = \mathrm{e}^{\pm j\psi} F^{(n)} . \qquad (8.61)$$

Hence the solution of Eq. (8.59) has the form $F^{(n)} = C_V \exp(jn\psi)$, where C_V is a constant column vector. The system of equations can thus be written:

$$\left[A_{(n)}(k) - 2A_{(n-1)}(k)\cos\psi\right]C_V = 0 \qquad (8.62)$$

Since it is homogeneous, it has a non-trivial solution only when its determinant is vanishes:

$$|A_{(n)}(k) - 2A_{(n-1)}(k)\cos\psi| = 0 . \qquad (8.63)$$

The solutions $k(\psi)$ of this equation, which correspond to *real values* of the phase advance ψ, define the *pass-bands* of the periodic structure.

For a large but finite number N of cavities, the determinant is also large, but after diagonalisation it can be decomposed into a product of eigenvalues of the form $A_{(n)}(s,k) - 2A_{(n-1)}(s,k)\cos\psi$ for the s-th pass-band. The above determinant is zero if any one of these factors vanishes individually, which yields the simple equations:

$$A_{(n)}(s,k) - 2A_{(n-1)}(s,k)\cos\psi = 0 , \quad n, s = 1, 2, \dots N . \qquad (8.64)$$

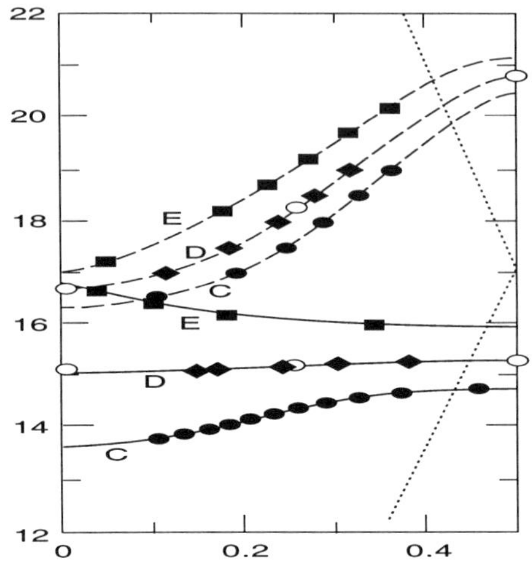

Figure 8.7: Brillouin diagram for two lowest pass bands: monopole (solid lines) and dipole (dashed lines). Mode frequency f [GHz] versus phase advance/cell $\psi/2\pi$ [rad] for three different disk loaded waveguides: equal cell lengths 8.75 mm, iris thickness 1.46 mm for all 3, C) b=5.25, d=11.12mm D) b=4.625, d= 10.83mm E) b=4.00, d=10.58 mm.

Their solutions $k_s(\psi)$ define the *dispersion curves* for the s-th passband in the *Brillouin diagram.* They are periodic functions of the phase advance per cell ψ, with period 2π. Moreover, they are symmetric with respect to the sign of the phase $k(-\psi) = k(\psi)$. For a given phase advance per cell, Eqs. (8.60) also yield the vector C_V, which defines the *EM* fields in the periodic structure up to an arbitrary constant. The Brillouin diagrams for 3 periodic structures with slightly different beam tube and cavity radii b and d are shown in Fig. 8.7. The intersections with the speed of light line (dotted) give the frequencies and phase advances of these modes in the two lowest pass bands.

Chapter 9

Integral Equation Methods and Diffraction Theory

The *EM* fields excited in structures by a charged particle beam can be formulated in terms of integral equations. They are essentially equivalent to the Maxwell equations in the differential form.

Integral equations are particularly useful for finding approximate solutions at high frequencies. The asymptotic behaviour of the impedance with frequency is shown to approach $\omega^{-1/2}$ for a single, and $\omega^{-3/2}$ for a large number of cavities, which can be explained by the interference of their diffracted fields. Furthermore, the energy loss increases with energy as $\gamma^{1/2}$ for a point charge, but becomes independent of energy for bunches of non-zero length. For Gaussian bunches of length $\sigma = \sigma_s/\beta c$ it is proportional to $\sigma^{-1/2}$. For an array of a large number cavities, the impedance and the energy loss increase only with the square root of their number. The criterion for transition between different regimes is considered below.

Several methods of solution will be demonstrated here, mainly for axially symmetric geometries. Integral equations can be applied to a rather broad class of structures, such as cavities, both single and in arrays, collimators, tapers etc. From the solutions for the fields, the coupling impedances of the structures can be obtained.

9.1 Integral Equation for Axially Symmetric Fields

We consider an axially symmetric cavity of arbitrary cross section in the r, z plane, with a concentric beam pipe of radius b on either side extending to infinity as shown in Fig. 9.1.

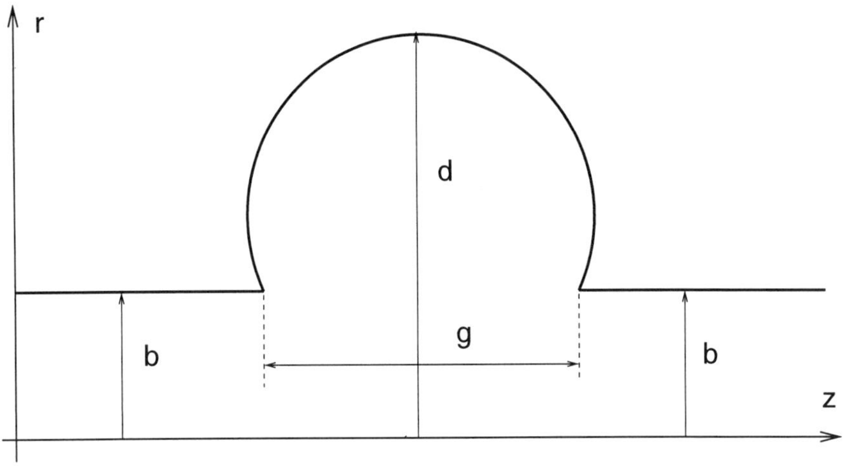

Figure 9.1: Geometry of an axially symmetric cavity with beam tubes.

The $m = 0$ Fourier harmonics of the longitudinal electric field are excited by an ultra-relativistic beam moving on the axis with a Fourier harmonic of the current density $\tilde{J}_z = q\delta(r)\exp(-jkz)/2\pi r$. In the region $r < b$ they can be represented as an expansion into the eigenfunctions $J_0(\nu r)\exp(-jpz)$ of the wave equation in a pipe, with the unknown coefficients $A(p)$:

$$\widetilde{E}_z(r,z) = \int_{-\infty}^{\infty} dp\ A(p)\frac{J_0(\nu r)}{J_0(\nu b)}\exp(-jpz)\ , \tag{9.1}$$

where $\nu^2 = k^2 - p^2$, and $k = \omega/c$. The denominator of the integrand vanishes where $J_0(\nu b) = 0$. To ensure that the radiated field contains only outgoing waves, the contour of integration in the complex plane p should pass above the poles $J_0(\nu b) = 0$ on the positive half-axis $\mathrm{Im}\, p = 0, \mathrm{Re}\, p < 0$, and below them on the negative side. In the ultra-relativistic limit, $\gamma \to \infty$, there is no longitudinal electric field of the source in free space, and the particular solution of the Maxwell equation is zero.

The unknown function $A(p)$ is given by the inverse Fourier transformation of $\widetilde{E}_z(b,z)$:

$$A(p) = \frac{1}{2\pi} \int_{-\infty}^{\infty} dz \ \exp(jpz) \ \widetilde{E}_z(b,z) \ . \tag{9.2}$$

The azimuthal component of the magnetic field strength at $r = b$ becomes:

$$Z_0 \widetilde{H}_\theta(b,z) = \frac{Z_0 q}{2\pi b} \exp(-jkz) + jkb \int_{-\infty}^{\infty} dp \ A(p) \ \frac{J_1(\nu b)}{\nu b J_0(\nu b)} \exp(-jpz) \ . \tag{9.3}$$

The first term on the r.h.s. of this equation is the magnetic field at $r = b$ of a beam in free space, cf. Eq. (7.11), i.e. a particular solution of the Maxwell equations with the source term included. The second term is the general solution at $r = b$ of the homogeneous equation, containing the unknown function $A(p)$ which must be chosen to fulfill the boundary conditions.

The ratio of Bessel functions can be represented as a series in the zeros j_{0s} of the Bessel function J_0 [Abramowitz65]:

$$\frac{J_1(x)}{xJ_0(x)} = \sum_{s=0}^{\infty} \frac{2}{j_{0s}^2 - x^2} = \frac{2}{b^2} \sum_{s=1}^{\infty} \frac{1}{p^2 - \lambda_s^2} \ , \tag{9.4}$$

where $\lambda_s^2 = k^2 - j_{0s}^2/b^2$. The second form is obtained by putting $x = \nu b$, and shows explicitly the location and the residues of the poles in the integrand of Eq. (9.3) at $p = \pm\lambda_s$ in the complex p-plane.

After substituting $A(p)$ and this expansion, the integration over p in Eq. (9.3) can be performed by closing the contour of integration with a semi-circle of infinitely large radius in the lower (upper) complex half-plane p for $z > z'$ $(z < z')$. The result is:

$$Z_0 \widetilde{H}_\theta(b,z) = \frac{Z_0 q}{2\pi b} \exp(-jkz) - \frac{jkb}{2\pi} \int_{-\infty}^{\infty} dz' \ \widetilde{E}_z(b,z') \ K_b(z,z') \ , \tag{9.5}$$

with the following *kernel for the pipe region*:

$$K_b(z,z') = \frac{2\pi j}{b^2} \sum_{s=1}^{\infty} \frac{1}{\lambda_s} \exp\Big(- j\lambda_s |z - z'|\Big) \ . \tag{9.6}$$

The fields in the cavity region $r > b,\ 0 < z < g$ can be represented as an expansion in the eigenmodes of the closed cavity. For the azimuthal magnetic field component one obtains

$$Z_0 \widetilde{H}_\theta(b, z) = \frac{jkb}{2\pi} \int_0^g dz'\ \widetilde{E}_z(b, z')\ K_c(z, z')\ . \tag{9.7}$$

The integration limits can be extended to $\pm\infty$ since the $\widetilde{E}_z(b, z') = 0$ at the beam pipe walls which are assumed to be perfectly conducting. As shown in Chapter 2, the *kernel for the cavity region* is given by:

$$K_c(z, z') = \sum_{s=1}^{\infty} \frac{h_s(b, z) h_s(b, z')}{k^2 - k_s^2}\ . \tag{9.8}$$

The functions $h_s(b, z)$ are the Fourier harmonics of the normalised azimuthal magnetic eigenfields[1] at $r = b$ of the s-th cavity mode with eigenfrequencies $\omega_s = k_s c$. Eqs. (9.5) and (9.7) describe the same magnetic field at the boundary between the cavity and the pipe regions. Equating them produces the desired integral equation for the longitudinal component of the electric field:

$$\int_{-\infty}^{\infty} dz' \widetilde{E}_z(b, z') \Big[K_b(z, z') + K_c(z, z') \Big] = -j \frac{Z_0 q}{kb^2} \mathrm{e}^{-jkz}\ . \tag{9.9}$$

It is convenient to redefine the kernels by including the exponential factors:

$$\widetilde{K}_{b,c}(z, z') = K_{b,c}(z, z')\ \exp\left[jk(z - z')\right]\ , \tag{9.10}$$

and to introduce the dimensionless function $G(z)$ by the definition:

$$G(z) = \frac{kb^2}{qZ_0}\ \widetilde{E}_z(b, z)\ \exp(jkz)\ . \tag{9.11}$$

[1] The expansion of a partially open cavity into eigenmodes of a closed cavity often causes problem of understanding: indeed, each eigenmode of the closed cavity has vanishing tangential electric field at the opening to fulfill the boundary conditions on a perfectly conducting wall, but their sum is expected to have a non-vanishing value to match the fields in the pipe region. This is indeed possible for the sum of an infinite series [Teichmann75, Gluckstern89b], as is known e.g. from the Fourier expansion of a (periodic) step function. Such a sum converges to a non-zero value in the limit from one side, and to another one from the opposite side, while it takes the mean value in between. Thus the fields can converge to the correct value on the cavity side of the opening, be equal to zero at the opening itself, but would take the opposite values on the other side of the opening. However, in the beam region the fields are in any case described by a different expansion.

The integral equation then becomes:

$$\int_{-\infty}^{\infty} dz'\ G(z') \left[\widetilde{K}_b(z,z') + \widetilde{K}_c(z,z')\right] = -j\ . \tag{9.12}$$

In terms of the electric field, the longitudinal impedance is given by:

$$Z_{\|}(k) = -\frac{1}{q}\int_{-\infty}^{\infty} dz\ \tilde{E}_z(b,z)\exp(jkz)\ , \tag{9.13}$$

and thus one obtains:

$$Z_{\|} = -\frac{Z_0}{kb^2}\int_0^g dz\ G(z)\ . \tag{9.14}$$

If the total kernel $\widetilde{K}_{tot} = \widetilde{K}_b + \widetilde{K}_c$ were independent of z and z', the admittance would be proportional to it:

$$Z_0 Y_{\|}(k) = -jkb^2\widetilde{K}_{tot}\ . \tag{9.15}$$

9.1.1 Cavity Impedance at High Frequencies

Eq. (9.12) is particularly useful for the evaluation of the impedance of a cavity with beam pipes in the high frequency region $kb \gg 1$. At frequencies below or near the cut-off frequency, the impedance of the pipe has already been calculated in Chapter 8.

At very high frequencies, the resonant peaks of an impedance with finite quality factors overlap strongly. The effect of such an impedance on a bunched beam is mainly determined by the average of the impedance function over a frequency interval corresponding to the bunch length. This interval is usually large compared to the typical resonance width, but small compared to the frequency itself. Under these conditions the impedance becomes a smooth function of frequency and can be approximated with the averaged part of the kernels $\langle\widetilde{K}\rangle$ defined in Eq. (9.10).

For the beam pipe kernel we can rewrite Eq. (9.6) to get:

$$\widetilde{K}_b(z-z') = j\frac{2\pi}{b^2}\sum_{s=1}^{\infty}\frac{e^{j\psi_s}}{\lambda_s}\ , \tag{9.16}$$

where the phase is given by:

$$\psi_s = k\ (z-z') - \lambda_s\ |z-z'|\ . \tag{9.17}$$

When the frequency is high, the phase ψ_s is large almost everywhere, thereby producing rapid oscillations of the kernel $\widetilde{K}_b(z-z')$. By averaging over a frequency interval, the rapidly oscillating terms are eliminated and do not contribute significantly to the value of the smoothed kernel $\langle\widetilde{K}_b(z-z')\rangle$. The phase ψ_s is small only where the two terms nearly cancel, i.e. for $k \approx \lambda_s$ when $z > z'$, and never when $z < z'$. Since $\lambda_s = \sqrt{k^2 - j_{0s}^2/b^2}$, this happens when $k \gg j_{0s}/b$, where $\lambda_s \approx k - j_{0s}^2/kb^2$. In the vicinity of this region, the phase can be approximated by $\psi_s \approx (z-z')j_{0s}^2/k^2b$. Since $j_{0s} \approx \pi(s-1/4)$ - the summation can be limited to terms $s < \hat{s} = kb/\pi$

$$\widetilde{K}_b(z'-z) \approx \begin{cases} 0 & \text{for} \quad z < z' \;, \\ \dfrac{2\pi j}{kb^2} \displaystyle\sum_{s=1}^{\hat{s}} \exp\left[\frac{j(z-z')j_{0s}^2}{2kb^2}\right] & \text{for} \quad z > z' \;. \end{cases} \tag{9.18}$$

Again, terms with values of s for which $j_{0s}^2(z-z') > kb^2$ oscillate rapidly and can be neglected. For the other terms, the summation can be replaced by an integration, with the result:

$$\left\langle\widetilde{K}_b(z-z')\right\rangle \approx \begin{cases} 0 & \text{for} \quad z' > z \;, \\ \dfrac{(j-1)}{b}\sqrt{\dfrac{\pi}{k(z-z')}} & \text{for} \quad z' < z \;. \end{cases} \tag{9.19}$$

The geometry of the cavity must be specified in order to evaluate the cavity kernel $K_c(z,z')$. For a pillbox cavity, the calculation has been performed in ref.[Gluckstern89b]. The eigenfunctions for the magnetic field strength are given as products of Bessel functions, describing their radial dependence, and trigonometric functions, describing their dependence on the longitudinal coordinate z. For sufficiently high frequencies the Bessel functions can be replaced by their asymptotic values, and summation over the radial eigenvalues and the longitudinal eigenvalues can be replaced by integration. The result for the cavity kernel averaged over a small interval of frequencies turns out to be exactly the same as given in Eq. (9.19):

$$\left\langle\widetilde{K}_c(z-z')\right\rangle \approx \begin{cases} 0 & \text{for} \quad z' > z \;, \\ \dfrac{(j-1)}{b}\sqrt{\dfrac{\pi}{k(z-z')}} & \text{for} \quad z' < z \;. \end{cases} \tag{9.20}$$

The same result was also obtained for a cavity of a different, triangular shape [Gluckstern89a], suggesting that the high-frequency behaviour of the kernel is independent of the cavity shape. The same conclusion follows also from the diffraction model described later in this Chapter.

The approximate expressions for the kernels given in Eqs. (9.19) and (9.20) lead to an integral equation for $G(z)$:

$$\int_0^z dz' \frac{G(z')}{\sqrt{z-z'}} = (1-j)\frac{b}{4}\sqrt{\frac{k}{\pi}} \,. \tag{9.21}$$

The solution can be obtained with the help of a *Laplace transformation*:

$$G(z) = (1-j)\frac{b}{4\pi}\sqrt{\frac{k}{\pi z}} \,. \tag{9.22}$$

The high-frequency longitudinal impedance is then obtained from Eq. (9.14):

$$Z_{\|}(k) = Z_0 \frac{(1-j)}{2\pi b}\sqrt{\frac{g}{\pi k}} \,. \tag{9.23}$$

The same expression has also been found by a different method [Dôme85]. It furthermore agrees with another result for the real part of the impedance [Heifets89a]. The $\omega^{-1/2}$ dependence of the impedance at high frequencies was predicted already much earlier with an argument based on diffraction theory [Lawson68].

9.1.2 Periodic Chain of Cavities

Often the rf cavities in particle accelerators are built in units consisting of several closely spaced resonator cells. The impedance of such a structure can not be approximated by the sum of the impedances of each single cell due to the strong coupling between them. The interference of the *EM* fields of different cells changes in particular the high frequency behaviour of the impedance. As we will see, for a large number N of cavities, the real part of the longitudinal impedance decreases with frequency as $\omega^{-3/2}$ instead of $\omega^{-1/2}$ which was obtained above for a single cavity.

We consider a finite, periodic chain of identical, circular cylindrical cavities, connected to each other by coaxial beam pipes of radius b as illustrated in Fig. 9.2. We will assume that the period P of the cavities is close to their length g, otherwise the coupling between them would be small and each cavity could be considered independently.

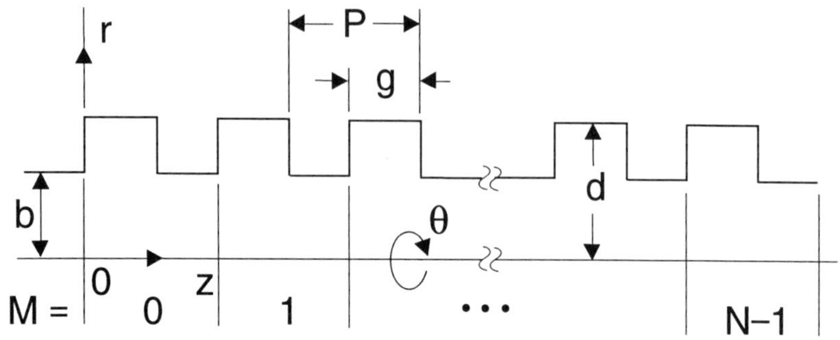

Figure 9.2: Geometry of finite periodic array of cavities.

The field in a chain of cavities can be described by a system of coupled integral equations, each of which is a generalisation of Eq. (9.9):

$$\sum_{n=0}^{N-1} \int dz'_n \; G(z'_n)\Big[\widetilde{K}_b(z_m, z'_n) + \delta_{nm}\widetilde{K}_c(z_m, z'_n)\Big] = -j \qquad (9.24)$$

for $m = 0, 1, \ldots N-1$. Here z'_n and z_m, with $0 \le n, m \le N-1$, denote the local longitudinal coordinates inside the n-th and m-th cavities. The coupling between cavities is described by the dependence of the pipe kernels on the variables belonging to different cavities. On the other hand, the cavity kernels are diagonal, i.e. each one of them depends only on its own variables.

In the high frequency region we can use the same approximations for the kernel as for a single cavity. Let us define the variables for the n-th cavity by $z'_n = nP + z'$ and $z_m = mP + z$. Using Eqs. (9.19) and (9.20) for the diagonal terms, and Eq. (9.18) for the non-diagonal ones, we obtain the m-th integral equation:

$$(1+j)\sqrt{\frac{\pi}{k}} \int_0^z dz' \frac{G_s(z')}{\sqrt{z-z'}} + \frac{\pi}{kb} \sum_{s=1}^{\infty} \sum_{n=0}^{m-1} \exp\left[\frac{j(m-n)Pj_{0s}^2}{2kb^2}\right] \int_0^g dz' G_n(z') = \frac{b}{2} \; . \qquad (9.25)$$

for $m = 0, 1 \ldots M$. In the last term, the difference $z_m - z'_n$ was replaced by its approximate value $(m-n)P$.

In terms of the solutions G_n of this system, the longitudinal impedance is given by:

$$Z_{\|}(k) = -\frac{Z_0}{kb^2} \sum_{n=0}^{N-1} \int_0^g dz G_n(z) \; . \qquad (9.26)$$

and can thus be found once the functions $G_n(z)$ are known.

Using Eq. (9.22), the functions G_n can be approximated by:

$$G_n(z) = (1-j)\frac{b\sqrt{k}}{4\pi\sqrt{\pi z}}H_n \ . \tag{9.27}$$

where unknown coefficients H_n do not depend on z. These coefficients satisfy the system

$$H_m + \frac{(1+j)}{b}\sqrt{\frac{g}{\pi k}}\sum_{s=1}^{\infty}\sum_{n=0}^{m-1} H_n \exp\left[j(m-n)\frac{Pj_{0s}^2}{2kb^2}\right] = 1 \ . \tag{9.28}$$

In terms of the coefficients H_n, the impedance can be written:

$$Z_{\parallel}(k) = Z_0\frac{(1+j)}{2\pi b}\sqrt{\frac{g}{\pi k}}\sum_{n=0}^{N-1} H_n \ . \tag{9.29}$$

The system of Eqs. (9.28) can be solved approximately [Gluckstern89b]. For an infinite periodic array, $N \to \infty$, the following expression for the longitudinal admittance per period is given:

$$Z_0 Y(k) = (1-j)\pi b\sqrt{\frac{\pi k}{g}} - \frac{j\pi k b^2}{P} \ . \tag{9.30}$$

On the other hand, for large frequencies $kb^2/P \gg N \gg 1$, the solution leads to the admittance per cell

$$Z_0 Y(k) = (1-j)\frac{\pi b}{2}\sqrt{\frac{\pi k}{P}}\left[2+\sqrt{\frac{Ng}{P}}\right] \ . \tag{9.31}$$

Since the admittance per cell is thus proportional to $N^{1/2}$, the impedance per cell falls off as $N^{-1/2}$, while the total impedance increases as $N^{1/2}$. The transition to the regime of an infinite periodic array, given in Eq. (9.28), takes place when $N \approx kb^2/P$. Note, that the first term in both formulae is the admittance of a single cavity, cf. Eq. (9.23).

The behaviour of the admittance in both limits and the transition from one regime to another is quite well described by the following empirical formula [Gluckstern89b]:

$$Z_0 Y(k) = (1-j)\pi b\sqrt{\frac{\pi k}{g}} - \alpha\sqrt{N-1}\ \arctan\left(\frac{\alpha}{2\sqrt{N}}\right) , \tag{9.32}$$

where

$$\alpha = (1-j)\sqrt{\frac{\pi k b^2}{P}} \,. \tag{9.33}$$

This expression gives the correct limits for $N \to \infty$ also when $kb^2 \gg NP$. In the latter case, the real part of the impedance tends towards

$$\operatorname{Re} Z_{\parallel}(k) = Z_0 \, \frac{P^2}{\sqrt{\pi k g} k b^3} \,, \tag{9.34}$$

which can be obtained from Eq. (9.30). For a periodic array of cavities, the leading term of the imaginary part of the impedance is $g/\pi k b^2$, i.e./ it is $\propto \omega^{-1}$. However, the real part decreases as $\omega^{-3/2}$ at high frequencies.

For a periodic array of *small obstacles*, i.e., which satisfies the inequality $kg \leq 1$, while $kP \gg 1$, Eq. (9.32) remains unchanged, except that the first term in it now describes the actual impedance of a single obstacle. For very small obstacles of surface area S and for frequencies $kg \ll 1$ the result can be written [Gluckstern89c]:

$$Z_0 Y(k) = 2\pi k b \left[-\frac{j}{k^2 S} + \sum_{s=1}^{\infty} \frac{1}{\lambda_s b} \exp\left(-j\lambda_s g\right) + j\frac{2\ln 2}{\pi} \right] \,. \tag{9.35}$$

which agrees with the results discussed in Chapter 7.

9.2 Kirchhoff Diffraction Theory

In this Section we develop another method which is suitable for the study of the high frequency behaviour of impedances. The *vectorial Kirchhoff integrals* [Jackson75], which are equivalent to the Maxwell equations, express the *EM* fields inside a closed volume by the fields at the boundary. When these fields are unknown, the Kirchhoff integrals constitute a system of integral equations for them. They usually can be solved only approximately using appropriate techniques, such as the *method of iteration* or *expansion into infinite series.* Often it is more convenient to use approximate boundary conditions of the unperturbed problem, such as the *Kirchhoff approximation.* This approach has been used to calculate e.g. the impedance of cavities and tapers [Heifets89a] or of collimators [Novokhatskii88].

The approach presented below is a different description of diffraction presented in Appendix 9A [Lawson68, Faltens75, Palmer89]. Our description here closely follows that in ref.[Heifets91a].

9.2.1 Method of Iteration

Consider a perfectly conducting closed surface S of arbitrary shape enclosing a volume V. The total *EM* field, excited by a (relativistic) particle traversing this volume, is the sum of its direct space charge field and of the radiated one. The radiated field needs to satisfy only the homogeneous wave equation, i.e. without source terms. Then its values anywhere inside the volume are given by the *Kirchhoff integrals* in terms of the fields at the inner surface [Stratton41, Jackson75]. In frequency domain, these equations for the Fourier transforms of the *EM* fields can be written:

$$\begin{aligned}\widetilde{\mathbf{E}} &= \int dS'\Big[-jk\mu(\mathbf{n}'\times\widetilde{\mathbf{H}}')\widetilde{G}+(\mathbf{n}'\cdot\widetilde{\mathbf{E}}')\cdot\nabla'\widetilde{G}+(\mathbf{n}'\times\widetilde{\mathbf{E}}')\times\nabla'\widetilde{G}\Big]\,,\\ \widetilde{\mathbf{H}} &= \int dS'\Big[\frac{jk}{\mu}(\mathbf{n}'\times\widetilde{\mathbf{E}}')\widetilde{G}+(\mathbf{n}'\cdot\widetilde{\mathbf{H}}')\cdot\nabla'\widetilde{G}+(\mathbf{n}'\times\widetilde{\mathbf{H}}')\times\nabla'\widetilde{G}\Big]\,. \qquad (9.36)\end{aligned}$$

Here $\mathbf{n}'$ is the unit vector at position $\mathbf{r}'$, normal to the surface pointing to the inside of the volume, and $\widetilde{G}$ is the Fourier transform of the *Green function* discussed in Chapter 2. Its integral representation is, cf. Eq. (2.40):

$$\widetilde{G}(\mathbf{r},\mathbf{r}') = -\frac{j}{8\pi}\int_{-\infty}^{\infty} dp\widetilde{G}_p(r,r')\exp\Big[-jp(z-z')\Big]\,, \qquad (9.37)$$

where $\widetilde{G}_p(r,r')$ was given in Eq. (2.41). The derivatives $\nabla'\widetilde{G}$ are taken with respect to $\mathbf{r}'$.

The system of integral equations (9.36) can be solved approximately by the *method of iteration.* The fields at the boundary should be chosen appropriately for the first iteration, and influence the fields in all successive steps. Their choice is crucial for the convergence of the iteration.

We consider a particle traveling along the z axis of an azimuthally symmetric structure with a velocity close to that of light, hence $\gamma \gg 1$.In this case, only the monopole mode $m = 0$ is excited. However, the method can also be extended to the general case $m \neq 0$.

In the frequency domain, the electric field of a relativistic particle in free space has a Fourier transform $\widetilde{\mathbf{E}}^{(0)}$ with a large radial component. Its longitudinal component, which is tangential to a cylindrical surface $b = constant$, is then very small. It excites an induced current in such a surface which produces an opposite tangential component $\widetilde{E}_t^{(1)}$ of the radiated electric field which compensates the unperturbed one. The induced current excites also small corrections to the normal component of the radiated electric field $\widetilde{E}_n^{(1)}$, and the tangential (azimuthal) component of the radiated magnetic field $\widetilde{H}_t^{(1)}$. The fields in an unperturbed cylindrical pipe are the sum of the fields in free space and those of the corrections.

Eq. (9.36), together with the radiation condition at infinity, gives the component $\widetilde{E}_z^{(1)}$ inside the pipe as the surface integral over the *metallic* walls of the pipe:

$$\widetilde{E}_z^{(1)} = \int dS' \, \frac{\partial \widetilde{G}}{\partial r'} \, \widetilde{E}_t^{(0)}(r', z') \; . \tag{9.38}$$

For a smooth pipe $b(z) = b$, $dS = 2\pi b \, dz$, and the electric field of a particle in free space is given at $r = b$ by $\widetilde{E}_t^{(0)} = -jE_\gamma I_0(\tau b)\, K_0(\tau r)\, \mathrm{e}^{-jkz}$, where $\tau = k/\gamma$ and

$$E_\gamma = \frac{q\tau^2}{2\pi\varepsilon_0\omega} = \frac{\mu_0\omega q}{2\beta^2\gamma^2} \; . \tag{9.39}$$

In this case Eq. (9.38) gives:

$$\widetilde{E}_z^{(1)} = jE_\gamma \; I_0(\tau r) \; K_0(\tau b)\, \mathrm{e}^{-jkz} \quad \text{for} \quad \beta \approx 1 \; , \tag{9.40}$$

which agrees with the exact solution for a straight pipe, cf. the space charge fields in Eqs. (8.29).

We now evaluate the field excited by a beam in a perfectly conducting structure described by the boundary function $r = b(z)$, in which the fields are considered as perturbations of those in a smooth pipe $b =$ constant. In first approximation, the boundary conditions for the *radiated field* are similar to those for a smooth pipe.

The sum of the tangential component of the direct space charge field of a particle $\widetilde{E}_t^{(0)}$, and that of the radiated field $\widetilde{E}_t^{(1)}$ on the perfectly conducting surface have to be zero:

$$\widetilde{E}_t^{(1)} = -\widetilde{E}_t^{(0)}, \quad \text{for} \;\; r = b(z) \; . \tag{9.41}$$

The condition that the normal component of the *radiated electric field* be zero on a conducting surface is approximately the same as for a smooth pipe:

$$\widetilde{E}_n^{(1)} = 0 \quad \text{for} \quad r = b(z) \; . \tag{9.42}$$

Hence the components of the radiated field at a *conducting boundary* at $r = b$ are given by:

$$\begin{aligned} \widetilde{E}_r^{(1)} &= -\widetilde{E}_t^{(0)} \sin\alpha \; , \qquad \widetilde{E}_z^{(1)} = -\widetilde{E}_t^{(0)} \cos\alpha \; , \\ \widetilde{E}_t^{(0)} &= \widetilde{E}_r^{(0)} \sin\alpha + \widetilde{E}_z^{(0)} \cos\alpha \; . \end{aligned} \tag{9.43}$$

Here $\widetilde{E}_r^{(0)} = (Z_0 q/2\pi b)\exp(-jkz)$ and $\widetilde{E}_z^{(0)} \propto 1/\gamma^2 \approx 0$, while the angle α is defined by $\tan\alpha = db\ (z)/dz$. The magnetic field in the axially symmetric case has only an azimuthal component $\widetilde{H}_\theta$. It can be shown [Heifets89a] that $\widetilde{H}_\theta^{(1)} \approx 0$ at $r = b$ follows from Eq. (9.42). The values of the fields at the boundary yield the first iteration of the approximate solution of Eq. (9.38). The first order approximation for the fields may then be taken as initial function for the next iteration.

The series obtained in this way are analogous to the Born series of scattering theory. Its expansion parameter is the ratio of the amplitude of the tangential component of the radiated field $\widetilde{E}_t^{(1)}$ to that of the particle field. Already the first approximation gives a good estimate of this parameter.

This method can be used to evaluate the electric field in a pill-box cavity with beam pipes from which an estimate of the impedance can be obtained.

9.2.2 Transition from Cavity to Step Regime

Here we consider an axially symmetric pill-box cavity of length g, radius d, and concentric beam pipes of radius b as sketched in Fig. 9.3. The surface integral Eq. (9.38) for this geometry is the sum of two contributions. The first one, producing the main part of the sum, is the integral over the side walls of the cavity at $z = \pm g/2$ for $b < r < d$, which can be evaluated with the space charge terms of Eq. (8.29).

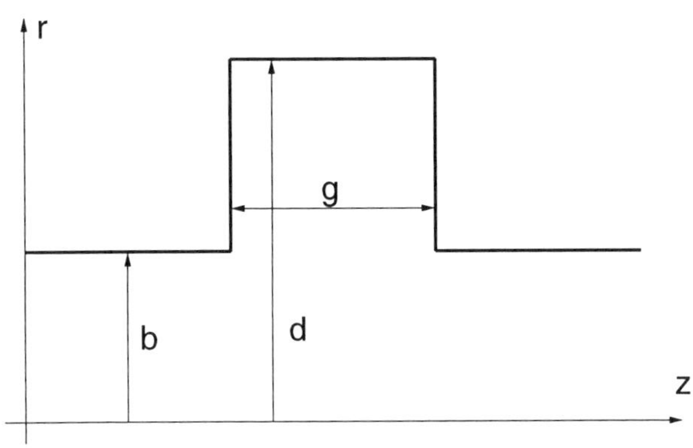

Figure 9.3: Geometry of a pillbox cavity with beam tubes.

For $0 < r < b$ in the beam pipe:

$$\begin{aligned}\widetilde{E}_z^{(1)}(r,z) &= -\frac{j}{4}\int_{-\infty}^{\infty} dp \int_b^d r'dr' K_0(r,r')\,\mathrm{e}^{-jpz}\,,\\ K_0(r,r') &= J_0(\nu r)\frac{dH_0^{(1)}}{dr'}\left[\widetilde{E}_z^{(0)}\left(r',-\frac{g}{2}\right)\mathrm{e}^{-jpg/2} - \widetilde{E}_z^{(0)}\left(r',\frac{g}{2}\right)\mathrm{e}^{jpg/2}\right]. \quad (9.44)\end{aligned}$$

The second integral is over the cylindrical wall $r = d$ for $-g/2 < z < g/2$, which gives a negligible small contribution of the order $1/\gamma^2$ since it contains the longitudinal field component. Similar considerations have been used to obtain the narrow-band, longitudinal impedance above the cutoff frequency of the beam pipe [Gluckstern85]. Performing the integration over r' we get:

$$\widetilde{E}_z^{(1)}(r,z) = \frac{-jZ_0 q}{2}\int_{-\infty}^{\infty} dp\,\mathrm{e}^{-jpz}\left[1-\mathrm{e}^{-j(k-p)g}\right] J_0(\nu r)\left[H_0^{(1)}(\nu d) - H_0^{(1)}(\nu b)\right]. \quad (9.45)$$

This expression gives the first approximation for the diffracted field in the cavity. To calculate the impedance we choose a path of integration parallel to the axis at $r = b$, in accordance with the discussion in Chapter 3.

Since $\widetilde{E}_{\omega z}(b,z) = 0$ on the (perfectly conducting) pipe wall, the range of integration is reduced to $-g/2 < z < g/2$. For $k > 0$ we obtain:

$$Z_{\parallel}(k) = \frac{Z_0}{2\pi}\int_{-\infty}^{\infty}\frac{dp}{k-p}\,J_0(\nu b)\left[H_0^{(1)}(\nu b) - H_0^{(1)}(\nu d)\right]\sin^2\frac{g}{2}\,(k-p)\,. \quad (9.46)$$

Estimates with this expression for the impedance of two different structures, *cavities* and *steps*, give the correct dependence on all physical parameters [Heifets90c, Bisognano88]. This was verified by numerical calculations using the code *TBCI* [Weiland82] over a wide range of parameters. The results are also valid for very short bunches, for which direct numerical calculations require excessive computing time and computer memory.

For the region of parameters where $g \ll kb^2$, called *cavity regime*, Eq. (9.46) reproduces the result given in Eq. (9.23) [Dôme85]:

$$Z_{\parallel}^{\mathrm{cav}}(k) = (1-j)\,\frac{Z_0}{2\pi b}\sqrt{\frac{g}{k\pi}}\,. \quad (9.47)$$

In the region of parameters where $kb^2 < g < kd^2$, called the *transition regime*:

$$Z_{\parallel}^{\mathrm{trans}}(k) = \frac{Z_0}{2\pi}\,\ln\left(\frac{g}{kb^2}\right), \quad (9.48)$$

and in the region of parameters where $g \gg kd^2$, called *step regime*, one obtains the impedance of two opposite steps: [Kheifets87]

$$Z_{\|}{}^{step}(k) = \frac{Z_0}{\pi} \ln \frac{d}{b} . \tag{9.49}$$

The difference between the impedances of a cavity, Eq. (9.47), and a step, Eq. (9.49), corresponds to different types of diffraction. For $g \ll kb^2$, the radius of the area illuminated by the diffracted wave increases as $(g/k)^{1/2}$ with distance g, which characterises the *Fresnel diffraction*, cf. Appendix A to this Chapter. For larger g, this radius becomes proportional to g/kb, which corresponds to the *Fraunhofer diffraction*.

In the integral in Eq. (9.46), the real part of the impedance is mainly determined by values of p in the range $-k < p < k$ [Heifets89a]:

$$\mathrm{Re}\, Z_{\|}(k) = \frac{Z_0}{2\pi} \int_{-k}^{k} \frac{dp}{k-p} J_0(\nu b)\left[J_0(\nu b) - J_0(\nu d)\right] \sin^2 \frac{g}{2}(k-p) . \tag{9.50}$$

The results of numerical integration of this expression, together with the approximations Eqs. (9.47) and (9.49) are shown in Fig. 9.4. The transition from the cavity to the step regime occurs for values of the cavity parameters such that

$$\eta \equiv \frac{g}{k(d-b)^2} \approx 1 . \tag{9.51}$$

We turn now to the evaluation of the correction term for the radial component $\widetilde{E}_r^{(1)}(r,z)$ of the radiated field. It can be derived from Eq. (9.38), where the surface integral is taken over the side walls. In the high-frequency region $kb \gg 1$, neglecting terms of order of $1/kb$, we obtain

$$\widetilde{E}_r^{(1)} = \frac{E_\gamma}{2\pi} \sum_{n=1}^{2} (z - z_n) \int_0^{2\pi} d\theta \left[\frac{\mathrm{e}^{-jk[R_n(b)+z_n]}}{[b - r\cos\theta] R_n(b)} - \frac{\mathrm{e}^{-jk[R_n(d)+z_n]}}{(d - r\cos\theta) R_n(d)} \right] . \tag{9.52}$$

The constant E_γ was defined in Eq. (9.39), and

$$R_n^2(x) = r^2 + x^2 - 2rx\cos\theta + (z - z_n)^2 \quad \text{for} \quad n = 1, 2 . \tag{9.53}$$

One can see that the wave radiated from the surface $z_1 = -g/2$ does not contribute to $\widetilde{E}_r^{(1)}$ at $z_2 = g/2$. Only the wave scattered from the other surface changes the field in the this approximation. The integrand has a weak singularity $(r-b)^{-1/2}$ at $r = b$ as discussed in ref.[Meixner49]. Note that at least in this approximation, a similar singularity appears also at $r = d$. Since these singularities leave the field energy finite, they are not important.

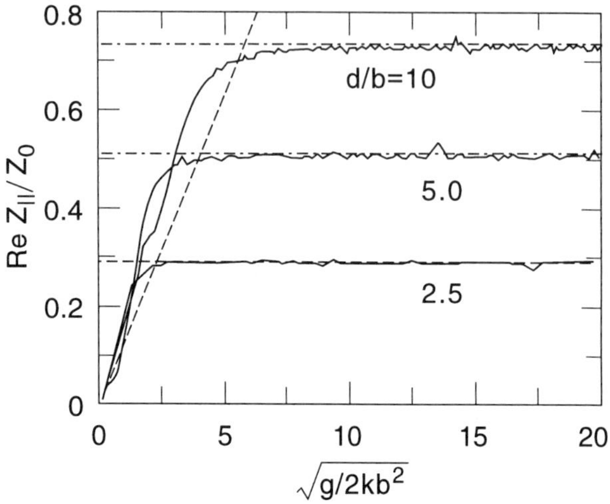

Figure 9.4: Transition of cavity to step regime: real part of impedance versus parameter $\sqrt{g/2kb^2}$ for various values of d/b.

9.2.3 Loss Factors of a Cavity

The high-frequency longitudinal impedance of a cavity is given by Eq. (9.47). Since it falls off only as $k^{-1/2}$, the high-frequency tail gives the main contribution to the energy loss of a short bunch whose spectrum extends to large frequencies:

$$\kappa_{\parallel}^{cav}(\sigma_s) = \frac{\Gamma(1/4)}{4\pi^2\varepsilon_0 b}\sqrt{\frac{g}{\pi\sigma_s}} \,. \tag{9.54}$$

This equation has been checked by *TBCI* computations for three different sets of parameters for *CEBAF* rf structures:

(a) the fundamental power coupler with $b = 3.5$, $g = 2.5$, $d = 5.5$ cm;

(b) the higher order mode coupler with $b = 3.75$, $g = 3.75$, $d = 5.5$ cm;

(c) the gate valve with $b = 1.75$, $g = 2$, $d = 3.5$ cm.

The rms length of the bunch σ_s was varied over the range $0.75 - 1.5$ mm. The observed agreement is within 10 % [Bisognano88].

For a very long cavity, this equation is not applicable. The longitudinal impedance of a step is given by formula Eq. (9.49). The impedance becomes independent of frequency, which corresponds to a wake proportional to a δ-function. The longitudinal loss factor for a single step is:

$$\kappa_{\parallel}^{step}(\sigma_s) = \frac{1}{2\pi\varepsilon_0\sqrt{\pi}\sigma_s} \ln\frac{d}{b} \ . \tag{9.55}$$

The transverse impedance can be calculated from the longitudinal one using the Panofsky-Wenzel theorem (see Chapter 4). The transverse loss factor of a cavity is then given approximately by:

$$\kappa_{\perp}^{cav}(\sigma_s) = \frac{1}{4\varepsilon_0 b^3}\sqrt{\frac{g\sigma_s}{\pi}} \ . \tag{9.56}$$

This formula also agrees well with numerical results obtained with the code *TBCI*.

The exact expression for the transverse loss factor of a step is not known. The following estimate has been given [Bisognano88]:

$$\kappa_{\perp}^{step}(\sigma_s) = \frac{1}{2\pi\varepsilon_0\sqrt{\pi}b^2} \ln\frac{d}{b}\ln\frac{d}{\sigma_s} \ . \tag{9.57}$$

The loss factors of a step, Eqs. (9.55) and (9.57), contain both radii b and d, in contrast to Eqs. (9.47) and (9.54) for a cavity. This is due to the fact that the signal from a step has always enough time to reach the backside of a bunch traveling with $v \to c$. If the parameter η_σ, obtained from Eq. (9.51) for $k \approx 1/\sigma_s$, is small:

$$\eta_\sigma \equiv \frac{2g\sigma_s}{(d-b)^2} \ll 1 \ , \tag{9.58}$$

the formulae for a cavity are valid. In the opposite case, when $\eta_\sigma \gg 1$, the *step regime* is pertinent. Eqs. (9.54)-(9.58) are convenient for a fast and reasonably accurate estimate of the energy loss or kick in an accelerator. However, Eq. (9.57) should be considered only as an empirical estimate.

9.3 Finite Array of Identical Cavities

The *EM* waves generated in one element of an accelerator propagate into the elements downstream of the system. There the waves interfere with the locally radiated field. We here derive the impedance of a finite array of identical cavities. When the number of cavities N is large, the interference can drastically change the impedance per cavity from that of a single one. In the high-frequency region, the iteration method can again be used

We consider an rf structure of a linear accelerator such as a disk-loaded waveguide, where the disks or *irises*, have negligible thickness. They have circular beam holes of radius b and are separated by cylindrical *cavities* of length g. The signal reflected from the outer cavity wall is assumed not to arrive in time to influence the short bunches investigated, even when they are moving close to the velocity of light along the accelerator axis. We may then assume that the outer cavity radius is infinite.

The radial component of the *total EM field* at location z of the accelerator is , to good approximation, the sum of the *direct space charge field* , and the *radiated field* $F(k,r)$ diffracted on the upstream irises:

$$\widetilde{E}_r^{(0)}(r,z) = \frac{qZ_0}{2\pi r}\Big[1 + rF(k,r)\Big]\,\mathrm{e}^{-jkz}\;. \tag{9.59}$$

We use this expression as the first approximation to obtain $\widetilde{E}_z$ from Eq. (9.44). For the longitudinal field between the irises, $0 < z < g$, this equation yields:

$$\begin{aligned}\widetilde{E}_z^{(1)}(r,z) &= \frac{-jqZ_0}{8\pi}\int_{-\infty}^{\infty} dp\,\mathrm{e}^{-jpz}\,\Big[1 - \mathrm{e}^{-j(k-p)g}\Big] \\ &\times \int_b^{\infty} dr'\,\Big[1 + r'F(k,r')\,\Big]\frac{\partial}{\partial r'}J_0(\nu r)\;H_0^{(1)}(\nu r')\;, \end{aligned} \tag{9.60}$$

where $\nu^2 = k^2 - p^2$. Integrating over z from 0 to g, and neglecting a small term containing the radial derivative of the function $F(k,r)$, the impedance per cell is approximately given by:

$$Z_{\parallel}(k) \;=\; \Big[1 + bF(k,b)\Big]\frac{Z_0}{2\pi}\int_{-\infty}^{\infty}\frac{dp}{k-p}\;J_0(\nu b)\;H_0^{(1)}(\nu b)\;\sin^2\frac{g}{2}(k-p)\;. \tag{9.61}$$

Except for the factor $1 + bF(k,b)$, this expression is the same as that for a single cavity, Eq. (9.46). As shown in ref.[Heifets89b], this factor is of order $1/kb$. Since the high-frequency dependence for a single cavity was found to be $k^{-1/2}$, that of a periodic cavity array becomes $\propto k^{-3/2}$. The real part of the impedance per cavity can be approximated by

$$\mathrm{Re}\,Z_{\parallel}(k) = \frac{2Z_0}{\sqrt{\pi}}\left[\frac{g}{kb^2}\right]^{3/2}\;, \tag{9.62}$$

in agreement with the first Section of this Chapter, apart from a numerical factor.

9.4 Transition from Cavity to Array Regime

In the previous Section we presented the transition from the cavity to the step regime. Here we show another approach to analyse the transition to the array regime.

When the structure under consideration can be separated into regions for which the solutions of the Maxwell equations are known or can be found, a natural method for obtaining the solution for the whole structure is the field matching technique discussed in the Chapter 8.

In the present section, using this method, we derive an exact system of equations that are suitable for the high-frequency region. An approximate solution of the system is obtained for cylindrically symmetric structures. The impedance is calculated for a cavity and for a periodic array of a finite number of cavities. A criterion for the transition from the *cavity regime* to the *array regime*, which is governed by the interference of the *EM* waves diffracted from different beam hole edges, is given. The results agree with those obtained above.

9.4.1 The Basic System of Algebraic Equations

The starting point of the calculations is a system of linear algebraic equations for the unknown coefficients of the field expansion. We derive it for the general case of a periodic array of N equal, circular cylindrical cavities [Heifets88b, Heifets89b] of radius d,(gap) length g, and period P, with concentric, infinitely long beam pipe of radius b. A point particle with charge q is assumed to move along the axis of the system $r = 0$ with velocity $v \sim c$ $(\gamma \gg 1)$. We choose the plane $z = 0$ to coincide with the beginning of the first cavity. Fig. 9.2 shows the layout of the geometry considered and the coordinate system used.

A particular case of a single cavity can be obtained by assuming $N = 1$. Likewise, a particular case of a periodic array of cavities can be obtained in the limit as $N \to \infty$.

The radiation field satisfies the homogeneous wave equation, and has to be finite at $r = 0$. It can be represented as a superposition of cylindrical eigenfunctions with unknown coefficients $A(p)$, similar to Eq. (9.3). For the region inside the pipe, $r \leq b$, the electric field components can be written:

$$\frac{\widetilde{E}_z}{E_\gamma} = jG_0(b,r)\, \mathrm{e}^{-jkz} + b\int_{-\infty}^{\infty} dpA(p)\; J_0(\nu r)\; \mathrm{e}^{-jpz} \;,$$

$$\frac{\widetilde{E}_r}{E_\gamma} = \gamma\, G_1(b,r)\mathrm{e}^{-jkz} + jb\int_{-\infty}^{\infty} dp\; A(p)\,\frac{p}{\nu}\; J_1(\nu r)\; \mathrm{e}^{-jpz}\;, \qquad (9.63)$$

where the constant E_γ was defined in Eq. (9.39) and

$$G_{0,1}(b,r) = I_0(\tau b)K_{0,1}(\tau r) \mp K_0(\tau b)I_{0,1}(\tau r)\;, \qquad (9.64)$$

with $\tau = k/\gamma$ and $\nu^2 = k^2 - p^2$ as before.

Introducing local coordinates $z^{(n)} = z - nP$, the radiation field components in the n-th cavity, $b < r < d,\; nP \le z^{(n)} < nP + g$ can be written, cf. Section 8.2:

$$\begin{aligned}
\widetilde{E}_z^{(n)} &= b\sum_{s=0}^{\infty} \lambda_s\, D_s^{(n)}\; F_s^{(0)}(d,r)\; \cos(\mu_s z^{(n)})\;, \\
\widetilde{E}_r^{(n)} &= b\sum_{s=0}^{\infty} \mu_s\, D_s^{(n)}\; F_s^{(1)}(d,r)\; \sin(\mu_s z^{(n)})\;,
\end{aligned} \qquad (9.65)$$

with the Bessel function combinations

$$F_s^{(0,1)}(d,r) = J_{0,1}(\lambda_s r)\; Y_0(\lambda_s d) - Y_{0,1}(\lambda_s r)\; J_0(\lambda_s d)\;. \qquad (9.66)$$

Here $\mu_s = \pi s/g$, $\lambda_s^2 = k^2 - \mu_s^2$, and $D_s^{(n)}$ are the unknown expansion coefficients for the n-th cavity. The fields in Eqs. (9.65) are constructed in such a way that their tangential components are zero at the outer walls $r = d$ of the all cavities, for $nP \le z^{(n)} < nP + g$, and at the side walls for $b < r < d$.

To find the unknown expansion coefficients $D_s^{(n)}$, we match the radial and longitudinal electric field components in the n-th cavity to those in the beam pipe at $r = b$, $0 \le z^{(n)} < g$. In addition, the longitudinal component $\widetilde{E}_z$ must be zero at the beam pipe surface between the cavities $r = b,\; g \le z^{(n)} < P$. This produces the following integral equation for the function $A(p)$:

$$\begin{aligned}
A(p) &= \frac{1}{\pi}\sum_{n=0}^{N-1}\sum_{s=0}^{\infty} C_s(k)\,\frac{V_s(p)}{J_0(\nu b)}\; \exp\Big[-j(k-p)nP\Big]\Big\{-\frac{j}{kb}\; V_s^*(k) \\
&+\; b\int_{-\infty}^{\infty} dp'\; p' A(p')\;\frac{J_1(\nu' b)}{\nu'}\; V_s^*(p')\exp\Big[-j(p'-k)nP\Big]\Big\}\;, \qquad (9.67)
\end{aligned}$$

where the following notations are introduced:

$$\begin{aligned}
C_s(k) &= \lambda_s b\; \frac{F_s^{(0)}(d,b)}{F_s^{(1)}(d,b)}\;, \\
V_s(p) &= \frac{1}{g}\int_0^g dz\; \exp(jpz)\; \cos(\mu_s z)\;.
\end{aligned} \qquad (9.68)$$

For $kg \gg \pi s$ one can use the asymptotic expansions for the Bessel functions to get approximately $C_s(k) = \lambda_s \tan[(d-b)\lambda_s]$.

The longitudinal impedance is given by the coefficient $A(k)$: $Z_{\parallel}(k) = -Z_0 A(k)$. For a smooth pipe $d = b$, all $C_s(k) = 0$, hence $A(p) = 0$ and therefore $Z_{\parallel}(k) = 0$, i.e. the impedance of a (perfectly conducting) smooth pipe is zero.

We seek the solution of Eq. (9.67) in the form

$$A(p) = \sum_{n=0}^{N-1} \sum_{s=0}^{\infty} B_s^{(n)} \frac{V_s(p)}{J_0(\nu b)} \exp\Big[-j(k-p)nP\Big] , \tag{9.69}$$

which gives the following system of linear algebraic equations for the coefficients $B_s^{(n)}$:

$$B_s^{(n)} = \frac{C_s(k)}{\pi} \left[-\frac{jV_s^*}{kb} + \sum_{n'=0}^{N-1} \sum_{t=0}^{\infty} \Gamma_{st}^{(n-n')} B_t^{(n')} \right] . \tag{9.70}$$

Here the following notation for the matrix elements is introduced:

$$\Gamma_{st}^{(n-n')} = \int_{-\infty}^{\infty} \frac{dp}{\nu} \frac{J_1(\nu b)}{J_0(\nu b)} V_s^*(p)\, V_t(p) \exp\Big[-j(p-k)(n-n')P\Big] . \tag{9.71}$$

Taking into account the radiation condition, the path of integration in the complex p plane should go below the negative real axis, and above the positive one. Closing the path by a circle of large radius either in the upper half-plane (for $n < n'$), or in the lower half-plane (for $n > n'$), one can see that the integral is equal to the sum of residues at zeros j_{0s} of the Bessel function J_0. The longitudinal impedance in terms of the coefficients $B_s^{(n)}$ then becomes

$$Z_{\parallel}(k) = -Z_0 \sum_{n=0}^{N-1} \sum_{s=0}^{\infty} V_s(k)\, B_s^{(n)}(k) . \tag{9.72}$$

The system Eq. (9.70) constitutes an exact system of equations for the radiation fields of a relativistic particle. We now search for approximate solutions.

9.4.2 The Impedance of a Cavity

In the high-frequency limit, the system of Eqs. (9.70) can be solved by the method of iteration. In the lowest approximation, we neglect the double sum to get simply:

$$B_s^{(n)} = -j \frac{C_s(k)}{\pi k b} V_s^*(k) . \tag{9.73}$$

Then the impedance per cell is

$$\frac{Z_{\|}}{N} = j\,\frac{Z_0}{\pi k b}\,\sum_{s=0}^{\infty} |V_s(k)|^2\,C_s(k)\;. \tag{9.74}$$

In this approximation, the impedance per cell does not depend on the number of cells in the array and equals the impedance of a single cavity, Eq. (9.23):

$$\frac{Z_{\|}}{N} = (1-j)\,\frac{Z_0}{2\pi}\,\sqrt{\frac{g}{\pi b}}\,\frac{1}{\sqrt{kb}}\;. \tag{9.75}$$

This means that neither the interference of radiation from different cavities, nor the energy escaping into the beam pipes have been included. To take these effects into account, we have to go to the next approximation. For a single cavity, $N = 1$, only the energy flow into the beam pipes must be taken into account. Eq. (9.70) takes the form:

$$B_s = \frac{C_s(k)}{\pi}\left[-\frac{jV_s^*(k)}{kb} + \sum_{t=0}^{\infty}\Gamma_{st}^0 B_t\right]\;. \tag{9.76}$$

We retain only the diagonal terms $t = s$, since all other terms give only small corrections. The main contribution to the sum comes from the term $t = t_0 = [kg/\pi]$. In this *diagonal approximation* [Heifets88b, Heifets89b] we find that the resonances of the impedance are no longer delta-functions, as in the lowest approximation, but have finite widths, although they are often quite narrow.

The integral over a resonance curve does not depend on its width, hence the real part of the impedance is the same as given by Eq. (9.75). The diagonal approximation gives an estimate of the correction due to the next iteration, and shows that they are small in the high-frequency limit [Heifets89b].

9.4.3 Impedance of an Array of Cavities

Consider now an array consisting of N identical cells. In this case the interference of waves generated in different cells must be taken into account. We solve Eq. (9.70) again in the diagonal approximation for the term $t = t_0$, but retain the contributions from all previous cavities. This gives the expression:

$$B_s^{(n)} Y(k) = -\frac{jV_s^*}{kb} + \sum_{n'=0}^{n-1}\Gamma_{ss}^{(n-n')} B_s^{(n')}\;, \tag{9.77}$$

where $n = 0, 1, \ldots, N-1$ and

$$Y(k) = \frac{\pi}{C_s(k)} - \Gamma_{ss}^{(0)} . \tag{9.78}$$

By omitting the terms with $n' > n$ we neglect the interaction of a particle with waves traveling in the opposite direction. Eq. (9.77) are recurrence relations for the coefficients $B_s^{(n)}$, which can be found sequentially starting with the lowest one :

$$B_s^0 = -\frac{jV_s^*}{kbY(k)} , \tag{9.79}$$

which is again the impedance of a single cavity.

Performing these substitutions and averaging over the small frequency interval $\pi/2g$, the result becomes:

$$\left\langle \mathrm{Re}\, \frac{Z_\parallel}{N} \right\rangle = \frac{2Z_0}{(kb)^{3/2}} \left(\frac{2P}{\pi b}\right)^2 \sqrt{\frac{\pi b}{g}}\, \Phi(k,N) , \tag{9.80}$$

where the function $\Phi(k, N)$ is shown in Fig. 9.5.

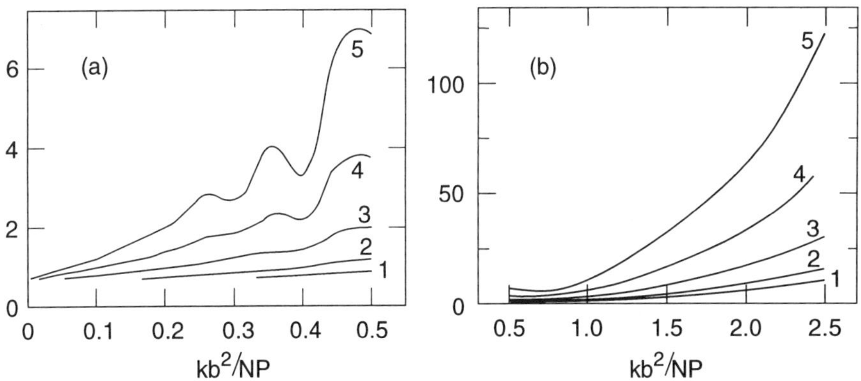

Figure 9.5: The function $\Phi(k, N)$ versus kb^2/NP a) from 0 to 0.5, b) 0.5 to 2.5, for various values of N: Curve 1) 500, 2) 1000, 3) 3000, 4) 10.000, 5) 30.000.

We introduce the parameter $\xi^2 = kb^2/P$ and investigate two limiting cases:
(a) For $\xi^2 \ll N$, the function $\Phi(k, N) \approx 1$ and we obtain:

$$\left\langle \mathrm{Re}\, \frac{Z_\parallel}{N} \right\rangle = \frac{2Z_0}{(kb)^{3/2}} \left(\frac{2P}{\pi b}\right)^2 \sqrt{\frac{\pi b}{g}} . \tag{9.81}$$

In this case, the real part of the impedance decreases with frequency as $\omega^{-3/2}$.

(b) For $\xi^2 \gg N \gg 1$, the function $\Phi(k, N)$ becomes proportional to k, and we get:

$$\left\langle \mathrm{Re}\,\frac{Z_\parallel}{N} \right\rangle = \frac{Z_0}{2\pi}\left(\frac{g}{\pi k b^2}\right)^{1/2}\left(\frac{2P}{gN}\right)^{1/2} . \tag{9.82}$$

The real part of the impedance decreases as $\omega^{-1/2}$. This result agrees with the results obtained above in Section 9.1.

For intermediate parameter values $\xi^2 \approx 1/N$, the impedance is in the *transition regime*, where it changes from the cavity to the array regime.

Let us summarise the results on the *real part of the impedance per cell*: for a *small number of cavities* it decreases with frequency as $\omega^{-1/2}$. In a *large array* of N cavities, we can distinguish two cases: for very high frequencies, it depends on frequency in a way similar to that of a single cavity, i.e. as $\omega^{-1/2}$, but falls off as $N^{-1/2}$ due to interference of the radiated waves emitted from different cavities. The interference also takes place at moderate - but still high - frequencies satisfying the criterion $\xi^2 \gg N \gg 1$. However, this results in a much faster decrease of the impedance $\sim \omega^{-3/2}$. These results agree both with numerical calculations performed for a small number of cavities [Bane90b] and with the optical resonator model described below [Brandt82].

The fast decrease of the real part of the impedance as $\omega^{-3/2}$ has a direct implication on the design of short bunch accelerators. Indeed, had the asymptotic decrease of the longitudinal impedance followed the law $\omega^{-1/2}$, the main contribution to the total energy loss would be given by the high-frequency tail of the impedance and the total energy loss would depend on the longitudinal rms size of the bunch σ_s as $\sigma_s^{-1/2}$. The situation is quite different when the impedance falls off as $\omega^{-3/2}$. In this case, the total energy loss is defined by the low-frequency range of the impedance and is smaller than in the first case.

9.5 Periodic Array of Half-Planes

We consider an infinite periodic disk-loaded structure with period equal to the gap length $P = g$ and with iris radius b. According to the Fresnel theory, a diffracted wave with frequency $\omega = kc$ has a radial extension $\Delta(g) \approx (g/k)^{1/2}$ from the edge of an iris. In the high-frequency region, of interest for short bunches, the extension is much smaller than the iris radius. Also the *EM* fields reflected from the cylindrical outer wall of the structure cannot reach a short enough bunch and thus can be ignored. Therefore an infinite periodic array of cavities can be approximated by an infinite periodic stack of perfectly conducting half-planes. Then the theory of *plane wave diffraction*, which is much better understood than that of cylindrical waves, can be used to evaluate the impedance of such a structure.

9.5.1 Diffraction of a Plane Wave

The diffraction of a plane wave, incident on an infinite array of periodic half planes, has been calculated in ref.[Vainshtein69]. The geometry consist of a stack of perfectly conducting half-planes, spaced at distance g from each other, in the region $y < 0$, as shown schematically in Fig. 9.6 [Stupakov95b].

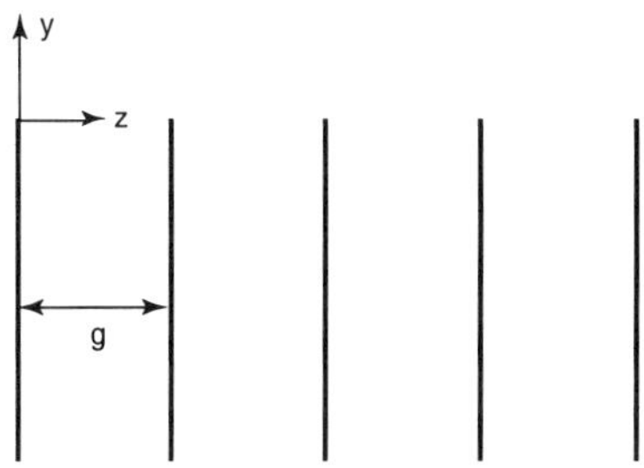

Figure 9.6: Geometry for plane wave diffraction.

A plane *EM* wave is incident at an arbitrary angle ϕ_i measured from the positive y axis, such that grazing incidence corresponds to $\phi_i = \pi/2$. The polarisation of the wave is assumed to have its magnetic field directed along the x-axis:

$$\widetilde{H}_x^i = A \exp\Big[- jk(zs_i - yc_i)\Big] \, . \tag{9.83}$$

where we write for short $s_i = \sin\phi_i$ and $c_i = \cos\phi_i$. The solution of the diffraction problem for the incident wave is represented by a sum of eigenmodes propagating between the plates at $y < 0$. For the space between the m-th and $(m+1)$-th plate:

$$\widetilde{H}_x^{(n)} = A\mathrm{e}^{-jkmgs_i} \left[T_0 \exp(jky) + \sum_{p=1}^{\infty} T_p \mathrm{e}^{j\lambda_p y} \cos\left(\frac{\pi p z}{g}\right)\right] \, , \tag{9.84}$$

where $\lambda_p^2 = k^2 - (\pi p/g)^2$. The sign of the imaginary part of $\lambda_p > 0$ is chosen in such a way that all eigenmodes decay exponentially as $y \to \infty$. The coefficients T_p are the amplitudes of the expansion of the incident wave into eigenmodes of the structure.

The field in the upper half plane $y > 0$ is given by

$$\widetilde{H}_x^d = A\Big[\mathrm{e}^{-jk(zs_i - yc_i)} + \sum_{p=-\infty}^{\infty} R_p \mathrm{e}^{-jk(zs_p + yc_p)}\Big] , \qquad (9.85)$$

where

$$c_p = \sqrt{1 - \frac{(p + kgs_i/2\pi)^2}{(kg/2\pi)^2}} , \qquad s_p = \sqrt{1 - c_p^2} . \qquad (9.86)$$

It is assumed that $\mathrm{Im}\, c_p < 0$.

The coefficients R_p are the amplitudes of the p-th diffracted wave. Their general expression, as well as those for the coefficients T_p, can be found in ref.[Vainshtein69]. In particular, the coefficient R_0 describes the field diffracted at the angle ϕ_i, i.e.in the direction of the incident wave.

9.5.2 Impedance of a Periodic Array of Irises

To evaluate the longitudinal impedance one needs to consider the grazing incidence of the wave $\phi_i = \pi/2$. Then both the incident wave and the diffracted wave with index $p = 0$ propagate parallel to the z-axis. Their sum can be identified with the *EM* field of the bunch at the edge of the half plane. Hence only the coefficient R_0 in the limit $c_i \to 0$ is needed.

The Maxwell equation $\partial \widetilde{H}_x/\partial y = -j\omega\varepsilon_0 \widetilde{E}_z$ can be used to obtain $\widetilde{E}_z$ from $\widetilde{H}_x$ at the boundary $y = 0$ [Stupakov95b]:

$$\widetilde{E}_z(y=0) = -\frac{Z_0}{S(kg)} \widetilde{H}_x . \qquad (9.87)$$

Here we introduced the function:

$$S(kg) = \frac{1}{2}\left\{1 + \frac{1+j}{\pi}\sqrt{\frac{kg}{2}}\Big[(\sqrt{2}-1)F(2kg) + C_S\Big]\right\} , \qquad (9.88)$$

where

$$F(x) = \int_{-\infty}^{\infty} \frac{dt}{\exp(t^2/2 + jx) - 1} , \qquad (9.89)$$

and the constant

$$C_S = 2\int_{-\infty}^{\infty} dt \frac{\exp(t^2/2) - t^2/2 - 1}{t^2[\exp(t^2/2) - 1]} \approx 3.658 \ . \tag{9.90}$$

The periodic function $F(x)$ can be found in ref.[Stupakov95b]. For a cylindrical geometry, Eq. (9.87) should be replaced by:

$$\widetilde{E}_z(r = b) = -\frac{Z_0}{S(kg)}\widetilde{H}_\theta \ . \tag{9.91}$$

This equation is similar to the relation for a pipe with finite conductivity σ_c, for which $\widetilde{E}_z = -\zeta\widetilde{H}_\theta$, where $\zeta = (j\mu\omega/\sigma_c)^{1/2}$ is the *surface impedance*, cf. Eq. (6.28). This similarity suggests the assignment of an *effective complex conductivity* σ_c^{eff} to the system of irises:

$$\sigma_c^{eff} = j\omega\mu S^2(kg) \ . \tag{9.92}$$

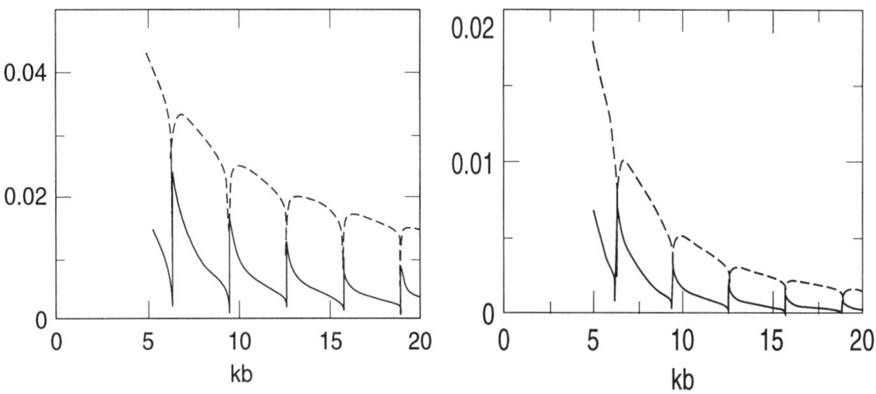

Figure 9.7: Real (solid lines) and imaginary (dashed lines) parts of a) the longitudinal impedance per unit length bZ_L/Z_0 and b) the transverse impedance b^2Z_T/Z_0 per unit length versus kb.

In analogy to the longitudinal electric field for a wall with finite conductivity (see Chapter 6) and using the boundary condition Eq. (2.34) one finds:

$$\widetilde{E}_z = -\frac{Z_0 I_0}{\pi b}\frac{1}{jkb + 2S(kg)} \ , \tag{9.93}$$

which then yields the longitudinal impedance per unit length (indicated by the prime):

$$Z'_{\parallel}(k) = \frac{Z_0}{2\pi b} \frac{1}{S(kg) + jkb/2} \, . \tag{9.94}$$

The real and imaginary parts of the longitudinal impedance become:

$$\operatorname{Re} Z'_{\parallel}(k) \approx \frac{2Z_0}{\pi k^2 b^3} \operatorname{Re} S(kg) \, , \quad \operatorname{Im} Z'_{\parallel}(k) \approx \frac{Z_0}{\pi k b^2} \, , \tag{9.95}$$

and are shown in Fig. 9.7a for the case $b = g$. This plot is very similar to that obtained by the method of auxiliary functions [Dôme92] (cf. next Subsection). Apart from the resonance peaks at $kg = m\pi$, the impedance falls off with frequency as $\omega^{-3/2}$ in agreement with all other calculations presented in this Chapter.

A similar calculation for the transverse impedance gives the result:

$$Z'_{\perp}(k) = \frac{Z_0}{\pi k b^3} \frac{1}{S(kg) + jkb/2 - j/(kb)} \, . \tag{9.96}$$

For very large k one finds asymptotically:

$$\operatorname{Re} Z'_{\perp}(k) \approx \frac{4Z_0}{\pi k^3 b^5} \operatorname{Re} S(kg) \, , \quad \operatorname{Im} Z'_{\perp}(k) \approx \frac{2Z_0}{\pi k^2 b^4} \, . \tag{9.97}$$

Fig. 9.7b shows the real and imaginary parts of the transverse impedance for the case $b = g$.

9.5.3 Method of Auxiliary Functions

We describe here the method of *auxiliary functions*, which was developed in refs.[Dôme91, Dôme92]. It is used to calculate the longitudinal impedance of a disk-loaded structure. The structure is modeled by a sequence of circular irises with radius b, spaced by distances g. The reference plane $z = 0$ is chosen at one of these irises.

The current induced on the iris surface by an ultra-relativistic point charge q has only a radial component. The Hankel transform $F(u)$ of its density distribution $\tilde{J}_r(r)$ is treated as the unknown function of the problem. From the radiation condition at infinity and the boundary condition for the radial component of the electric field the following dual integral equations are derived:

$$\begin{aligned}\int_0^\infty du\ uF(u)J_1(ur) &= 0 && \text{for} \quad 0 < r < b\ , \\ \int_0^\infty du\ uF(u)S(u,g)J_1(ur) &= jAK_1(\tau r) && \text{for} \quad b < r < \infty\ , \end{aligned} \tag{9.98}$$

where $\tau = k/(\beta\gamma)$, $A = q\gamma\tau^2/\pi$ and

$$S(u,g) = \frac{\sinh(g\sqrt{u^2-k^2})}{\cosh(g\sqrt{u^2-k^2}) - \cos(kg/\beta)}\ . \tag{9.99}$$

The wave number $k = \omega/c$ is assumed to have a small positive imaginary part $\operatorname{Im} k > 0$. In the case of a single screen one finds then:

$$S(u,\infty) = 1 \quad \text{for any } u\ . \tag{9.100}$$

The function $F(u)$ is the Hankel transform of the unknown radial component of the current density $\tilde{J}_r(r)$:

$$F(u) \equiv \int_0^\infty dr\ rJ_1(ur)\tilde{J}_r(r)\ . \tag{9.101}$$

The first Eq. (9.98) is simply the condition that the current density in the hole must be zero.

The system of Eqs. (9.98) cannot be solved analytically, and numerical integration is limited by slow convergence and numerical instability, especially for high frequencies. However, the system can be converted into a single *Fredholm integral equation* of the second kind for a properly chosen auxiliary function $p(r)$. One possibility is to have it satisfy the following equation:

$$F(u)\sqrt{u^2-k^2}S(u,g) = -q\frac{jk}{\pi\beta}\left[\frac{u}{u^2+k^2} - \int_0^b dr\ p(r)\sin(ur)\right]\ . \tag{9.102}$$

Such a choice of the function $p(r)$ is not unique. Nevertheless, the solution obtained with its help is unique since it satisfies all the radiation and boundary conditions including the *edge condition* [Meixner49] for E_r at $r = b$. The second Eq. (9.98) then is satisfied automatically while the first one becomes:

$$p(r) = T(r) + \frac{1}{2}\int_0^b dx\ p(x)\big[G(|x-r|) - G(x+r)\big] \quad \text{for } 0 < r < b\ , \tag{9.103}$$

where the following notations are used for the kernel

$$G(r) = \frac{2}{\pi} \int_0^\infty du \left[1 - N(u)\right] \cos(ur) \ , \tag{9.104}$$

for the term outside the integral

$$T(r) = \mathrm{e}^{-\tau r} - \frac{2}{\pi} \int_0^\infty udu \, \frac{\left[1 - N(u)\right]}{u^2 + \tau^2} \sin(ur) \ , \tag{9.105}$$

and the function

$$N(u) = \frac{u}{g} \sum_{n=0}^{\infty} \epsilon_{0n} \frac{1 - (-1)^n \cos(kg/\beta)}{u^2 - k^2 + (n\pi/g)^2} \ . \tag{9.106}$$

Here $\epsilon_{0n} = 1$ for $n = 0$, $\epsilon_{0n} = 2$ for $n \neq 0$ is the *Neumann symbol.*

In terms of the function $p(r)$, the longitudinal impedance per cell is given by:

$$Z_{\|}(k) = \frac{-jkZ_0}{2\beta^2} \left[\frac{2}{\pi} \int_0^\infty du \frac{u^2 N(u)}{(u^2 + \tau^2)^2} - \int_0^a drp(r)T(r)\right] \ . \tag{9.107}$$

The results of numerical integration of Eqs. (9.103) and (9.107) for $g = b, \beta = 1$ agree quite well with those of Fig. 9.7. An approximation for the impedance can be written:

$$\left[\frac{2\pi b Z_{\|}(k)}{gZ_0}\right]^{-1} = \frac{1 - j}{2}\sqrt{\pi kg} - \frac{jkb}{2} + j\frac{H_1^{(2)}(kb)}{H_0^{(2)}(kb)} \ , \tag{9.108}$$

which has some similarity to Eq. (9.30).

9.5.4 Optical Resonator Model

A quantitative description of the high-frequency behaviour of the coupling impedance can be based on the theory of diffraction of the *EM* fields on two parallel, circular mirrors [Kogelnik68]. Such mirrors are widely used for *lasers.* They constitute open resonators for optical frequencies, and therefore the method based on this approach is called *optical resonator model.*

In the high-frequency region, an infinite periodic array of cavities with period P can be approximated by parallel plates with circular openings for a beam. According to the *Babinet principle*, the *EM* fields diffracted by a screen are the same as those diffracted on an opening with the same shape and size in an infinite screen, except for a phase shift of π. The analogy between a pair of infinite screens with circular apertures of radius b, and a pair of circular mirrors of the same radius, has been invoked to apply the results known for the latter [Sessler68].

The fraction of energy of an *EM* wave lost per passage through an optical resonator, due to diffraction of the dominant first radial mode is [Vainshtein63]:

$$\alpha = 2j_{01}^2 \frac{C_{opt}(\xi + C_{opt})}{[C_{opt}^2 + (\xi + C_{opt})^2]^2} , \tag{9.109}$$

where j_{01} is the first zero of the Bessel function J_0. The constant can be expressed in terms of the *zeta function* $C_{opt} = -\zeta(1/2)/\sqrt{4\pi} \approx 0.412$. The parameter $\xi = \sqrt{kb^2/P}$, the same as introduced earlier in this Chapter, is proportional to the square-root of frequency.

The energy flux in the frequency interval dk of the field of a point charge q is [Sessler68]:

$$dU = \frac{q^2 b^2 k^2}{4\pi^2 \varepsilon_0 \beta^3 \gamma^2} K_1^2 \left(\frac{kb}{\beta\gamma} \right) dk , \tag{9.110}$$

where $K_1(x)$ is the modified Bessel function of the second kind.

The asymptotic energy loss U by a particle in one period can be obtained by integration of αdU over the high-frequency region:

$$U = \frac{j_{01}^2 C_{opt} q^2 b^2}{2\pi^2 \varepsilon_0 \beta^3 \gamma^2} \int_{2\pi/P}^{\infty} dk \; k^2 \frac{(\xi + C_{opt}) K_1^2(kb/\beta\gamma)}{[(\xi + C_{opt})^2 + C_{opt}^2]^2} . \tag{9.111}$$

This expression has been evaluated numerically [Keil72a]. For $\gamma \gg 1$, $K_1(x/\gamma) \approx \gamma/x$, and the above integral can be evaluated analytically:

$$U = \frac{j_{01}^2 P q^2}{2\pi^2 \varepsilon_0 b^2} \left[\frac{\xi' C_{opt}}{(\xi' + C_{opt})^2 + C_{opt}^2} + \arctan\left(\frac{C_{opt}}{\xi' + C_{opt}}\right) \right] , \tag{9.112}$$

where we introduced the parameter $\xi' = \sqrt{2\pi} b/P$.

This formula can be simplified in two limiting cases:

1) For a short period and large aperture $\xi' \gg C_{opt}$. Then

$$U = \frac{j_{01}^2 C_{opt} P^2 q^2}{\sqrt{2\pi}\pi^2 \varepsilon_0 b^3} . \tag{9.113}$$

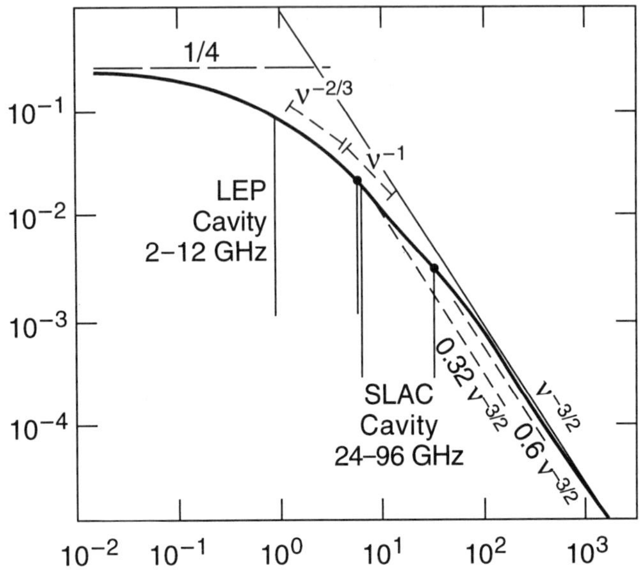

Figure 9.8: Frequency dependence of the real part of the longitudinal impedance in the optical resonator model versus parameter ν.

2) In the opposite case, for a long period and small aperture, $\xi' \ll C_{opt}$

$$U = \frac{j_{01}^2 P q^2}{4\pi\varepsilon_0 b^2} \; . \tag{9.114}$$

Hence in the optical resonator model, the energy loss of an ultra-relativistic particle does not depend on γ. This is in agreement with the physical picture considered in Appendix A to this Chapter.

The real part of the longitudinal impedance of an array of irises, again for one period, can be obtained from Eq. (9.111). For an ultra-relativistic particle:

$$\mathrm{Re} Z_{\|} = \frac{Z_0 j_{01}^2}{2\pi^2} \frac{C_{opt}(\xi + C_{opt})}{[(\xi + C_{opt})^2 + C_{opt}^2]^2} \; . \tag{9.115}$$

In the high-frequency region, $\xi \gg C_{opt}$, this yields:

$$\mathrm{Re} Z_{\|} = Z_0 \frac{j_{01}^2}{2\pi^2} \frac{P^3}{b} \frac{C_{opt}}{(kP)^{3/2}} \; , \tag{9.116}$$

in agreement with previous results for the frequency dependence of the impedance of an array of cavities. The wake function of an optical resonator has been calculated using a similar approach [Bane80b]. The estimates of the asymptotic high-frequency behaviour of both the wake function and impedance of rf cavities were improved in ref.[Brandt82]. The real part of the impedance in the optical resonator model, divided by $Z_0 j_{01}^2/2\pi^2 C_{opt}^2$, is plotted versus the dimensionless parameter $\nu = \xi^2/C_{opt}^2$ in Fig.9.8.

9.6 Wiener-Hopf Technique

A method which has been applied mainly to calculate radiation patterns of antennas [Vainshtein69] can also be used to solve the Maxwell equations for *EM* fields excited by charged particles traversing some idealised structures. The technique has been applied to a point charge passing through a circular cylindrical, semi-infinite beam pipe [Kheifets85b] or through an infinite one with a gap [Palumbo84], as well as to a sheet beam passing close to an infinite periodic set of half planes [Sessler68].

The *Wiener-Hopf Technique* used for this purpose is based on a pair of *coupled integral equations*, which can be solved by *factorisation* of the integral kernel by continuation into the complex plane. Although these solutions are exact, they are not always convenient to compute as they contain infinite series or products. Nevertheless, they can provide insight into the parameter dependence, and in some cases yield simple approximate expressions. The same method has been used before [Levine49] to obtain an explicit solution to the problem of the radiation of sound from the end of a pipe.

In this section we describe two examples of the application of the Wiener-Hopf factorisation method [Wiener31, Vainshtein69]. One is used for finding the longitudinal impedance of an axially symmetric *semi-infinite beam pipe.* Contrary to the usual symmetry properties,in this case the impedance does depend on the direction of charge motion. Another example is the calculation of the *EM* field diffracted by a planar waveguide.

9.6.1 Semi-Infinite Circular Waveguide

We consider a charge *entering* a beam pipe of radius b positioned at $z > 0$ whose open end is located at $z = 0$. A similar calculation can be done for a charge *exiting* from the beam pipe. Due to the axial symmetry of the problem, the current density has only a z component, and can be expressed as the sum of the current densities of the source and the induced charges.

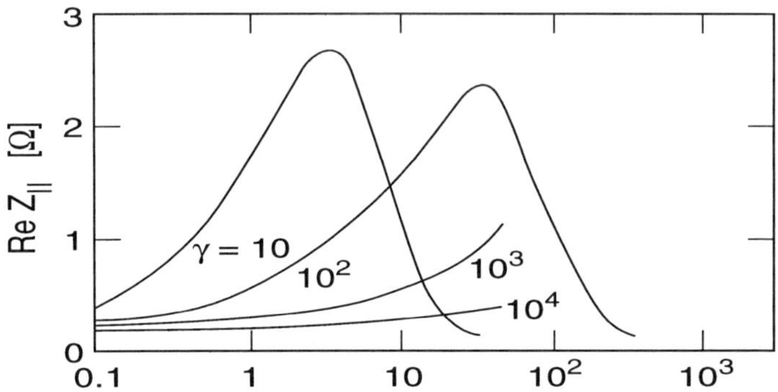

Figure 9.9: Real part of longitudinal impedance of semi-infinite pipe, as function of normalised frequency $\omega b/c$ for various values of the energy factor γ.

We calculate the longitudinal current density distribution induced in the wall of the pipe. The first of Eqs. (9.63) for the longitudinal electric field component can be rewritten in the following form:

$$\frac{\widetilde{E}_z}{E_\gamma} = jI_0(\tau b)K_0(\tau r)\ \mathrm{e}^{-jkz} + b\int_{-\infty}^{\infty} dpF(p)\ \nu^2 J_0(\nu r)H_0^{(1)}(\nu b)\ \mathrm{e}^{-jpz}\ , \quad (9.117)$$

where $\nu^2 = k^2 - p^2$ as before. In this equation, the function $F(p)$ was introduced instead of $A(p) = -(\pi\gamma^2\nu^2/2k^2)\ H_0^{(1)}(\nu b)\ F(p)$. The function $F(p)$ defined in this way can be interpreted as the Fourier component of the induced current density [Kheifets85b] . We have also replaced $G_0(r,b)$ by the free-space radial dependence $I_0(\tau b)K_0(\tau r)$, so that the induced current is contained entirely in the second term of Eq. (9.117).

We define a function of p which will become the *kernel* of the integral equation:

$$L(p) = \pi(\nu b)^2 J_0(\nu b)H_0^{(1)}(\nu b)\ . \quad (9.118)$$

Then the boundary conditions $\widetilde{E}_z(b,z) = 0$ for $z > 0$ can be written:

$$\int_{-\infty}^{\infty} dpF(p)L(p)\exp(-jpz) = \exp(-jkz) \quad \text{for} \quad z > 0\ . \quad (9.119)$$

Since there are no metallic surfaces at $z < 0$, the induced current density there must be zero:

$$\int_{-\infty}^{\infty} dpF(p)\exp(-jpz) = 0\ , \quad \text{for} \quad z < 0\ . \tag{9.120}$$

Eqs.(9.119) and (9.120) constitute a system of linear integral equations for the function $F(p)$. Once the function $F(p)$ is found, the longitudinal impedance can be obtained by integrating the electric field strength using Eq. (9.13). The solution of this system of equations can be obtained by *factorisation* of the kernel $L(p)$ in such a way as to satisfy the following requirements [Vainshtein69]:

a) In the lower half-plane of the complex variable p, the product $F(p)L(p)$ has only a single pole at $p_0 = k$. The value of the residue at this pole is $1/2\pi j$. In all other points of the lower half-plane this product is an analytic function which tends to zero when $|p| \to \infty$.

b) In the upper half-plane, the function $F(p)$ is analytic and also tends to zero when $|p| \to \infty$.

The analytic behaviour of $F(p)$ and $F(p)L(p)$ listed under points a) and b) guarantees that $F(p)$ is a solution of the system of equations. The function $F(p)$ which satisfies both of the above requirements can be written [Vainshtein69]:

$$F(p) = \frac{j}{2\pi b^2 (k-p)^{3/2}} \frac{\Gamma_-(pb)}{\Gamma_+(kb)\sqrt{k}}\ , \tag{9.121}$$

where the functions $\Gamma_\pm$ are given by:

$$\begin{aligned} \Gamma_\pm(pb) &= [2\Sigma I_0(\Sigma)K_0(\Sigma)]^{\pm 1/2} \exp\left\{ \frac{pb}{j\pi} \int_0^{kb} dt\, \frac{\ln\left[\pi\Sigma_1 J_0(\Sigma_1)H_0(\Sigma_1)\right]}{p^2b^2 - t^2} + \right. \\ &+ \left. \frac{jpb}{\pi}\, PV \int_{kb}^{\infty} dt\, \frac{\ln\left[2\Sigma_2 I_0(\Sigma_2)K_0(\Sigma_2)\right]}{t^2 - p^2b^2} \right\}\ , \end{aligned} \tag{9.122}$$

with $\Sigma^2 = b^2(p^2 - k^2)$, $\Sigma_1^2 = k^2b^2 - t^2$, and $\Sigma_2^2 = t^2 - k^2b^2$. Note that in the above equations there are no singularities.

In terms of these functions, the impedance produced by the radiation from the open end of the waveguide for $\gamma \gg 1$ is [Kheifets85b]:

$$Z_\parallel(k) = \frac{Z_0\, kb\, K_0(\tau b)}{2\pi\, \gamma^2\, I_0(\tau b)} \left[\frac{\gamma I_1(\tau b)}{I_0(\tau b)} - \frac{\Gamma'_+(kb)}{\Gamma_+(kb)} - \frac{1}{4kb} \right]\ . \tag{9.123}$$

Figure 9.9 presents the real part of the longitudinal impedance Re $Z(k)$ for several values of the Lorentz factor γ.

In the asymptotic region $kb \gg 1$ the longitudinal impedance is:

$$Z_{\|}(k) = \frac{Z_0}{2\pi} \ln \frac{2\gamma}{kb} \ . \tag{9.124}$$

This result is similar to the impedance of a step from radius b to an infinitely large one.

9.6.2 Diffraction on a Planar Waveguide

Another example of the application of the Wiener-Hopf technique is the calculation of the diffraction of an *EM* wave on an open end of a planar waveguide [Vainshtein63]. Consider two semi-infinite perfectly conducting plates positioned at $y = \pm g$, $z > 0$ at the open end of which arrives an *EM* wave $\widetilde{H}_x(y, z)$. The field is assumed to be independent on the coordinate x and has the indicated polarisation. Similar solution can be found for the other polarisation.

In the high frequency region where the wave number $k = \omega/c$ satisfies conditions:

$$kg = \pi(\frac{q}{2} + s), \quad q \gg 1, \quad |s| < \frac{1}{2} \tag{9.125}$$

the diffracted field at the plate $z = g$ is represented by an integral:

$$\widetilde{H}_x(g, z) = \int_C dpF(p)e^{-jpz} \ , \tag{9.126}$$

where the path C of integration goes mainly along the real axis in the complex plane p avoiding from above the point $p = -p_0$ (see below). This wave number corresponds to the wave arriving at the end of the waveguide. Function $F(p)$ is proportional to the surface density of the current induced in the upper plate. It is determined by the coupled integral equations, cf. previous Subsection:

$$\begin{aligned} \int_C dpF(p)e^{-jpz} &= 0 \quad \text{for} \quad z < 0 \ , \\ \int_C dpL(p)F(p)e^{-jpz} &= 0 \quad \text{for} \quad z > 0 \ . \end{aligned} \tag{9.127}$$

Under conditions

$$|p| \ll k, \quad 2gk^{-3}|p|^4 \ll 1 \tag{9.128}$$

the function $L(p)$ is approximated by

$$L(p) = 1 - e^{-j(2\pi s - \frac{g}{k}p^2)} . \tag{9.129}$$

Function $L(p)$ vanishes at points $p = \pm p_n$, where

$$p_n = \sqrt{\frac{k}{2g}}\bar{p} , \quad \bar{p} = \sqrt{4\pi(n+s)} . \tag{9.130}$$

These wave numbers define the diffracted waves in the planar waveguide.

The system of Eqs. (9.127) is solved by representing $L(p)$ as the product of two functions:

$$L(p) = L_1(p)L_2(p), \quad L_2(p) = -L_1(-p) , \tag{9.131}$$

where $L_1(p)$ is function analytic in the lower half-plane $\operatorname{Im} p \leq 0$ and which satisfies there condition $L_1(p) \to 1$ for $|p| \to \infty$. It is given by the formulas:

$$\begin{aligned} L_1(p) &= \mathrm{e}^{U(\bar{p},s)}, \quad \bar{p} = \sqrt{\frac{2g}{k}}p , \\ U(\bar{p},s) &= \frac{j}{2\pi}\int_{-\infty}^{\infty}\frac{dt}{t - \bar{p}\exp(-j\pi/4)}\ln\left[1 - \mathrm{e}^{-2j\pi s - t^2/2}\right] . \end{aligned} \tag{9.132}$$

Properties of function $U(\bar{p}, s)$ and tables of its values can be found in the book [Vainshtein63].

In terms of functions $L_1(p), L_2(p)$ solution of Eqs. (9.127) is obtained in the form:

$$F(p) = -\frac{jA}{2\pi}\frac{L_1(p_0)}{(p+p_0)L_2(p)} , \tag{9.133}$$

while the field then is:

$$\widetilde{H}_x(y,z) = A(e^{jp_0 z} + \sum_{n=-\infty}^{n=\infty} R_n e^{-jp_n z}\cos\frac{2\pi}{g}y) , \tag{9.134}$$

where A is the amplitude of the incident wave, $p_0 = \sqrt{2\pi ks/g}$ and R_n is the relative amplitude of the n-th diffracted wave:

$$R_n = \frac{j}{(\bar{p}_0 + \bar{p}_n)\bar{p}_n}e^{U(\bar{p}_0,s)+U(\bar{p}_n,s)} . \tag{9.135}$$

In particular, R_0 is the reflection coefficient. The values $n = 1, 2, ...$ correspond to waves propagating in the waveguide, while the values $n = -1, -2, ...$ give evanescent waves.

Similar calculations employing the same method can be performed for an array of semi-infinite perfectly conducting plates. The results thus obtained [Vainshtein63] are used in Section 9.4 to derive the impedance of a periodic array of irises.

Appendix 9.A Simplified Model of Diffraction

Here we present a qualitative description, in the time domain, of the diffraction of *EM* fields excited by a bunch passing through one or more axially symmetric cavities[Palmer89]. The results obtained in this Appendix are supported by the more rigorous calculations presented in the previous Sections of this Chapter.

9.A.1 Diffraction in a Single Cavity

Consider a bunch of length L, consisting of charged ultra-relativistic particles, moving along the axis of an axially symmetric beam pipe of radius b. The field of the bunch is confined to a disk of thickness $w(r)$, given approximately by

$$w(r) \approx L + \frac{2r}{\gamma} \ . \tag{9.136}$$

The denominator γ arises as the result of the Lorentz contraction of the *EM* field. The radial component of the electric field at radius r is

$$E_r(r) \approx \frac{q}{2\pi\varepsilon_0 r w(r)} \ . \tag{9.137}$$

A similar equation holds also for the azimuthal component of the magnetic field. The longitudinal component of the electric field of a relativistic beam is very small and will be neglected. The field energy contained in the small interval dr is of order

$$dU_0(r) \approx \frac{q^2 dr}{2\pi\varepsilon_0 r w(r)} \ . \tag{9.138}$$

We consider what happens when the bunch arrives at a position where a beam pipe suddenly opens up into the cavity. The field of the bunch will be modified to fulfill the boundary conditions there. When the field lines reach the edge at $r = b$ and $z = 0$, the induced charges in the wall must change the direction of their velocity. Therefore they will radiate in a forward cone, spreading the field to the radii $b \pm \Delta_r(z)$, where $\Delta_r(z)$ is its radial extension. The front of this field moves with light velocity c, and reaches the back-end of the direct field moving with the beam velocity βc when

$$\sqrt{z^2 + \Delta_r^2(z)} = \frac{z + w(r)}{\beta} \ . \tag{9.139}$$

Assuming a bunch length L, short compared to the axial distance z, and that the spread of the direct field $w(r) \approx w(b)$, one can solve this equation to get

$$\Delta_r(z) \approx \sqrt{2zw(b)}\ . \qquad (9.140)$$

The field energy at radius r should be that given by Eq. (9.138) up to radius $b - \Delta_r$, and then decrease to zero at the radius $b + \Delta_r$. Assuming the linear decrease, the field energy at position z, outside the radius b, is approximately:

$$U_{out}(z) \approx \frac{q^2}{2\pi\varepsilon_0 b w(b)} \frac{\Delta_r(z)}{4}\ . \qquad (9.141)$$

This result is valid only when $\Delta_r < b$, hence the gap g of the cavity should not be very large or, conversely, the cavity should not be too shallow. We will also assume that the outer wall of the cavity is far enough that the field reflected by it can be ignored.

In reality, the field lines cannot suddenly stop but have to bend backwards towards the cavity side wall as shown e.g. in Fig. 2.1. The field energy behind the bunch is not taken into account in Eq. (9.141), but is assumed to be small compared to the other contributions.

When the field reaches the other cavity side wall at $z = g$, the energy $U_{out}(g)$ outside the radius b will be reflected and lost for the bunch. In addition, an approximately equal amount of energy inside the radius b can not catch up with the bunch and is also lost [Lawson90].

Introducing a numerical factor F_1 which should accommodate for all the approximations made, the energy lost by the bunch to the cavity is:

$$U_{loss} \approx F_1 \frac{q^2}{2\pi\varepsilon_0 b} \sqrt{\frac{g}{2w(b)}}\ . \qquad (9.142)$$

In the frequency domain, the essential frequencies for a point charge lie below $k \approx 1/w(b)$, where the pulse length is $w(b) = b/\gamma$ due to the relativistic contraction of the field. This corresponds to a wave number $k \approx \gamma/b$. From Eq. (9.140) one then finds $\Delta_r \approx (2g/k)^{1/2}$. This estimate agrees essentially with the results obtained from the Kirchhoff equations, cf. Section 9.2.

Point Charge

For a point charge $w(b) = b/\gamma$, Eq. (9.142) yields

$$U_{loss} \approx F_1 \frac{q^2}{2\pi\varepsilon_0} \sqrt{\frac{g\gamma}{2b^3}}\ . \qquad (9.143)$$

which agrees, up to a factor of order unity, with an earlier result [Lawson90]. There it was obtained by expanding the bunch field into a series of plane waves with frequencies ω. The formulae for Fresnel diffraction of each of these waves on a straight, 90° edge was used. Integrating over all frequencies then gave the total energy diffracted at the cavity opening and reflected by the exit wall.

Bunch of Finite Length

For infinite $\gamma \to \infty$, the width $w(b)$ can be replaced by the bunch length L. For a Gaussian bunch, it is approximately $2.5\sigma_s$. Calculating the loss factor from $U_{loss} \equiv q^2 \kappa_{\shortparallel}(\sigma_s)$, we get:

$$\kappa_{\shortparallel}(\sigma_s) \approx \frac{F_1}{2\pi\varepsilon_0 b}\sqrt{\frac{g}{5\sigma_s}} \ . \tag{9.144}$$

This result agrees quite well with numerical calculation using the program *TBCI* [Bane90b]. Expressed in terms of the real part of the longitudinal coupling impedance, this result is also consistent with analytic perturbation calculations [Heifets90d] (cf. Section 9.3).

9.A.2 Diffraction in a Periodic Array of Cavities

A similar calculation can be applied to an infinite periodic array of irises [Sessler68]. We can model them as a sequence of thin irises periodically spaced at distance g from each other. The *outward* extension Δ_r of the field, before it is reflected by the next iris at $z = g$, is the same as for a single cavity. But the field depletion inside the iris radius $r < b$ will extend farther than for a single cavity. The N-th iris is at $z = Ng$, and since the spread extension grows as the square root of distance, its value is:

$$\Delta_r(Ng) \approx \sqrt{N}\Delta_r(g) \ . \tag{9.145}$$

The full radial extension over which the field energy is depleted is then

$$\Delta_r(Ng) + \Delta_r(g) \approx (1+\sqrt{N})\Delta_r(g) \ . \tag{9.146}$$

Assuming a linear decrease of the field energy between $b - \Delta_r(Ng)$ and $b + \Delta_r(g)$, the energy reflected by the n-th iris is

$$U_N = \frac{F_2}{2\pi\varepsilon_0}\,\frac{q2}{bw(b)}\,\frac{\Delta_r^2(g)}{\Delta_r(g)+\Delta_r(Ng)} \ , \tag{9.147}$$

where F_2 is again a factor of order unity. In terms of the energy loss of a single cavity, U_{loss} given in Eq. (9.142), one obtains:

$$U_N \approx U_{loss} \frac{2}{(1+\sqrt{N})}\,, \tag{9.148}$$

where the factor two has been added to make the last term equal to one for a single cavity.

Note that a condition of validity for these formulae is that all fields diffracted from different irises have the same phase.

The loss per cavity decreases when the number of cavities in the array increases. That cannot hold for an arbitrary large N since our assumption that $\Delta_r(Ng) < b$ breaks down at some value of N. An approximate value of N where this is no longer valid is

$$N \approx \frac{b^2}{2gw(b)}\,. \tag{9.149}$$

At this point the field pattern reaches an equilibrium at which the field energy is continually scraped off by the irises, being replenished on the axis by energy taken from the beam.

An estimate of the magnitude of this equilibrium energy loss per cavity is obtained by substituting in Eq. (9.147) b instead of $\Delta_r(Ng)$, and $2w(b)g$ instead of $\Delta_r^2(g)$. For $\Delta_r(g) \ll b$ the functions $w(b)$ cancel, and the average energy loss per unit length of the array becomes:

$$\frac{dU_n}{dz} = \frac{F_2}{2\pi\varepsilon_0}\frac{q^2}{b^2}\,. \tag{9.150}$$

Point Charge

As for a single cavity, the case of a point charge is obtained by substituting $w(b)$ by b/γ. In this case, it follows from the two preceding equations, that the energy loss per unit length becomes independent of energy [Keil72a] when

$$\gamma \gg \gamma_{cr} = \frac{2g}{b}\,. \tag{9.151}$$

Bunch of Finite Length and Infinite γ

When $\gamma \to \infty$ it follows from the same two equations that the energy loss per unit length is:

$$\frac{dU_n}{dz} = \frac{F_2}{2\pi\varepsilon_0} \frac{q^2}{b^2} \ . \tag{9.152}$$

This approximation is valid so long as

$$L \ll \frac{b^2}{2g} \ . \tag{9.153}$$

Hence the energy loss depends neither on the bunch length σ_s, nor on the gap length g. This result is similar to that for the energy loss in a pipe with a resistive wall.

Appendix 9.B Generalised S-Matrix Formalism

The concept of a *generalised scattering matrix* S is an extension of the conventional scattering (or S)-matrix of a transmission line. While the matrix elements of the latter describe the reflection and transmission of a *single mode* at a *single junction*, those of the generalised scattering matrix take into account the scattering not only of the fundamental mode, but also of all higher-order modes (*HOMs*), including evanescent ones, at any number of junctions.

When there are more than one junctions, the generalised scattering matrix describes the scattering process including the interaction between different junctions. Since all higher-order modes are included in the overall matrix S, the method of finding the fields by means of this matrix is - at least in principle - exact [Itoh89]. In practice, the solutions can only be obtained by replacing the exact matrix by an approximate one. Therefore the convergence of the solutions of the approximate matrices to that of the exact matrix has to be verified.

Here we will illustrate the formalism by describing the generalised scattering matrix for the case of two junctions, which we call A and B. When a mode with index n is incident on the first junction A, its field is partially reflected and partially transmitted. Both of these fields components can be expanded into series of eigenmodes, the order of which we designate by the index m. The complex expansion coefficients of these series contain the information on the (relative) amplitudes and phases of each mode.

We call *region 1* the transmission line ahead of junction A, *region 2* that between the two junctions, and *region 3* that beyond the second junction B. The complex amplitude of the m-th mode in region 1, which was reflected on junction A, will be called $S_{11}^A(mn)$, even if it is not propagating. The complex amplitude of the m-th mode transmitted (but not necessarily propagating) through junction A into region 2 is called $S_{12}^A(mn)$. Similar notations are used for the junction B. The fields in each region can be described as a *mode vector* $\mathbf{X}$ such that X_m is the complex amplitude of the m-th mode. In particular, the incident field is given by the mode vector $\mathbf{X}_1$.

Consider now the cascaded scattering processes on both junctions. The field transmitted into region 2 is $S_{21}^A\mathbf{X}_1$. Part of it is reflected back from the junction B as $S_{22}^B S_{21}^A\mathbf{X}_1$, and part is transmitted into region 3 as $S_{32}^B S_{21}^A\mathbf{X}_1$.

The field in region 3 is described by the vector $\mathbf{X}_3$:

$$\begin{aligned}\mathbf{X}_3 &= S_{32}^B S_{21}^A \mathbf{X}_1 + S_{32}^B S_{22}^A S_{22}^B S_{21}^A \mathbf{X}_1 \\ &+ S_{32}^B (S_{22}^A S_{22}^B)^2 S_{21}^A \mathbf{X}_1 + S_{32}^B (S_{22}^A S_{22}^B)^3 S_{21}^A \mathbf{X}_1 \\ &= \sum_{l=0}^{\infty} S_{32}^B (S_{22}^A S_{22}^B)^l S_{21}^A \mathbf{X}_1 = S_{32}^B (I - S_{22}^A S_{22}^B)^{-1} S_{21}^A \mathbf{X}_1 \ , \qquad (9.154)\end{aligned}$$

where I is the identity matrix. All matrices in this equation have infinite size. The field at the exit of the system is given by:

$$\mathbf{X}_3 = S_{31}^{AB} \mathbf{X}_1 \ , \qquad (9.155)$$

and the generalised scattering matrix S for the junctions A and B is:

$$S_{31}^{AB} = S_{32}^B (I - S_{22}^A S_{22}^B)^{-1} S_{21}^A \ . \qquad (9.156)$$

The same formalism can also be applied to an arbitrary number of consecutive junctions. The generalized S-matrix is remarkably similar to the matrices used in the field matching technique described in Chapter 8. However, there are differences between the two methods. The most obvious one is that field matching allows also to consider the excitation by a beam, and not only by an incident wave. The second difference concerns the convergence of the numerical solutions in the two methods. While field matching for a disc-loaded structure seems to have no restrictions on the number of modes retained in the truncated matrices [Kheifets95], above the obvious one stemming from desire to achieve a certain accuracy, it was discovered [Mitra63] that in order for the solution of the scattering matrix method to converge to the correct one, the numbers of modes retained in the cavity region and in the iris region should be in a certain ratio related to the geometric dimensions.

Appendix 9.C Equivalent Circuit Formalism

We have given examples of the application of equivalent circuits before: e.g. it was used to describe eigenmodes of a cavity, or for the fields excited by a small obstacle in the beam pipe.

As another example of its application, we will here evaluate the transverse wake field of an accelerating section with a large number N of cells (cavities and irises), each with slightly different dimensions, also called a *detuned structure* [Thomson94] for linear accelerators. Such a structure is modeled by a chain of resonant circuits with different levels of coupling.

9.C.1 Periodic Array of Uncoupled Cells

For an infinite periodic array of axially symmetric, identical cavities, the characteristic parameters of dipole modes of the first passband - i.e. the *synchronous frequencies* $f_n^{syn} = \omega_n^{syn}/2\pi$, the *kick factors* $\kappa_{\perp n}$, the frequencies of the π - and 0 modes $f_n^{(\pi)}$ and $f_n^{(0)}$ for the n-th cell, can be calculated e.g. with the computer code *TRANSVRS* [Zotter79b].

The $\pi/2$ mode frequency $f_n^{(\pi/2)}$, and its (relative) coupling factor η_n, can be expressed in terms of $f_n^{(0)}$ and $f_n^{(\pi)}$ in the following way:

$$f_n^{(\pi/2)} = \sqrt{\frac{2\left[f_n^{(\pi)}\right]^2\left[f_n^{(0)}\right]^2}{\left[f_n^{(\pi)}\right]^2+\left[f_n^{(0)}\right]^2}}\,, \qquad \eta_n = \frac{\left[f_n^{(\pi)}\right]^2-\left[f_n^{(0)}\right]^2}{\left[f_n^{(\pi)}\right]^2+\left[f_n^{(0)}\right]^2}\,. \tag{9.157}$$

The wake field for $s > 0$ in an array of N identical cells is approximated by:

$$W_\perp(s) = \frac{2}{N}\sum_{n=1}^{N}\kappa_{\perp n}\sin\frac{\omega_n^{syn}s}{c}\,. \tag{9.158}$$

This simple expression does not include realistic coupling between cells, and is valid only for not too large distances s behind the bunch.

9.C.2 Chain of Coupled Resonant Circuits

On the next level of approximation, the structure is modeled by a chain of resonant circuits which are inductively coupled to their nearest neighbours.

The Fourier transform of the current $\tilde{I}_n(f)$ in the n-th circuit must satisfy the following set of n difference equations:

$$\left[1 - \frac{\left[f_n^{(\pi/2)}\right]^2}{f^2}\right] \tilde{I}_n + \frac{M_{n,n+1}}{2L_n} \tilde{I}_{n+1} + \frac{M_{n,n-1}}{2L_n} \tilde{I}_{n-1} = 0 \; . \tag{9.159}$$

The coefficients of the mutual inductances $M_{n,n\pm1}$, as well as the cell inductances L_n, are expressed in terms of the relative coupling coefficients η_n:

$$M_{n,n\pm1} \equiv \frac{1}{2} \left[\eta_n + \eta_{n\pm1}\right] \sqrt{L_n L_{n\pm1}} \; . \tag{9.160}$$

Eq. (9.159) must be solved numerically with the boundary conditions:

$$f_0 = f_1 \; , \quad f_{N+1} = f_N \; , \quad \eta_{1/2} = \eta_1 \; , \quad \eta_{N/2} = \eta_N \; . \tag{9.161}$$

The wake field can be constructed from the solution of these difference equations. A refinement is obtained by modeling the structure with two parallel chains, in which each cell is represented by two resonant loops inductively coupled to their neighbours [Bane92].

The difference equations for the currents in the equivalent circuit can also be solved in the time domain [Yamamoto91]. The results obtained by both methods are in good agreement. The advantage of doing the calculations in the frequency domain is that the integration of the set of difference equations allows one to define the wake function also for large distances s, while it is limited in the time domain approach.

Chapter 10

Radiation Impedance and Curvature Effects

Until now we have considered effects of the interaction of a bunch with its environment under the assumption that it moves with constant velocity on a straight trajectory. The interaction of relativistic particles with each other in the same bunch is strongly suppressed in this case since the electric part of the Lorentz force is compensated by the magnetic one with an accuracy $1/2\gamma^2$. Without this *space charge compensation* the stability of bunches in an accelerator would be restricted to much smaller bunch currents.

The assumption of a straight trajectory with constant velocity is clearly not valid when a charge moves in a circular accelerator. Only the absolute value of its velocity can still be considered constant, while its direction is continuously changing due to the magnetic field which creates a transverse acceleration. For the usual case of a constant magnetic field, the trajectory is essentially circular.

When a bunch moves on a curved trajectory, new effects arise. First of all, the compensation of the space-charge force mentioned above is strongly reduced. Such effects are particularly important for particles moving relativistically with a small radius of curvature, as was the case for beams in the *electron ring accelerators*. These were intensively studied some years ago with the intention to accelerate protons or heavy ions by ring beams of electrons. The effects of curvature on the *EM* forces were described in several papers [Laslett70, Koronev71].

On such curved trajectories, there also appears *synchrotron radiation* due to the acceleration which accompanies the motion, in particular of light particles with high energy factors $\gamma \gg 1$. Since the radiation travels on a chord, it can catch up with and even overtake a bunch in ultra-relativistic motion. Hence, the interaction of a particle with the radiated field is not restricted to the trailing particles. In fact, the *radiation wake* is actually much stronger in front of the bunch then behind it.

The properties of synchrotron radiation have been widely studied, both theoretically [Sokolov80] and experimentally [Winick80]. The main part of this radiation is produced by each particle in a bunch independently, i.e. *incoherently*. For an ultra-relativistic beam, the transverse acceleration has a much stronger effect than a longitudinal one - the power emitted by a particle accelerated in the direction of its velocity is γ^2 times smaller [Jackson75].

As a source of radiation with unique characteristics, synchrotron radiation is used in a wide variety of applications. For high-energy electrons, it is emitted in a very narrow cone $\sim 1/\gamma$ around the velocity vector of the charge motion. The emitted power increases with energy as γ^4, hence the power density or *brilliance* of the radiation can become extremely large. The *critical frequency* - the inverse of the *critical wave length* - divides the power spectrum into two equal parts. It is proportional to γ^3, and lies at very short wave lengths, usually in the ultraviolet or X-ray region. Several electron (or positron) storage rings in the multi-GeV region have been constructed to obtain very high brilliance at very short wave lengths.

Nevertheless, the spectrum of the synchrotron radiation extends all the way up to wave lengths comparable to the length of the orbit. In this range of wave lengths, the *EM* fields of different particles in the bunch can interfere constructively, enhancing the radiated power. Such *coherent radiation* is in many respect similar to the radiation produced by a bunch due to its interaction with the environment. In particular, the power of both types of radiation is proportional to the square of the bunch current, and they both can be described in terms of an effective impedance.

The power lost by a bunch in the form of coherent radiation can be viewed as lost by its current flowing through an effective impedance. The calculation of this impedance and an estimate of the radiation power is the goal of this Chapter.

10.1 Radiation Impedance

The space-charge forces on a curved trajectory produce both longitudinal and transverse effects. In this Section we consider the influence of synchrotron radiation on beam motion.

10.1.1 Incoherent Synchrotron Radiation

We start with a list of the most important formulae for incoherent synchrotron radiation which are useful as reference background. Fig. 10.1 shows the geometry and coordinate system used.

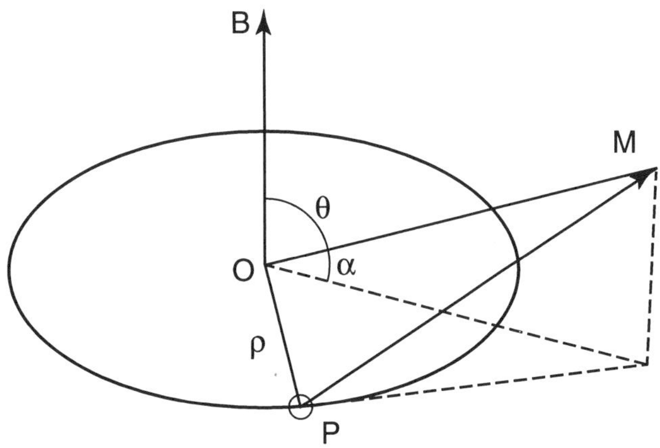

Figure 10.1: Geometry and coordinate system for the motion in magnetic field. The letter M marks an observation point for the radiation emitted by a charge at point P in the direction of vector PM.

We assume that a point charge q of the mass m moves in the horizontal plane $z = 0$, in a magnetic field with vertical component B. The radius ρ of its circular orbit, its velocity $v = \beta c$, its revolution frequency ω_0, and its energy $\mathcal{E}$ are connected in the following way:

$$\omega_0 = \frac{v}{\rho} = \frac{qB}{m\gamma}, \quad \mathcal{E} = qvB\rho \, . \tag{10.1}$$

Synchrotron radiation is emitted at discrete angular frequencies $\omega = n\omega_0$, where n is called the *harmonic number*. The angular and spectral distribution of the radiated power $P_n(\theta, \phi)$ is given by the expression [Schott07]:

$$dP_n(\theta,\phi) = \frac{n^2 r_0 \mathcal{E}_0 \omega_0}{2\pi\rho}\Big[J_n^2(n\beta\cos\theta)\tan^2\theta + \beta^2 J_n'(n\beta\cos\theta)\Big]\sin\theta d\theta d\phi\ , \quad (10.2)$$

where θ and ϕ are the polar and the azimuthal angles of the unit vector in the direction of the radiation, $r_0 = e^2/4\pi\varepsilon_0 mc^2$ is the classical radius of an electron, and $\mathcal{E}_0 = mc^2$ is the particle rest energy.

We introduce the more convenient variables $\alpha = \pi/2 - \theta$ and $\varepsilon = \alpha^2 + \gamma^{-2}$. When γ is large, both these quantities are very small. Using the asymptotic expansions for Bessel functions of large index and arguments of the same order of magnitude, one gets:

$$dP_n(\alpha,\phi) = \frac{n^2 r_0 \mathcal{E}_0 \omega_0}{6\pi^3\rho}\Big[\varepsilon^2 K_{2/3}^2(\frac{n}{3}\varepsilon^{3/2}) + \varepsilon\alpha^2 K_{1/3}^2(\frac{n}{3}\varepsilon^{3/2})\Big]d\alpha d\phi\ . \quad (10.3)$$

The modified Bessel functions of the second kind of orders 1/3 and 2/3 decay exponentially when their argument is larger than one:

$$K_{1/3}^2(x) \approx \frac{\pi}{2}\frac{\mathrm{e}^{-2x}}{x}\ , \quad K_{2/3}^2(x) \approx \frac{\pi}{2}\left(\frac{2}{3}\right)^{2/3}\frac{\mathrm{e}^{-2x}}{x^{1/3}} \quad \text{for } x \gg 1\ . \quad (10.4)$$

It follows that the angular distribution of the radiation of an ultra-relativistic particle is concentrated in a narrow cone around the instantaneous particle velocity with a width $\delta\theta \cong 1/\gamma$. When $\alpha < 1/\gamma$, the harmonic number n is limited by the condition $n < n_c$, where $n_c \cong 3\gamma^3/2$ is the *critical harmonic number*. The harmonic number is large for the major part of the radiation. This explains the fact that classical dynamics is sufficient to describe this radiation over most of its spectrum. When $\alpha > 1/\gamma$, the radiation at harmonics $n \ll n_c$ is emitted mainly into angles smaller then $\alpha \cong n^{-1/3}$.

The total radiated power is obtained by integration of Eq. (10.2) or (10.3) over all angles and summing over all harmonic numbers n. The result is:

$$P_{tot} = \frac{2}{3}\frac{r_0 \mathcal{E}_0 \omega_0}{\rho}\gamma^4\ . \quad (10.5)$$

A comprehensive but succinct description of incoherent synchrotron radiation can be found in ref.[Hofmann89], which also contains some useful integrals which are relevant for the evaluation of its characteristics.

10.1.2 Coherent Synchrotron Radiation

Charges radiate coherently at the wave length λ if their relative distances are smaller than half of it. The *EM* fields of such particles interfere constructively, and the resulting field is the same as that radiated by a single particle with the total charge equal to the sum of all charges. If the characteristic distance between charges is σ_c, then the condition for coherence of synchrotron radiation in free space is:

$$\sigma_c \leq \frac{\lambda}{2} \ . \tag{10.6}$$

Unlike the case of incoherent radiation, for which the radiation power is proportional to the particle number N in the bunch, the power of coherent radiation is proportional to N^2. To understand how this comes about, consider two particles moving in a magnetic field on a circular trajectory. Their positions have azimuthal angles ϕ_1 and ϕ_2 with respect to some arbitrary azimuth taken as zero. Correspondingly, their currents have the phase factors $\exp(-i\phi_1)$ and $\exp(-i\phi_2)$. The n-th harmonics of the fields excited by the particles have the phase factors $\exp(-in\phi_1)$ and $\exp(-in\phi_2)$.

To find the total field created by all N particles of a bunch, we need to sum their individual fields taking into account the corresponding phase factors. For the radiation power averaged over the longitudinal particle distribution $S(\phi)$ we find:

$$P_n|\Sigma_k^N e^{-in\phi_k}|^2 = NP_n + N(N-1)P_n f_n \ , \tag{10.7}$$

where P_n is the incoherent power emitted by each particle, and the form factor for coherent radiation f_n is:

$$f_n = \Big[\int d\phi \ \cos(n\phi) S(\phi)\Big]^2 \ . \tag{10.8}$$

For example, for a Gaussian distribution with the rms bunch length σ_s:

$$\begin{aligned} S(\phi) &= \frac{\rho}{\sqrt{2\pi}\sigma_s} \exp\left(-\frac{\phi^2\rho^2}{2\sigma_s^2}\right) \\ f_n &= \exp\left(-\frac{n^2\sigma_s^2}{\rho^2}\right) \ . \end{aligned} \tag{10.9}$$

Since $\lambda \approx \rho/n$, from these equations the coherence condition Eq. (10.6) follows once more in a more formal way.

The first term on the right hand side of Eq. (10.7) gives the power of incoherent radiation of the bunch, which is proportional to the number of particles in the bunch or to its current. The second term gives the power of coherent radiation, which for $N \gg 1$ is proportional to the square of the number of particles or the bunch current I. Such a signature of coherent radiation makes it suitable to describe it by an effective impedance Z_n which can be defined in accordance with Ohm's law $P_n = Z_n I_n^2$.

For all practical particle distributions, the condition Eq. (10.6) limits coherent synchrotron radiation to microwave frequencies. Hence, the wave lengths of the radiated power are usually smaller than, or comparable to, the transverse size of the vacuum chamber and/or the distance to the poles of the guiding magnets. Under these conditions the presence of walls and magnets can substantially change the radiation and its characteristics. When the walls are conducting, the induced charges tend to decrease the *EM* fields created by the charge itself. Poles of magnets have similar effect due to induced currents. This phenomenon is referred to as *radiation shielding* and is the stronger, the closer the induced charges are to the beam. From this consideration it is clear that the shielding must depend strongly on the size of the vacuum chamber or the distance to the poles.

10.1.3 Coherence Length, Absorption Length, and Radiation Shielding

To take into account the presence of conducting walls around a bunch, and to evaluate their *shielding effect* for coherent radiation, we need to modify the coherence condition Eq. (10.6).

Consider, for example, a case of two parallel conducting plates placed above and below the plane of motion at $z = \pm h/2$. The tangential components of the electric field should satisfy the proper boundary conditions on the plates. For perfectly conducting plates these components should be zero. That means that the *EM* field must contain factors which make appropriate components of the field vanish on the plates. The radiation field can be expanded into a double sum over two integers: the radial harmonic number n, and the vertical harmonic number p. Accordingly, the wave vector $\mathbf{k}, |k| = \omega/c$ of the *EM* field acquires a vertical component k_z which cannot be smaller than π/h - the lowest value which is consistent with the boundary condition. As discussed in the previous subsection, the opening angle α into which radiation is emitted is smaller than $n^{-1/3}$, which means that $\pi/h < k\alpha = \omega\alpha/c = n\alpha/\rho < n^{2/3}/\rho$, where ρ is the local radius of curvature of the trajectory. Therefore only harmonics with $n > (\pi\rho/h)^{3/2}$ can be emitted coherently. As we will show below, a more exact longitudinal *coherence condition* is:

$$n > n_{th} = \frac{1}{\sqrt{3}} \left(\frac{\pi\rho}{h} \right)^{3/2} . \tag{10.10}$$

Usually ρ is much bigger then h. Then the radiation can be emitted coherently only at harmonics with rather large $n \gg 1$.

That means that the *condition of coherence* in the presence of conducting plates is more stringent than Eq. (10.6) in free space:

$$\sigma_s < \frac{\lambda_{th}}{2} \approx \frac{\rho}{n_{th}} = \frac{h}{\pi} \sqrt{\frac{h}{\pi\rho}} . \tag{10.11}$$

The finite conductivity of the plates can be taken into account by introducing an effective *absorption length* L_{ab} defined as the length in the longitudinal direction in which the field decays by a factor e due to absorption in the walls. For radiation at a wave length λ it can be estimated as $L_{ab} \approx Q\lambda/2\pi$ [Heifets91b]. The quality factor Q is the ratio of the field energy stored in a vacuum chamber with typical size h (proportional to the cross section h^2) to the field absorbed in the wall within the skin depth of thickness δ_ω (proportional to the absorption cross section $h\delta_\omega$):

$$L_{ab} \approx \frac{\rho h}{n\delta_\omega} \approx \frac{\rho h}{n^{1/2}\delta_{\omega 0}} . \tag{10.12}$$

The effect of finite conductivity is usually small, since for most metals used for vacuum chambers the absorption length L_{ab} is much larger than the effective (longitudinal) shielding length

$$L_{sh} \approx \frac{\rho}{n^{1/3}} . \tag{10.13}$$

The latter is defined as the length in which the phase of the field slips with respect to the phase of the bunch current by π:

$$\omega t - k_{\parallel} s = \left[\frac{\omega}{c} - k(1 - \frac{\alpha^2}{2}) \right] L_{sh} \approx \pi . \tag{10.14}$$

As we will show, the power spectrum of coherent radiation increases toward shorter wave lengths. This increase is limited by the bunch length since the radiation ceases to be coherent when the wave length becomes smaller than the bunch length.

10.1.4 Radiation Wake and Impedance

The *EM* field of a point charge q moving on a circular trajectory has been derived already at the beginning of this century [Schott07]. For an ultra-relativistic particle, the ϕ component of the radiated electric field simplifies, and can be written [Murphy95]:

$$E_\phi(\phi) = -\frac{U_0}{\pi q \rho} G_{\|}(\phi) \ , \tag{10.15}$$

where $U_0 = 4\pi r_0 \mathcal{E}_0 \gamma^4 / 3\rho$ is the energy loss per revolution, and

$$G_{\|}(x) = \begin{cases} 0 & \text{for} \quad x < 0 \ , \\ \frac{1}{2} & \text{for} \quad x = 0 \ , \\ W(x) & \text{for} \quad x > 0 \ . \end{cases} \tag{10.16}$$

Here the *wake function* $W(x)$ is expressed as function of the dimensionless quantity $x = 3\gamma^3\phi/2$. The sign convention used is such that $\phi > 0$, and hence *$x > 0$ in front of the particle*:

$$W(x) = \frac{dV(x)}{dx} \ , \qquad V(x) = \frac{9}{4} \frac{\cosh(\frac{5}{3}\text{arsinh}x) - \cosh(\text{arsinh}x)}{\sinh(2\text{arsinh}x)} \ . \tag{10.17}$$

Directly in front of the particle $\phi = 0$ and $x = 0$, while directly behind the the particle $\phi = 2\pi$ and $x = 3\pi\gamma^3$. The total radiated power is $P_{tot} = qcE_\phi(0)$ and is given by Eq. (10.5). Fig. 10.2 shows the wake function $W(x)$ and its integral $V(x)$.

The *longitudinal impedance* Z_n due to synchrotron radiation at a harmonic number n is as usual the Fourier transform of the longitudinal electric field, normalised by the beam current: $Z_n = 2\pi\rho\widetilde{E}(n)/\widetilde{I}_n$, where

$$\widetilde{E}(n) = \frac{1}{2\pi} \int_{-\pi}^{\pi} d\phi \widetilde{E}_\phi(\phi) e^{jn\phi} \ , \tag{10.18}$$

and $\widetilde{I}_n = q\omega_0/2\pi$. For $n \ll \gamma^3$ the impedance then becomes:

$$Z_n = Z_0 \pi \beta n \Big[J'_{2n}(2n\beta) + j\mathbf{E}'_{2n}(2n\beta) \Big] \ , \tag{10.19}$$

where the Bessel and the Weber functions are defined in the usual way [Gradshteyn80]:

$$\begin{aligned} J_n(x) &= \frac{1}{\pi}\int_0^{\pi} dt \cos(nt - x\sin t) \ , \\ \mathbf{E}_n(x) &= \frac{1}{\pi}\int_0^{\pi} dt \ \sin(nt - x\sin t) \ . \end{aligned} \tag{10.20}$$

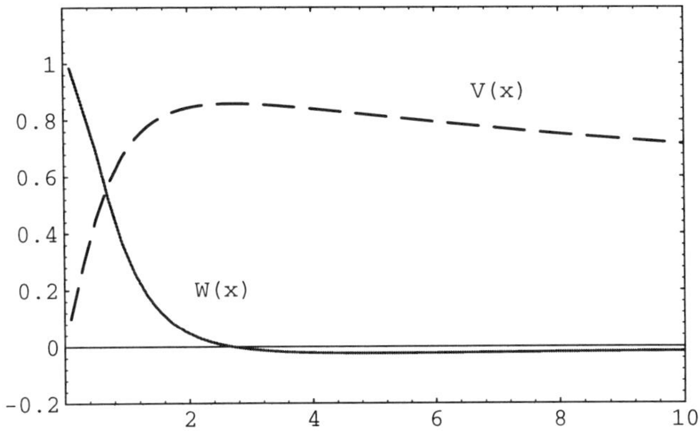

Figure 10.2: Wake potential $W(x)$ and its integral $V(x)$ for a circular orbit as function of dimensionless variable $x = 3\gamma^3\phi/2$.

Using the asymptotic expansions of the Bessel and Weber functions for a large index, in the region where their arguments are of the order of the index, one obtains the result [Faltens75]:

$$Z_{\|} = Z_0 \frac{\Gamma(2/3)}{2\,3^{1/3}}\left[\sqrt{3} - j\right] n^{1/3} \ . \tag{10.21}$$

Coherent radiation of the ultra-relativistic particle $\gamma \gg 1$ is produced mostly at harmonic numbers in the range $n_{th} \ll n_c$. For such n, the significant contribution to the integrals defining the Bessel and the Weber functions comes from the range of t which satisfies the inequality $nt^3 \leq 1$. Hence, $t \ll 1$ and therefore:

$$\begin{aligned} J_n'(2n) &\approx \frac{1}{\pi}\int_0^{\infty} dt \ t \sin\frac{nt^3}{3} \ , \\ \mathbf{E}'_n(2n) &\approx -\frac{1}{\pi}\int_0^{\infty} dt \ t \cos\frac{nt^3}{3} \ . \end{aligned} \tag{10.22}$$

The upper limit of integration can be extended to infinity without much error. From here the result Eq. (10.21) is recovered if one takes into account the expressions for the Γ function. For $0 < x < 1$:

$$\Gamma(x) = \frac{1}{\cos(\pi x/2)} \int_0^\infty dt\ t^{x-1} \cos t = \frac{1}{\sin(\pi x/2)} \int_0^\infty dt\ t^{x-1} \sin t\ . \tag{10.23}$$

For circular machines, the impedance divided by the mode number is more appropriate for beam stability:

$$\frac{Z_\parallel}{n} \approx \frac{Z_0}{n^{2/3}} \frac{\Gamma(2/3)}{2\, 3^{1/3}} \left[\sqrt{3} - j\right]\ . \tag{10.24}$$

This ratio decreases with higher mode numbers n. On the other hand, for $n < n_{th}$, the radiation impedance decreases exponentially. Therefore Re Z_n/n has a maximum at $n \approx n_{th}$:

$$\mathrm{Re} \left(\frac{Z_\parallel}{n}\right)_{max} \approx 300 \frac{h}{2\rho}\ [\Omega]\ . \tag{10.25}$$

This expression was first obtained in ref.[Faltens75].

To take into account the effect of shielding, we will consider the case when a charge moves between two conducting plates positioned at $z = \pm h/2$ above and below the plane of the orbit [Schwinger45, Nodvick54].

The n-th harmonic of the *EM* field of a point charge moving in the mid-plane between the plates can be found using the Green function which is zero at $z = \pm h/2$ [Morse53]:

$$\begin{aligned} \tilde{G}_n(r,\phi,z;r',\phi',z') &= -\frac{j}{2h} \sum_{p=1}^{\infty} \sum_{m=-\infty}^{\infty} \sin p\pi \left(\frac{z}{h} + \frac{1}{2}\right) \sin p\pi \left(\frac{z'}{h} + \frac{1}{2}\right) \\ &\quad \cdot \exp\left[-jm(\phi - \phi')\right] \tilde{G}_m(r,r')\ , \end{aligned} \tag{10.26}$$

where

$$\tilde{G}_m(r,r') = \begin{cases} J_m(\gamma_p r) H_m^{(1)}(\gamma_p r') & \text{for } r < r'\ , \\ J_m(\gamma_p r') H_m^{(1)}(\gamma_p r) & \text{for } r > r'\ , \end{cases} \tag{10.27}$$

and $\gamma_p^2 = (n\beta/\rho)^2 - (p\pi/h)^2$. The power of the radiation at the n-th harmonic can be found by integration over all angles:

$$P_n = \frac{4\pi n r_0 \mathcal{E}_0 \omega_0}{h} \sum_{p=1,3,\ldots}^{p<nh\beta/\pi\rho} \left[\beta^2 {J'_n}^2(x) + \frac{(p\pi\rho/h)^2}{(n\beta)^2 - (p\pi\rho/h)^2} J_n^2(x)\right]\ , \tag{10.28}$$

where the argument of the Bessel functions is $x = \gamma_p \rho$. Only propagating waves with mode numbers $p < nh\beta/\pi\rho$ contribute to the energy loss, since for evanescent modes the product $J_m H_m^{(1)}$ is purely imaginary.

From now on we assume that $\beta = 1$. The real part of the impedance due to synchrotron radiation can be found from the above expression by applying Ohm's law:

$$Z_n = nZ_0 \frac{\rho}{h\beta^2} \sum_{p=1,3,\dots} \left[\beta^2 {J'_n}^2 + \frac{(p\pi\rho/h)^2}{n^2 - (p\pi\rho/h)^2} J_n^2 \right] . \tag{10.29}$$

The same result can be obtained by employing another approach using the expressions for *EM* fields excited by a point charge in a perfectly conducting pill-box cavity of radius d and length g. Solutions which satisfy all boundary conditions can be written as expansions into cylindrical waves [Stratton41]. Such an approach has been used by several authors [Piwinski85, Warnock90, Ng89, Ng90, Warnock91].

The radiation in a closed cavity can only be absorbed in its walls, unlike in open structures where it can escape. In such a situation, the real part of the impedance has narrow peaks around the resonance frequencies of the cavity. In the idealised case of perfectly conducting walls, the resonance peaks become δ-functions. This means that the equilibrium between the emission and absorption of the radiation, can only be reached after an infinitely long time, but such a solution has no practical significance. For the purpose of evaluating the radiated *EM* field in an open structure, the desired result can be obtained by letting the radius d tend to infinity.

The double Fourier transform of the longitudinal component of the electric field E_ϕ, in time at frequency ω and in the azimuthal angle ϕ at harmonic number n, is proportional to a a *generalized impedance function* $Z(n, \omega)$. The radiation field interacts with a particle most effectively when its phase velocity is equal to the particle velocity. In terms of the function $Z(n, \omega)$ this implies that the impedance is obtained by taking its value at $\omega = n\omega_0$. For a pill-box cavity one finds [Warnock90]:

$$Z(n,\omega) = -jn\pi^2 Z_0 \frac{\rho}{h} \sum_{p=1,3,5,\dots} \left[\frac{\alpha_p^2}{\gamma_p^2} \frac{J_n(x)}{J_n(y)} F_n(x,y) + \frac{\omega\rho}{nc} \frac{J'_n(x)}{J'_n(y)} G_n(x,y) \right] , \tag{10.30}$$

where

$$\alpha_p = \pi p/h, \ \gamma_p = \sqrt{(\omega/c)^2 - \alpha_p^2} \ ,$$

$$F_n(x,y) = J_n(x)Y_n(y) - J_n(y)Y_n(x) \ ,$$

$$G_n(x,y) = J_n'(x)Y_n'(y) - J_n'(y)Y_n'(x) \ , \tag{10.31}$$

and $x = \gamma_p \rho, \quad y = \gamma_p d.$

To compare this expression with the previous result, we evaluate its limit for $d \to \infty$. The main contribution comes from the vicinity of the zeros of the denominators, which are defined by $\gamma_p d = j_{nm}$ and $\gamma_p d = j'_{nm}$, where j_{nm} and j'_{nm} are the m-th roots of the Bessel function J_n and of its derivative J'_n. These roots give the resonance frequencies:

$$\omega_{nm} = \sqrt{(\frac{j_{nm}}{d})^2 + (\frac{\pi p}{h})^2} \ , \quad \omega'_{nm} = \sqrt{(\frac{j'_{nm}}{d})^2 + (\frac{\pi p}{h})^2} \ . \tag{10.32}$$

Expanding the Bessel functions in the vicinity of these points, we get an expression for $Z(n,\omega)$ which is a double sum over n and m of two terms: one over the frequencies ω_{nm}, and another one the frequencies ω'_{nm}:

$$Z_n = j\pi Z_0 \frac{\rho}{h} n \Big[\Sigma_1 + \Sigma_2 \Big] \ , \tag{10.33}$$

where

$$\begin{aligned} \Sigma_1 &= \sum_{p,m} \frac{1}{\gamma_p d - j_{nm}} \left(\frac{\alpha_p}{\gamma_p} \right)^2 J_n^2 \left(\frac{\rho}{d} j_{nm} \right) \frac{Y_n(j_{nm})}{J_n'(j_{nm})} \ , \\ \Sigma_2 &= \sum_{p,m} \frac{1}{\gamma_p d - j'_{nm}} J_n'^2 \left(\frac{\rho}{d} j'_{nm} \right) \frac{Y_n'(j'_{nm})}{J_n''(j'_{nm})} \ . \end{aligned} \tag{10.34}$$

In the first sum, the main contribution comes from large values of m, and the sum over m can be evaluated replacing it by an integral. A real part of this integral arises from the residues of the poles at $\gamma_p d = j_{nm}$. The ratio $Y_n(d\gamma_p)/J_n'(d\gamma_p)$ is approximately 1 for $d \to \infty$. Hence the first sum over m is approximately $-jJ_n^2(\gamma_p\rho)/\pi$. Similarly, the second sum over m is $-jJ_n'^2(\gamma_p\rho)/\pi$. Calculating the function $Z(n,\omega)$ for $\omega = n\omega_0$ we obtain the expression for the impedance Eq. (10.29).

10.1.5 Spectrum of Coherent Radiation

The power of the coherent synchrotron radiation, P_n, emitted at the n-th harmonic, can be obtained by multiplying the impedance Z_n due to synchrotron radiation by the square of the current of the circulating charge $(Ne\omega_0/2\pi)^2$. We can use Eq. (10.29) for the impedance.

In the ultra-relativistic case, and when the condition $\pi\rho/h \gg 1$ is fulfilled, most of the radiation is produced at large harmonic numbers n, where the asymptotic expansions of the Bessel functions are valid:

$$\begin{aligned} J_n(\gamma_p\rho) &= \frac{1}{\sqrt{3}\pi}\left(\frac{p\pi\rho}{nh}\right) K_{1/3}\left[\frac{1}{3n^2}(\frac{p\pi\rho}{h})^3\right] , \\ J'_n(\gamma_p\rho) &= \frac{1}{\sqrt{3}\pi}\left(\frac{p\pi\rho}{nh}\right)^2 K_{2/3}\left[\frac{1}{3n^2}\left(\frac{p\pi\rho}{h}\right)^3\right] . \end{aligned} \tag{10.35}$$

Since for large argument the modified Bessel functions $K_\nu(x)$ are exponentially small, cf. Eq. (10.4), only harmonics with mode numbers $n > (p\pi\rho/h)^{3/2}$ contribute to the sum in Eq. (10.29). Hence $n \gg p\pi\rho/h$, and the second term in this equation can be simplified. The result is [Schwinger45, Nodvick54]:

$$P_n = \frac{N^2 r_0 \mathcal{E}_0 \omega_0}{\rho}\frac{4\rho}{3\pi h} n \sum_{p=1,3,\ldots}^{p<nh/\pi\rho} (\frac{p\pi\rho}{nh})^4 \left[K_{2/3}^2(x) + K_{1/3}^2(x)\right] , \tag{10.36}$$

where the argument of the modified Bessel functions is $x = (p\pi\rho/h)^3/3n^2$.

The main contribution to the sum comes from terms for which the argument of the modified Bessel functions is small. The condition $x < 1$ for $p = 1$ is the same as that given in Eq. (10.11). When $n_{th} \gg 1$ (cf. Eq. (10.10), the sum in Eq. (10.36) can be replaced by an integral over x. Hence,

$$P_n = \frac{2C}{3^{1/3}\pi^2}\frac{N^2 r_0 \mathcal{E}_0 \omega_0}{\rho} n^{1/3} \quad \text{for} \quad n_{th} < n < \rho/\sigma_s, \tag{10.37}$$

where [Hofmann89]

$$C = \int_0^\infty dx x^{2/3}\left[K_{2/3}^2(x) + K_{1/3}^2(x)\right] \approx 3.68 . \tag{10.38}$$

One can see that the spectrum of the coherent radiation is essentially the same as that of incoherent one, but for $\gamma \gg 1$ is effectively cut off at harmonics $n \ll n_c = 3\gamma^3/2$. The additional factor N increases the power by a large amount, making a short bunch moving in a magnetic field a very powerful source of *EM* radiation. Eq. (10.37) defines the radiation spectrum for $n_{th} < n$. To obtain the spectrum for the range $n < n_{th}$, we approximate the sum in Eq. (10.36) for $x > 1$. For this we can use the asymptotic expansions of the functions K_ν for $\nu = 1/3$ and $2/3$, Eq. (10.3). Since these expressions decrease exponentially with increasing x, it is sufficient to keep only the first term of the sum over p. Thus one gets:

$$P_n = \frac{2^{5/3}}{\pi}\frac{N^2 r_0 \mathcal{E}_0\, \omega_0}{\rho} n^{1/3} \left(\frac{n_{th}^2}{2n^2}\right)^2 \exp\left(-\frac{n_{th}^2}{2n^2}\right) \quad \text{for} \quad n < n_{th}\ , \tag{10.39}$$

which yields the shape of the spectrum below the threshold.

10.1.6 Power of Coherent Radiation

To find the total power of coherent synchrotron radiation, the power spectrum should be multiplied by the bunch spectrum and summed over all harmonic numbers. For the particular case of the Gaussian distribution, the spectrum is given by Eq. (10.9) and thus:

$$P_{coh} \approx \frac{N^2 r_0 \mathcal{E}_0 \omega_0}{\rho} \sum_{n_{th}}^{n_c} n^{1/3} \exp\left(-\frac{n^2\sigma_s^2}{\rho^2}\right) . \tag{10.40}$$

The main contribution to this sum comes from very large harmonic numbers, hence the summation can be replaced by an integral over $x = n^2\sigma_s^2/\rho^2$:

$$P_{coh} \approx \frac{N^2 r_0 \mathcal{E}_0 \omega_0}{2\rho} \left(\frac{\rho}{\sigma_s}\right)^{4/3} F(x_{th})\ , \tag{10.41}$$

where the form factor is given by the integral

$$F(x_{th}) = \int_{x_{th}}^{x_c} dx x^{-1/3} e^{-x}\ . \tag{10.42}$$

The limits of integration are

$$x_{th} = \frac{2}{3}\left(\frac{\pi\rho}{h}\right)^3 \left(\frac{\sigma_s}{\rho}\right)^2 , \quad x_c = \frac{9}{4}\gamma^6 \left(\frac{\sigma_s}{\rho}\right)^2 \gg 1\ . \tag{10.43}$$

Since the integrand in Eq. (10.42) is exponentially small for large x, the integration can be extended to ∞ without introducing much error. Then the form factor $F(x)$ reduces to the incomplete Γ-function of order $2/3$:

$$F(x) = \Gamma\left(\frac{2}{3}, x_{th}\right)\ . \tag{10.44}$$

Fig. 10.3 shows the function $F(x_{th})$. For very small or very large values of its argument, simpler expressions can be obtained by using appropriate expansions of this function.

1. When $x_{th} \ll 1$, $F(x_{th}) = \Gamma(2/3)$. The total power of the radiation is given by the formula Eq. (10.41) with this value of $F(x_{th})$ [Schiff46].

2. In the more usual case when $x_{th} \gg 1$, $F(x_{th}) \approx x_{th}^{-1/3} e^{-x_{th}}$, and the total power of the radiation is exponentially small.

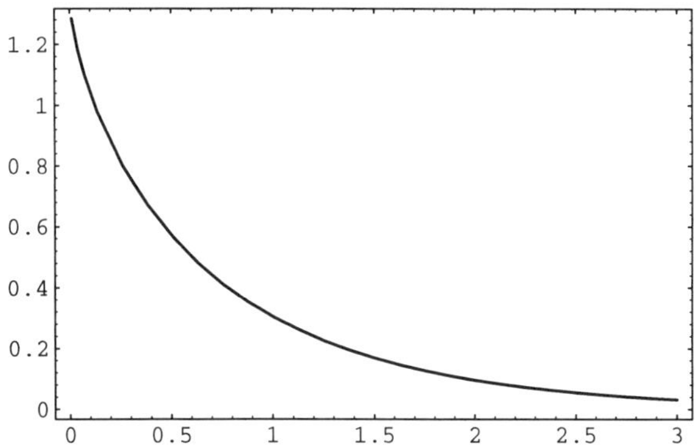

Figure 10.3: The reduction factor $F(x_{th})$ for shielding of coherent radiation, versus $x_{th} = (2/3)(\pi\rho/h)^3$ for chamber height h and bending radius ρ.

Similar results can be obtained for other particle distributions, e.g. for a uniform distribution $-\sigma_s < s < \sigma_s$. The density spectrum of such a distribution is $\lambda = \sin^2 x/x^2$, where $x = n\sigma_s/\rho$. Although it is different from the one for the Gaussian distribution, nevertheless, for $x \gg 1$ it is also a sharp function of its argument. In the limit $n \to \infty$,it tends to the δ-function: $\lambda \to \pi\delta(\sigma_s/\rho)/n$.

This was probably overlooked in the paper [Nodvick54]. In the derivation of the total power of radiation, the order of summation over the eigenmodes p and n was interchanged, and then $sin^2 x$ replaced by its average value 1/2. Therefore the total power of radiation given in this paper has no exponential term $\exp(-x_{th})$. Apart from this, their expression agrees with Eq. (10.41) if $F(x_{th})$ is approximated by $x_{th}^{-1/3}$.

The total power of coherent synchrotron radiation is thus mainly determined by the parameter x_{th} which depends on two independent effects:

a) *shielding*, characterised by the ratio ρ/h, and

b) *bunching*, characterised by the ratio σ/ρ.

In high-energy storage rings usually $x_{th} \gg 1$, and thus coherent synchrotron radiation is very weak. In order to observe this radiation, a storage ring with $x_{th} \approx 1$ would be required. This can be achieved only when the ratio ρ/h is not too large, and at the same time the ratio σ/ρ is not too small. In other words, the transverse size of the vacuum chamber should be large, while the effective length of the bunch should be small.

It is interesting to evaluate x_{th} for conditions in which coherent synchrotron radiation was observed with a linac beam [Ishi91]. For the parameters of the beam and apparatus given in the paper: $\rho = 2.44$ m, $h = 0.2$ m, $\sigma \approx 1$ mm , the value of $x_{th} \approx 6.3 \cdot 10^{-3}$ is very small. This indicates that the experimental conditions were not quite optimised for the observation.

In the considerations so far we were concerned with longitudinal effects such as energy loss, or the longitudinal impedance and wake function due to coherent synchrotron radiation emitted by charged particles moving with relativistic velocity on a curved trajectory. However, there are also transverse effects which might have to be taken into account, and caused a flurry of excitement some 10 years ago.

10.2 Transverse Curvature Effects

The transverse electric and magnetic forces due to the *space charge* of a beam almost cancel each other when acting on a charged particle which is moving in a straight line with relativistic velocity. The combined *EM* force is reduced by a factor γ^2 which is usually large compared to unity, in particular in storage rings for light particles such as electrons. But also in circular proton accelerators with energies above a few GeV the space-charge forces are very small.

A number of effects can destroy this nearly perfect cancellation. A quite common one is partial *charge neutralisation*, which may be caused by particles of opposite charge - e.g. ions in electron beams, or electrons in proton beams - being created near or even attracted towards the beam. Their electric field, proportional to the charge, tends to reduce that of the beam. However, since the particles of opposite charge usually move much slower than those of a high-energy beam, the corresponding current is low, and hence their magnetic field is smaller than that of the beam. In addition, if these particles move in the opposite direction, their magnetic field will add to that of the beam. These additional fields will disturb the delicate balance between electric and magnetic fields and the *EM* forces may be strongly increased enhancing the transverse *space-charge impedance.* In proton storage rings like the *ISR*, special *clearing electrodes* to remove electrons were used to reduce this effect. Electron machines are sometimes converted to operation with positrons to avoid ion effects.

Another effect perturbing this delicate equilibrium is due to particles moving on a curved trajectory. The electric and magnetic forces then cannot balance even when $v \to c$. In particular the so-called *centrifugal space-charge force*, was suspected to be important in storage rings [Talman86]. The effect of this force was the subject of a number of publications [Piwinski85, Bassetti86b, Sands86], until it was realised that it does not lead to an increase of the beam emittance both for coasting beams [Lee87] and for bunched beams [Derbenev96]. The reason for this is the near cancellation of the effect by that of particles with energy deviation under the influence of the transverse electric field of the space charge.

For a particle moving on a circular trajectory, one can find the *EM* fields at a field point P due to a charge q at source point P' from the *Liénard-Wiechert* formula [Jackson75]:

$$\mathbf{E}(\mathbf{r},t) = \frac{q}{4\pi\varepsilon_0}\left[\frac{\hat{\mathbf{n}}-\mathbf{v}'/c}{\gamma^2 D^3 L^2} + \frac{\hat{\mathbf{n}}\times\left[(\hat{\mathbf{n}}-\mathbf{v}'/c)\times\mathbf{a}'\right]}{D^3 L}\right], \tag{10.45}$$

where L is the distance PP', and $\hat{\mathbf{n}}$ the unit vector pointing in that direction. The velocity vector $\mathbf{v}'$ and the acceleration vector $\mathbf{a}'$ should be evaluated at the *retarded time* $t' = t - L/c$. The magnetic field can be found from the electric one by the relation $Z_0\mathbf{H} = \hat{\mathbf{n}}\times\mathbf{E}$. The derivative of time with respect to the retarded time is $D \equiv dt/dt' = 1 - \hat{\mathbf{n}}\cdot\mathbf{v}'/c$. For circular motion around an origin O, it can be expressed as $D = 1 - \beta\cos(\theta/2) \approx 1/2\gamma^2 + \theta^2/8$ when θ is the angle between OP and OP'.

While the first term in the above equation is the usual electric *Coulomb field* of a charge moving with constant velocity, the second term is proportional to the *acceleration* $\mathbf{a}'$. A longitudinal acceleration has only a very small effect, but the transverse one on a curved trajectory is an essential contribution. For a circular orbit with radius ρ, the centripetal acceleration $a' = v^2/\rho$ balancing the *centrifugal force.*

Evaluating the radial component of the *EM* force due to a uniform line charge density λ between the opening angles θ_{min} and θ_{max}, a short calculation yields approximately for the centrifugal force on a particle at the central orbit [Talman86]:

$$F_r = \frac{\lambda}{4\pi\varepsilon_0\rho}\left[\left(1-\frac{1}{2\gamma^2}\right)\ln\frac{\tan(\theta_{max}/4)}{\tan(\theta_{min}/4)}-1\right] . \tag{10.46}$$

This radial force falls off rapidly on both sides off the central orbit over distances comparable to the transverse beam size. The effects of screening due to the vacuum chamber walls has been ignored. Most of the force is due to particles close to the test charge, and one can obtain a better approximation using this fact [Bassetti86a].

For realistic particle distributions, the line density $\lambda(s)$ is a function of the distance s along the bunch, and hence this transverse force depends on the longitudinal coordinate. This could lead to coupling of longitudinal and transverse motion called *synchro-betatron coupling.*

A more consistent derivation of the space-charge forces on a curved trajectory is given in ref.[Derbenev96]. For small transverse dimensions of a bunch, $\sigma_\perp^{3/2} \ll \sigma_s\rho^{1/2} \ll h^{3/2}$, the *EM* potentials are approximated by:

$$\Phi(\mathbf{r}) = \int d\mathbf{r}_1\frac{\rho(\mathbf{r}_1, t-L/c)}{L} , \qquad \mathbf{A}(\mathbf{r}) = \int d\mathbf{r}_1\frac{\mathbf{v}_1\rho(\mathbf{r}_1, t-L/c)}{cL} , \tag{10.47}$$

where $L = |\mathbf{r}-\mathbf{r}_1|$ and ρ is the *space-charge density.* The *interaction Hamiltonian* is then:

$$V(s,r,z) = e\left[\Phi-\beta A_s\right] = U(s)\left[1+\frac{r}{\rho}\right] - F(s)r + \tfrac{1}{2}g(s)\left[3r^2+z^2\right] \tag{10.48}$$

up to terms quadratic in the transverse coordinates, with the functions:

$$\begin{aligned} U(s) &= \frac{2Ne^2}{(3\rho^2)^{1/3}}\int_0^\infty \frac{ds_1}{s_1^{1/3}}\lambda(s-s_1) , \\ F(s) &= \frac{2Ne^2}{\rho}\lambda(s) , \\ g(s) &= \frac{2Ne^2}{(3\rho^2)^{2/3}}\frac{d}{ds}\int_0^\infty \frac{ds_1}{s_1^{2/3}}\lambda(s-s_1) . \end{aligned} \tag{10.49}$$

The function dU/ds defines the power of the radiation from location s inside the bunch. The function $F(s)$ is the effective centripetal force, and the terms proportional to $g(s)$ describe the focusing in both transverse planes.

Chapter 11

Models and Measurements of Impedance

First we discuss a number of different models which have been proposed to imitate the impedances of real structures used in accelerators. They permit considerable simplifications in the analytic calculation of various quantities such as effective impedance, loss factor, wake functions or wake potentials. A number of free parameters can usually be chosen to give approximate agreement with the impedances of the actual structures over the range of interest. Next we describe methods of impedance measurements in the laboratory (*bench measurements*) and in existing accelerators (*machine measurements*).

At last we discuss briefly purely numerical methods for the calculation of impedances and wakes and list the most widely used computer programs for this purpose.

11.1 Broad-Band Equivalent Impedances

The coupling impedance of an accelerator is the result of contributions of many components with different characteristics. Each contribution is a complicated function of frequency which usually has many resonant peaks below or around the cut-off frequency of the beam pipe, as well as a characteristic high frequency tail. Many of the detailed features of the resulting impedance depend critically on the exact dimensions of the structures which may be variable with temperature or other external parameters. However, often these details are not required to predict the stability of beams correctly.

In particular, for single bunch stability one needs only the short-range wake field. In the frequency domain, this corresponds to a smoothed impedance obtained by an averaging procedure, i.e. there is limited resolution of the details of the impedance function. Furthermore, beyond a maximum frequency of order of $1/\sigma$ the bunch is no longer sensitive to any effects of the impedance.

Hence, the problem consists of constructing an equivalent impedance in a simple way to calculate the relevant machine parameters (wake potentials, loss factors, etc.) in order to be able to correctly describe the beam-wall interaction for a wide range of the beam characteristics (charge distribution, bunch length and displacement, etc.). The parameters of the equivalent impedance should be deduced from the results of numerical calculations, and checked by comparison with the bench or machine measurements.

11.1.1 The SPEAR Impedance Model

Historically, the first model was the so-called *SPEAR impedance* which was obtained from measurements of parameters of short bunches in a particular machine [Wilson77]. The frequency dependence of the real part of the longitudinal impedance was fitted with a simple power law ω^a. For very short bunches, the exponent was found to be approximately $a = -0.68 \approx 2/3$ from the variation of the bunch length with $\xi^{1/(2+a)}$. Here the *'scaling parameter'* is defined [Chao76] by $\xi = \alpha I_b e/\nu_s^2 \mathcal{E}$, where α is the momentum compaction factor, I_b the bunch current, ν_s the synchrotron tune, and $\mathcal{E}$ the energy of the beam.

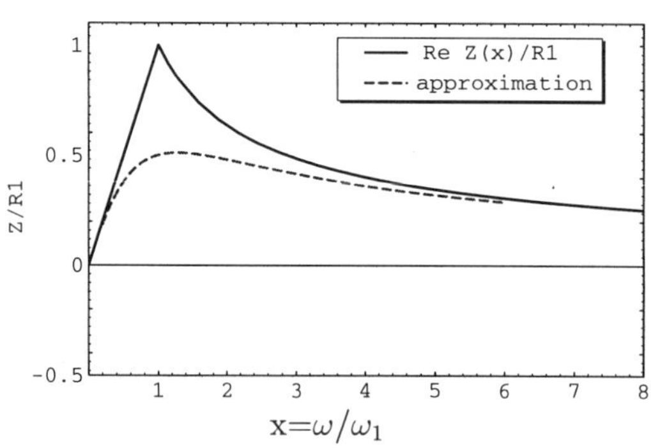

Figure 11.1: The *SPEAR* impedance model.

We can thus write the high frequency part of the (longitudinal) impedance as:

$$Z(\omega) = R_s \left(\frac{\omega}{\omega_1}\right)^a \quad \text{with} \quad a \approx -\frac{2}{3} \quad \text{for} \quad \omega > \omega_1 \ . \tag{11.1}$$

For long bunches, however, the impedance is inductive and the parameter a is close to unity. This yields the expected dependence of bunch length on current $\sigma \sim I_b^{1/3}$. The expression for the impedance at low frequencies, can be written $Z = R_s\omega/\omega_1$. In addition to the constant R_s one has to determine the frequency ω_1 where the two expressions change over. By measuring the dependence of the loss factor on bunch length at various energies and synchrotron tunes, it was found that the coupling impedance of *SPEAR II* was best described by $R_s = 9$ KΩ and $\omega_1 = 1.3$ GHz as shown in Fig. 11.1. Although this model could be adjusted for other machines, it was often used indiscriminately with the parameters valid only for *SPEAR*. The non-integer power law makes analytic calculations difficult.

11.1.2 The Broad-Band Resonator Model

More satisfying is the broad-band resonator model [Hofmann79b] described already in Chapter 4 and shown in Fig. 11.2, normalised by the shunt impedance R_s.

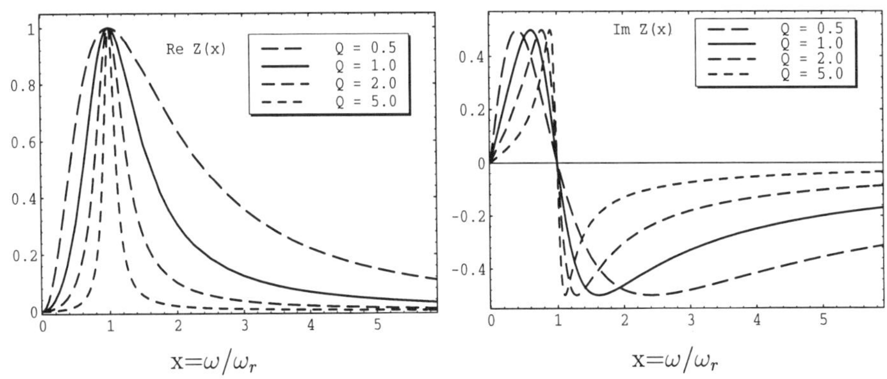

Figure 11.2: Normalised impedance of resonators with various quality factors.

It consists of a single expression, valid for all frequencies:

$$Z(\omega) = \frac{R_s}{1 + jQ_r(\omega/\omega_r - \omega_r/\omega)} \,. \tag{11.2}$$

The *reduced* longitudinal impedance Z/n, is proportional to the *transverse impedance*, which is shown in Fig. 11.3, normalised by $(\omega_0/\omega_r)(R_s/Q_r)$.

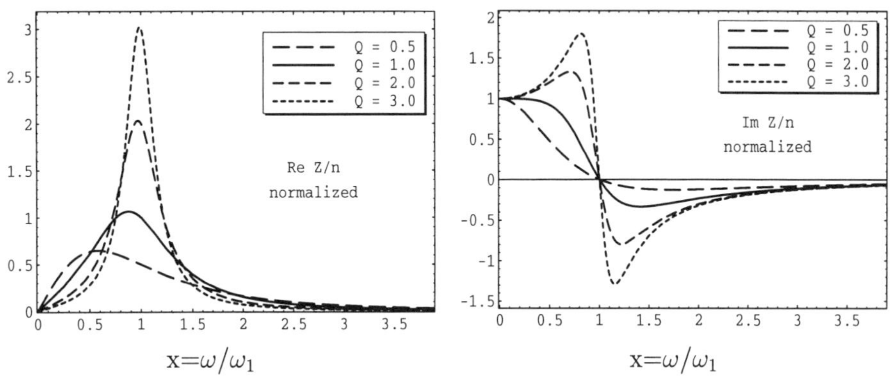

Figure 11.3: Reduced impedance of resonators with various quality factors.

At low frequencies, the real part of the resonator impedance depends quadratically on frequency, Re $Z \approx R_s\omega^2/Q_r^2\omega_r^2$, while the imaginary part is linear: Im $Z \approx R_s\omega/Q_r\omega_r$. This behaviour correspond to an inductance, and the quantity $|Z/n|$ ($n = \omega/\omega_0$) approaches the constant value $\omega_0 R_s/\omega_r Q_r$. On the other hand, at high frequencies the real part tends towards Re $Z \to R_s\omega_r^2/Q_r^2\omega^2$, i.e. it falls off with the inverse square of frequency, while the imaginary part falls off only with the inverse first power. This fast decrease of the real part does not describe correctly the asymptotic behaviour of a cavity impedance, whose real and imaginary parts fall only with the square root of frequency, and not even of a periodic structure, where the real part should fall off with a 3/2 power. For very short bunches, for which the high-frequency part of the impedance is important, one may thus obtain too low estimates of the stability limits and the energy loss.

However, the simple expression for the resonator model impedance permits analytic evaluation of many expressions in closed form. For the longitudinal wake function one finds:

$$G(\tau) = \frac{\omega_r R_s}{Q_r}\left[\cos(\omega_r'\tau) - \frac{1}{2Q_r'}\sin(\omega_r'\tau)\right]\exp\left(-\frac{\omega_r\tau}{2Q_r}\right), \qquad (11.3)$$

where $\omega_r' = \omega_r Q_r'/Q_r$ and $Q_r' = \sqrt{Q_r^2 - 1/4}$. The transverse wake function is even simpler and has only a *sine* term, cf. Eq. (3.24).Both the longitudinal and the transverse wake functions, normalised by $\omega_r R_s/Q_r$, are shown in Fig. 11.4.

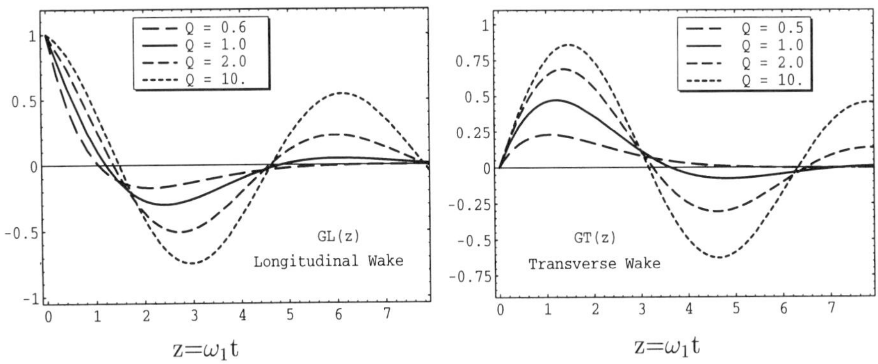

Figure 11.4: Normalised longitudinal and transverse wake functions of broad-band resonators with various quality factors.

This model also permits explicit evaluation of the loss factors for a Gaussian bunch. However, the resulting decrease with the third power of the bunch length does not agree with experiment or numerical calculations for (perfectly conducting) cavities, where the loss factor falls exponentially. To remedy these shortcomings, several other impedance models were therefore suggested. We describe some of them below [Palumbo94].

11.1.3 Improved Broad-band Model for Periodic Structures

It is also possible to find pairs of Fourier transforms which have an explicit expression for the wake function, and the desired asymptotic frequency dependence of the impedance [Hofmann88]. For a periodic structure, the real part of the longitudinal impedance is expected to decrease with frequency as $\omega^{-3/2}$. A suitable candidate for the wake function is its Fourier transform, i.e. the *complementary* error function erfc(z) = 1 - erf(z), with its argument proportional to the square root of time delay $z = \sqrt{\omega_1 \tau}$, where ω_1 is a free parameter. In order to fulfil the requirement that the real part of the impedance vanishes at zero frequency, we subtract a term which makes the integral over all z equal to zero. Choosing an exponential will not change the asymptotic behaviour of the impedance, as it corresponds to an impedance whose real part decreases as ω^{-2} and thus vanishes faster than the main term. With the second free parameter ω_2, or rather the ratio $\alpha = \omega_2/\omega_1$, the wake function for this model 1A can be written:

$$G(\tau) = \omega_1 R_s \Big[2 \ \mathrm{erfc}(\sqrt{\omega_1 \tau}) - \alpha \exp\left(-\alpha \sqrt{\omega_1 \tau}\right) \Big] \tag{11.4}$$

for $\tau > 0$, while $G(\tau)$ vanishes for $\tau < 0$. It is shown in Fig. 11.5, together with the wake function for model 1B, for various values of the parameter α.

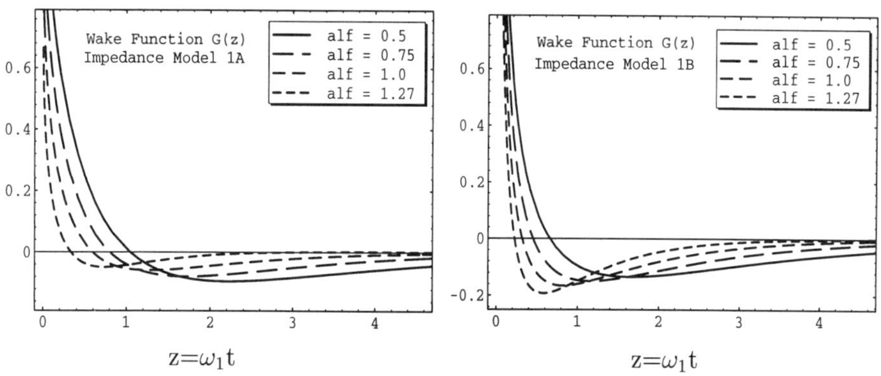

Figure 11.5: Normalised wake functions for model impedances 1A and 1B.

With $x = \omega/\omega_1$ and $u = \sqrt{x^2+1}$, the corresponding impedance becomes:

$$\begin{aligned} \mathrm{Re}\, Z(x) &= R_s \left[\frac{1}{u}\sqrt{\frac{2}{u+1}} - \frac{\alpha^2}{x^2+\alpha^2} \right] , \\ \mathrm{Im} Z(x) &= R_s \left[\frac{1}{u}\sqrt{\frac{2}{u-1}} + \frac{\alpha^2}{x^2+\alpha^2} - \frac{2}{x} \right] , \end{aligned} \tag{11.5}$$

and is shown in Fig. 11.6 for several values of $\alpha = \omega_2/\omega_1$. The frequencies ω_1, ω_2 and the shunt impedance R_s are free parameters which can be chosen to fit the impedance of a particular structure.

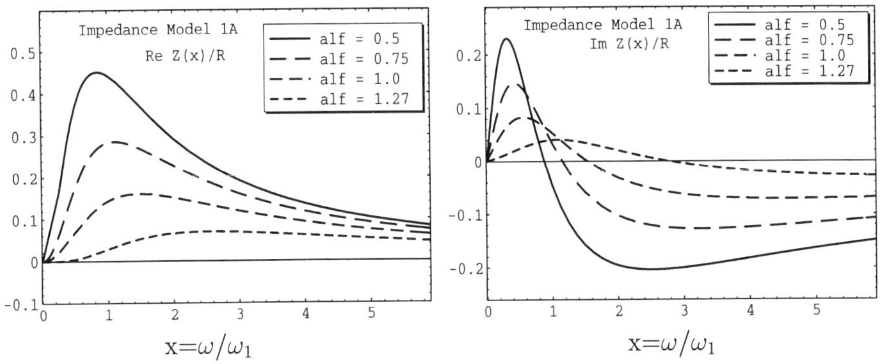

Figure 11.6: Real and imaginary parts for normalised impedance model 1A.

The asymptotic behaviour of the impedance is then given by:

$$\operatorname{Re} Z(x) \sim x^{-3/2}\,, \qquad \operatorname{Im} Z(x) \sim \frac{2-\alpha}{x}\,, \tag{11.6}$$

which agrees with the asymptotic behaviour of the impedance of an infinite periodic array of cavities.

The expansion of the impedance at low frequencies is

$$\operatorname{Re} Z(x) \sim \left(\frac{1}{\alpha^2} - \frac{5}{8}\right) x^2\,, \quad \operatorname{Im} Z(x) \sim \left(\frac{1}{\alpha^2} - \frac{3}{4}\right) x\,. \tag{11.7}$$

The impedance increases as ω^2, which gives a (too) slow decrease of the loss factor for long Gaussian bunches proportional to σ^{-3} as for the resonator model. This can be somewhat improved by choosing $\alpha = \sqrt{5/8}$, for which value the impedance increases as ω^4 and the loss factor becomes proportional to σ^{-5}.

11.1.4 Improved Broad-band Model for a Cavity

The expected asymptotic behaviour of the real part of the impedance of a cavity is $\omega^{-1/2}$. With $z = \sqrt{\omega_1 \tau}$, a suitable wake function for $\tau > 0$ is:

$$G(\tau) = R_s \left[\frac{\exp(-z)}{\sqrt{\pi z}} - \alpha \exp(-\alpha z) \right] . \tag{11.8}$$

With $x = \omega/\omega_1$ and $u = \sqrt{x^2+1}$ as before, the expression for the corresponding impedance becomes:

$$\begin{aligned} \operatorname{Re} Z(x) &= R_s \left[\frac{\sqrt{u+1}}{\sqrt{2}u} - \frac{\alpha^2}{x^2+\alpha^2} \right] , \\ \operatorname{Im} Z(x) &= -R_s \left[\frac{\sqrt{u-1}}{\sqrt{2}u} - \frac{\alpha x}{x^2+\alpha^2} \right] . \end{aligned} \tag{11.9}$$

This impedance and its wake function, for several values of the parameter α, are shown in Fig. 11.7 The expansion of this impedance at low frequencies is

$$\text{Re}Z(x) \sim \left(\frac{1}{\alpha^2} - \frac{3}{8} \right) x^2 , \quad \text{Im}Z(x) \sim \left(\frac{1}{\alpha^2} - \frac{1}{2} \right) x . \tag{11.10}$$

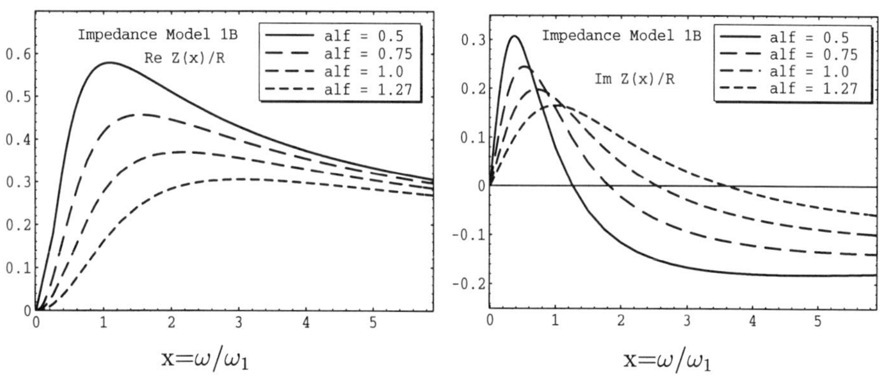

Figure 11.7: Real and imaginary parts of normalised impedance of model 1B.

The loss factor for a Gaussian bunch could be expressed analytically by parabolic cylinder functions or modified Bessel functions of order 1/4, but since these are rather unfamiliar functions, it is actually simpler to calculate them by numeric integration. The impedance and wake function for several values of the parameter α are shown in Fig. 11.7.

11.1.5 Second Broad-band Model for Periodic Structures

The two models 1A and 1B describe the asymptotic behaviour of the impedance correctly, but give a too slow decrease of the loss factor with bunch length for long Gaussian bunches. As shown in Fig. 11.8, the two models 2A and 2B improve this by introducing a cut-off frequency ω_1 as it occurs for more realistic structures. However, the impedance is then no longer given by a single expression, valid at all frequencies, and the wake function cannot be expressed simply by well-known functions.

For an asymptotic frequency dependence proportional to $\omega^{-3/2}$, such as for an infinite periodic array of cavities, one gets with $x = \omega/\omega_1$:

$$\operatorname{Re} Z(x) = R_s \begin{cases} 0 & \text{for } |x| < 1 , \\ \dfrac{\sqrt{|x|-1}}{x^2} & \text{for } |x| > 1 . \end{cases} \tag{11.11}$$

This function has a maximum at $x = \omega/\omega_1 = 4/3$, and decreases as $\omega^{-3/2}$ at high frequencies. Its Hilbert transform yields the imaginary part:

$$\operatorname{Im} Z(x) = \frac{R_s}{x^2} \begin{cases} \sqrt{1+x} - \sqrt{1-x} - x & \text{for } |x| < 1 , \\ \sqrt{1+x} - x & \text{for } |x| > 1 . \end{cases} \tag{11.12}$$

At low frequencies $x \ll 1$, the real part is identically zero, and the imaginary part tends towards Im $Z/R_s \approx 3x/4$, i.e. the impedance is purely inductive. The corresponding wake function could be expressed by Fresnel functions, but is just as easily found by evaluating directly the integral expression

$$G(y) = \frac{2}{\pi} \int_1^\infty dx \, \frac{\sqrt{x-1}}{x^2} \cos xy \, . \tag{11.13}$$

This impedance and its wake function, as well as the one described in the following subsection, are shown in Fig. 11.8.

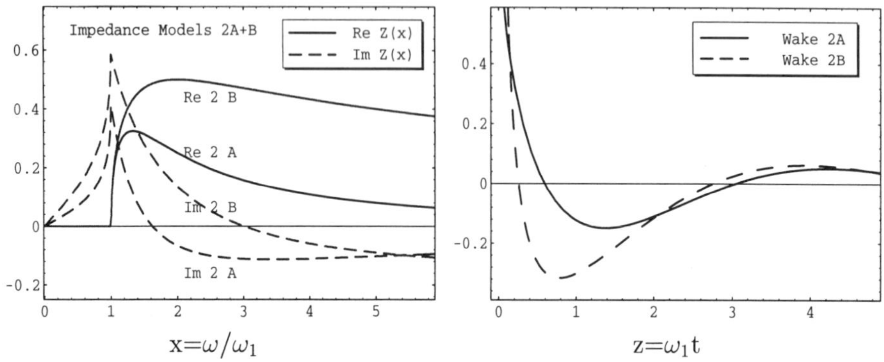

Figure 11.8: Normalised impedances and wake functions, models 2A+2B.

11.1.6 Second Broad-band Model for a Cavity

A similar model can be constructed for the impedance of a cavity, which tends asymptotically towards $\omega^{-1/2}$. For this case the impedance can be written:

$$\operatorname{Re} Z(x) = R_s \begin{cases} 0 & \text{for } |x| < 1 \;, \\ \dfrac{\sqrt{|x|-1}}{|x|} \;, & \text{for } |x| > 1 \;. \end{cases} \tag{11.14}$$

This function has a maximum at $x = 2$ and decreases as $x^{-1/2}$ at high frequencies. The expression for the imaginary part is:

$$\operatorname{Im} Z(x) = \frac{R_s}{x} \begin{cases} 2 - \sqrt{1+x} - \sqrt{1-x} & \text{for } |x| < 1 \;, \\ 2 - \sqrt{1+x} & \text{for } |x| > 1 \;. \end{cases} \tag{11.15}$$

The loss factor of these two models can be obtained by numerical integration of the product of the real part of the impedance and the charge distribution of the bunch. For long Gaussian bunches this yields an exponential decrease with bunch length, in agreement with numerical results.

11.1.7 Composite Model

An impedance model consisting simply of the sum of the expressions for its various elements has been used [Heifets92] in order to fit the results of calculations and measurements of wakes in storage rings [Bane89, Bane90a]. The model has two attractive features: firstly, each term of the impedance has a clear physical meaning; secondly, the expressions for the corresponding wake fields and loss factors can be derived analytically, which simplifies the fitting procedure to obtain the model parameters. However, the model diverges at low frequencies and needs to be supplemented by a cut-off.

The suggested expression for the longitudinal impedance is:

$$Z(\omega) = j\omega L + R_s + B\Big[1 - j\ \mathrm{sign}(\omega)\Big]\sqrt{|\omega|} + A\frac{1 + j\ \mathrm{sign}(\omega)}{\sqrt{|\omega|}} \ . \tag{11.16}$$

The model has 4 distinct terms: the first term represents a low-frequency inductive impedance which is typical for shallow structures on the accelerator vacuum chamber such as smooth tapers, bellows, vacuum ports, slots, collimators, etc. For small discontinuities the impedance has an inductive character up to rather high frequencies.

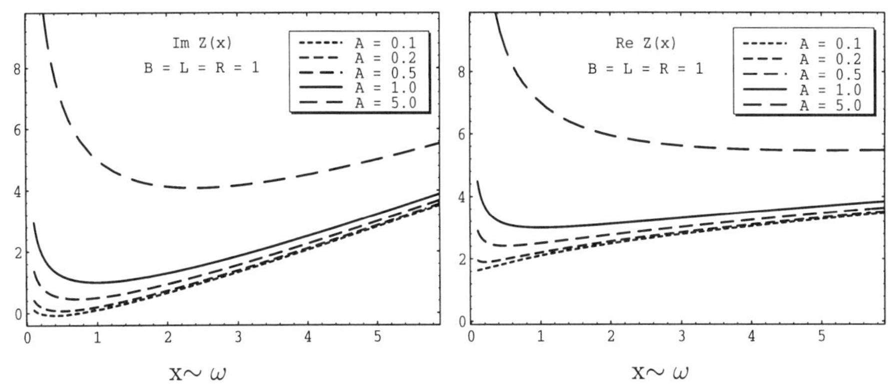

Figure 11.9: Real and imaginary parts of impedance for composite model.

For the inductive term, the longitudinal wake potential of a Gaussian bunch with rms length $\sigma = \sigma_s/c$ is given by:

$$W^{ind}(\tau) = \frac{L\tau}{\sqrt{2\pi}\sigma^3} \exp\left(-\frac{\tau^2}{2\sigma^2}\right) \ . \tag{11.17}$$

It is an odd function of the time delay τ, and hence the loss factor of a Gaussian bunch is identically zero. The wake function has extrema at the points $\tau = \pm\sigma$

of magnitude:

$$\hat{W}^{ind} = \frac{L}{\sqrt{2\pi\sigma^2}} \ . \tag{11.18}$$

The second term in Eq. (11.16) is resistive. For a Gaussian bunch it yields a wake potential given by:

$$W^R(\tau) = \frac{R_s}{\sqrt{2\pi}\sigma} \exp\left(-\frac{\tau^2}{2\sigma^2}\right) \ . \tag{11.19}$$

Such a wake is usually produced by cavities when the bunch length is comparable to the beam pipe radius. The corresponding loss factor is:

$$\kappa_{\parallel}(\sigma) = \frac{R_s}{2\sqrt{\pi}\sigma} \ . \tag{11.20}$$

The third term in Eq. (11.16) describes the impedance due to a non-zero wall resistivity. The corresponding wake potential can be expressed in terms of modified Bessel functions I_ν of fractional order ν and argument $\eta = (\tau/2\sigma)^2$:

$$W^{rw}(\tau) = B\frac{|\tau|^{3/2}}{4\sigma^3}\left[I_{-\frac{3}{4}}(\eta) - I_{\frac{1}{4}}(\eta)\, I_{-\frac{1}{4}}(\eta) + \mathrm{sign}(\tau) I_{\frac{3}{4}}(\eta)\right] \mathrm{e}^{-\eta} \ . \tag{11.21}$$

The loss factor of a Gaussian bunch of length σ becomes:

$$\kappa_{\parallel}(\sigma) = B\, \frac{\Gamma(3/4)}{2\pi\sigma^{3/2}} \ , \tag{11.22}$$

where $\Gamma(3/4) \approx 1.2254$.

The last term in Eq.(11.16) has the same high frequency dependence as the impedance of a single cavity. Its wake potential is found to be given by:

$$W^{cav}(\tau) = A\frac{\sqrt{|\tau|}}{2\sigma}\left[I_{\frac{1}{4}}(\eta) + \mathrm{sign}(\tau) I_{-\frac{1}{4}}(\eta)\right]\mathrm{e}^{-\eta} \ , \tag{11.23}$$

and the corresponding loss factor becomes:

$$\kappa_{\parallel}(\sigma) = A\, \frac{\Gamma(1/4)}{2\pi\sigma^{3/2}} \ , \tag{11.24}$$

where $\Gamma(1/4) \approx 3.6256$.

The four parameters of the model A, B, L, R_s can be obtained by fitting the wake function $W(\tau)$ or the loss factor $\kappa_{\parallel}(\sigma)$ to the values obtained either numerically or by measurements. The behaviour of this impedance is illustrated in Fig. 11.9 for several values of the parameter A, which is a coefficient of the cavity term, while all other parameters are set to unity.

11.1.8 Modified Inductance Model

Recently a model was suggested [Berg95, Heifets96] which is designed to represent the inductive impedance at low frequencies, but has non-vanishing loss factor there. The model has a $\omega^{-1/2}$ dependence at high-frequencies, and two free parameters L and T for fitting to numerically computed data:

$$Z(\omega) = \frac{j\omega L}{(1 + j\omega T)^{3/2}} \ . \tag{11.25}$$

The real and imaginary parts of this impedance, normalised by L/T are shown in Fig. 11.10. The corresponding wake function can be obtained explicitly by integration:

$$G(\tau) = \frac{L}{T} \begin{cases} 0 & \text{for } \tau < 0 \ , \\ \dfrac{\exp(-\tau/T)}{\sqrt{\pi \tau T}\ (1 - 2\tau/T)} & \text{for } \tau > 0 \ . \end{cases} \tag{11.26}$$

In the limit $T \to 0$ this function tends to the inductive wake $G(\tau) = L\partial\delta(\tau)/\partial\tau$. The parameter L has the meaning of the total ring inductance and can be found by summing the contributions of all elements. The inductance of each of them can be estimated numerically with the help of existing codes such as described in the Appendix.

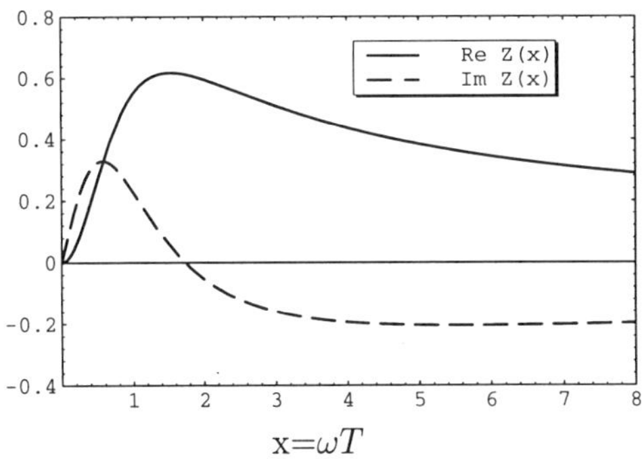

Figure 11.10: Normalised impedance of modified inductance model.

The second parameter, T, which has the meaning of the inverse of the effective cut-off frequency of the vacuum chamber, can be found from the loss factor. For a long bunch, $T \ll \sigma$, it can be written:

$$\kappa_{\parallel}(\sigma) = \frac{3LT}{8\sqrt{\pi}\sigma^3} \ . \tag{11.27}$$

One can also find explicitly the corresponding transverse impedances and wake functions for this model.

11.2 Narrow-Band Cavity Impedance

Next we describe a model for the cavity impedance in the vicinity of a resonance with frequency ω_r, based on a physical picture of the *EM* fields and currents which has been suggested recently [Heifets94]. Fig. 11.11 shows a schematic cross section of a cavity of arbitrary shape and its equivalent circuit.

The current I_c, induced by the Fourier harmonics $I_b \exp(j\omega t)$ of a beam travelling along its axis, flows through the cavity wall represented by its effective inductance L, the capacitance C of the cavity gap, and the resistance[1] R.

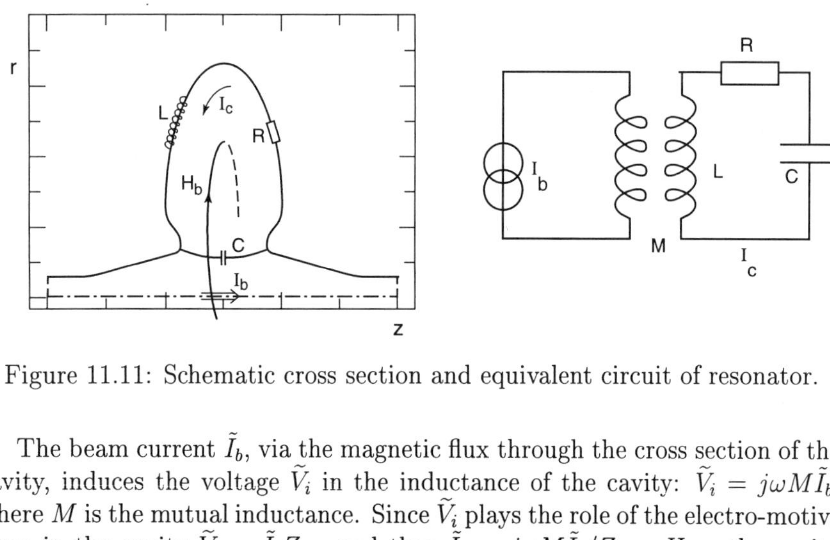

Figure 11.11: Schematic cross section and equivalent circuit of resonator.

The beam current $\tilde{I}_b$, via the magnetic flux through the cross section of the cavity, induces the voltage $\tilde{V}_i$ in the inductance of the cavity: $\tilde{V}_i = j\omega M \tilde{I}_b$, where M is the mutual inductance. Since $\tilde{V}_i$ plays the role of the electro-motive force in the cavity $\tilde{V}_i = \tilde{I}_c Z_{cav}$ and thus $\tilde{I}_c = j\omega M \tilde{I}_b / Z_{cav}$. Here the cavity impedance is:

$$Z_{cav} = R + j\omega L + \frac{1}{j\omega C} = \frac{jL}{\omega}\left[\omega^2 - \omega_r^2 - j\omega\Gamma\right] , \qquad (11.28)$$

[1]In this Section, R means the resistance in the equivalent circuit and *not* the machine radius as before.

where $\omega_r = 1/\sqrt{LC}$ and $\Gamma = R/L$. From conservation of energy one obtains $\widetilde{V}_i \tilde{I}_c = \widetilde{V}_b \tilde{I}_b$, where $\widetilde{V}_b = Z\tilde{I}_b$ is the change of beam voltage produced by the presence of the cavity. For the coupling impedance follows:

$$Z = \frac{\widetilde{V}_b}{\tilde{I}_b} = \frac{\widetilde{V}_i^2}{\tilde{I}_b^2 Z_{cav}} = -\frac{\omega^2 M^2}{Z_{cav}} = \frac{j\omega^3 M^2/L}{\omega^2 - \omega_r^2 - j\omega\Gamma} . \tag{11.29}$$

For a large *quality factor* $Q_r = \omega_r L/R = \omega_r/\Gamma \gg 1$, the last equation can be written:

$$Z = j\kappa_{\parallel} \left[\frac{1}{\omega - \omega_r - j\omega_r/2Q_r} + \frac{1}{\omega + \omega_r - j\omega_r/2Q_r} \right] , \tag{11.30}$$

where $\kappa_{\parallel} = \omega^2 M^2/2L$ can be identified with the *loss factor* $\kappa_{\parallel} = \omega_r R_s/2Q_r$, cf. Eq. (5.13), if we express the *shunt impedance* as:

$$R_s = \frac{\omega^2 M^2}{R} . \tag{11.31}$$

The last expression shows that the shunt impedance of a cavity increases quadratically with the coefficient of mutual inductance, which is proportional to the cross section of the cavity and inversely proportional to the wall resistance.

11.3 Impedance Measurements

We distinguish between *bench* and *machine measurements*, i.e. those where a device is tested in the laboratory by applying an *EM* excitation, and those where it is measured in an existing accelerator or storage ring. In addition, there exist *wake field test facilities* in a number of laboratories [Simpson86, Batchelor92], consisting of small linacs, which permit sending a short but strong bunch of charged particles through a structure, and to measure directly its influence on a weak *test* or *witness* bunch following at an adjustable distance.

Each of these methods has its advantages and its limitations, which have to be evaluated before undertaking the measurement. A general difficulty of measuring in the laboratory is the need for a proper termination of the structure under test, so it does not introduce too many effects caused by the closure itself. Beam pipes should in general be tapered gently on both ends, antennas and couplers should be small enough not to introduce new resonances, absorbing material should be installed in a manner to minimise reflections and should not depend on frequency. Furthermore, it is usually preferable to measure the difference between lines with and without a structure [Hahn78].

11.3.1 Bench Measurements

Measurements of the characteristics of a structure can be done in the frequency or time domain, either by applying a continuous rf excitation via coupling loops or antennas, or by sending a short pulse through it, usually on a centrally located wire. The first method is mainly used to measure higher-order mode resonances (*HOM*s) of cavities or cavity-like objects, such as e.g. kicker tanks, widened chambers for separators or pick-up plates etc.

Perturbation Methods

A common method to measure resonant *EM* fields - and hence impedances - in a cavity is to determine the *detuning* - i.e. the change of a resonant frequency - by introducing small metallic or dielectric objects, such as spherical beads, discs, or needles [Slater52]. They are usually pulled through the structure by a thread or wire, on or near the axis of the structure where the beam is expected to be located. The *EM* fields can be obtained from the frequency shifts with the help of form factors K_i, which are known analytically for various ellipsoidal shapes. For a bead on the axis, the only non-zero field component is E_z and can be obtained from the relation:

$$\frac{\Delta f}{f_0} = K_i \frac{\varepsilon_0 E_z^2}{4U} , \tag{11.32}$$

where U is the stored energy in the structure under test. For a spherical bead with radius a, the form factor is $K_1 = 4\pi a^3$, i.e. three times its volume. The measurement is more complicated off-axis, where other field components are present, but it can still be done, and is in particular required for measuring transverse fields.

An alternative method is to measure the phase shift at a fixed frequency, and to obtain the corresponding frequency shift from $\Delta f / f_0 = \tan(\Delta\phi)/Q_L$, where Q_L is the *loaded quality factor*, i.e. the quality factor modified by the effects of coupling devices.

At any particular resonance with frequency ω_r, shunt impedance R_s, and quality factor Q_r, the *geometric factor* R_s/Q_r is (nearly) independent of the wall resistivity. It is related to the modal loss factor[2] by $\kappa_{\parallel r} = \omega_r R_s / 2Q_r$. It can be found by integrating the electric field

$$\frac{R_s}{Q_r} = \frac{1}{2\omega_r U} \left[\left(\int dz E_z(z) \cos(k_r z) \right)^2 + \left(\int dz E_z(z) \sin(k_r z) \right)^2 \right] , \tag{11.33}$$

where $k_r = \omega_r / c$ is the wave number at the resonance.

This method has been described already in classical books on electrical measurements [Ginzton57], and nowadays is often automated such that the beads are moved through the structure by a stepping motor, and the impedances is calculated on-line by computer.

Coaxial Wire Methods

A method to estimate short-range wake fields or total loss factors of bunched beams [Sands74] is to send a short current pulse $I(t)$ along an inner conductor on the axis of the structure under test, and to compare it with the current through a reference line without such a structure. From the difference $\Delta I =$

[2]In the US literature, the *Linac Ohm* is often used as unit of the shunt impedance R_s. It equals half a conventional Ohm, and hence the factor 2 in the denominator of the expression for $\kappa_{\parallel}$ has to be replaced by 4.

$I(t) - I_{ref}(t)$ one obtains the loss factor by integration:

$$\kappa_{\parallel}(\sigma) = \frac{2Z_{ch}}{q^2} \int dt \; I(t) \Delta I(t) \; , \tag{11.34}$$

where $Z_{ch} = (Z_0/2\pi)\ln(b/a)$ is the *characteristic impedance* of a coaxial line, with radius a of the inner and b of the outer conductor, while $q = \int dt I(t)$ is the pulse charge. When the outer conductor is tapered in order to minimise reflections, also the inner one should be tapered to keep Z_{ch} constant. This method has been used widely, e.g. to measure the loss factors of components to be installed in the storage rings in $SLAC$ [Wilson91].

Similarly, one can measure the (complex) transmission coefficients of a structure, usually expressed as the element S_{21} of the scattering or *S-matrix* [Collin50], see Appendix 9B. The impedance can be obtained from:

$$Z(\omega) = 2Z_{ch} \frac{S_{21} - S_{21}^{ref}}{S_{21}^{ref}} \; , \tag{11.35}$$

where S_{21}^{ref} is the coefficient for the reference line.

Errors due to reflected power can be reduced by *gating* the signals in time. Pulses with high spectral power density can be simulated using fast Fourier transforms, which is also called *synthetic pulse technique* [Caspers85]. The introduction of a conducting wire on the structure axis constitutes a problem in itself, as it introduces additional losses and may short out electric fields just where they are most important. Therefore the wire should be chosen thin enough to cause as little perturbation as possible, but it also needs to be strong enough to withstand the tension required to avoid a large sagitta (hanging through by gravity) in structures of some length. This problem could be avoided by putting the device vertical, but often the space requirements then become prohibitive. Instead of exciting *EM* fields in a coaxial geometry, also wave guide modes can be used in order to reduce the resulting perturbation [Caspers92].

These methods are sufficiently well described in the literature [Caspers88, Karantzoulis91], and hence we shall concentrate here on measurements which can be performed if a circular accelerator or storage ring is available - and which are very valuable to optimise the performance of such a machine.

11.3.2 Machine Measurements

Once a structure is installed in an accelerator or storage ring, its effects can in principle be obtained by measuring appropriate beam parameters. The main difficulty here is the fact that there are a large number of other components, which also influence these parameters, and it is sometimes difficult to disentangle the desired effect from those influences. Furthermore, some of these measurements are at the limit of sensitivity of the beam diagnostic equipment available in a particular machine. With the installation of better instrumentation, measurements can be refined, and therefore machine parameters such as impedances and short-range wakes become known in some detail only after a few years of operation. Other interesting quantities, such as narrow-band resonances and long-range wakes, are so sensitive to particular operating conditions such as ambient temperature, air pressure, or structure alignment, that they vary continuously and only their average behaviour can be estimated.

11.3.3 Measurement of Longitudinal Impedance

A number of methods can be used to obtain the effective longitudinal impedance or the loss factor, which are both convolutions of the coupling impedance with the power density of the bunch spectrum.

Energy Loss Measurements

The real part of the longitudinal impedance can be estimated from measurements of the energy loss into the vacuum chamber. Its distribution around the machine circumference can be obtained e.g. by measuring the shift of the closed orbit in regions with non-zero dispersion as function of current. One should take the measurements with different currents in quick succession in order to minimise errors due to orbit drifts. If the orbits of individual bunches in a machine can be resolved by the pickups, it is even possible to make the measurements simultaneously, using bunches filled to different current levels. However, one must ascertain that the bunches do not influence each other by long-range wakes, e.g. by varying the distance between them. The perturbing effects of betatron oscillations induced by local energy losses in dispersive regions can be reduced by averaging the measured shifts over several betatron periods.

The loss factors can then be obtained from the expression

$$\kappa_{\|} = \frac{\mathcal{E} f_0}{e\langle D\rangle} \frac{\Delta x_{co}}{\Delta I_b} , \tag{11.36}$$

where $\mathcal{E}$ is the particle energy, $f_0 = \omega_0/2\pi$ the revolution frequency, $\langle D\rangle$ is the (horizontal) dispersion function averaged over all beam position monitors, Δx_{co} the change of the length of the closed orbit with a change of bunch current ΔI_b. Measurements on *LEP* gave good agreement with the computed loss factors of both the copper cavities and the more recently installed superconducting cavities.

An estimate of the loss factor can also be obtained by storing simultaneously two bunches with different charges in the machine, and measuring the shift of the synchrotron phase of the strong bunch relative to the weaker one. However, this measurement requires a rather high accuracy in the phase determination, and yields only the total loss factor of a machine.

Turbulent Threshold

A possibility to obtain the effective longitudinal impedance more directly is from measurements of the *turbulent threshold* for bunch lengthening, i.e. the current level at which the slope of bunch lengthening suddenly increases. In the broad-band resonator model - quite suitable for single bunch effects as discussed in Chapter 5 - the effective impedance for short bunches is given by these measurements. In practice, the accompanying increase of energy spread after the onset of *turbulence* is difficult to determine exactly.

Quadrupole Mode

All these measurements give only overlap integrals of the real part of the impedance with the bunch spectrum. The imaginary part of the (effective) impedance can be obtained from the shift of the *incoherent synchrotron frequency* with bunch current, i.e. the frequency of a single particle, usually chosen at the centre of the bunch. The incoherent frequency cannot easily be determined directly, but can be estimated from the beam current dependence of the difference of the frequencies of longitudinal quadrupole oscillations f_{s2} and twice the (coherent) synchrotron (dipole) frequency f_s. Both these frequencies are usually visible in the spectrum of a short bunch even without external excitation (which can be applied if necessary). The current dependence of the difference $\Delta f_{s2} = f_{s2} - 2f_s$ then yields:

$$\mathrm{Im}\frac{Z_{\|}}{n} = 4\pi^2 \frac{hV_{rf}\cos\phi_s}{f_0}\frac{\Delta f_{s2}}{\Delta I_b}\left(\frac{\sigma_s}{R}\right)^3 , \tag{11.37}$$

where V_{rf} is the accelerating rf voltage and h the harmonic number of its frequency. For storage rings operating above transition energy, such as all electron machine, $\cos\phi_s < 0$ and the imaginary part of the impedance is negative (capacitive) for positive shifts. This is usually the case with short electron bunches, while the shift may become negative for longer (proton) bunches, for which the effective impedance tends to become inductive.

11.3.4 Measurement of Transverse Impedance

The transverse impedance can in principle be obtained from the variation of the longitudinal impedance with transverse displacement of a bunch from the structure axis. It is possible to displace a particle beam locally with magnetic *bumps* or electrostatic separators, and thus to measure the effect of a single structure, but one should make sure that contributions from neighbouring structures remains negligible. By exciting a betatron oscillation around the whole machine, one can also measure the total transverse impedance. However, since the displacement is then not constant, one should repeat the measurement with a second excitation shifted by 90 degrees in betatron phase, and add the effects of the sinusoidal and cosinusoidal excitations quadratically. The result will still be weighted with the beta function in the plane of the excitation, but this quantity is needed anyhow in stability calculations.

Now we turn to more direct methods to measure the effective transverse impedance of the total machine, both in the horizontal and the vertical directions. From these data it is possible to obtain information on the relative contributions of structures with rotational symmetry - e.g. rf cavities with circular beam holes - and of structures in which the vertical impedance is dominant, such as cross-section variations in flat vacuum chambers.

Detuning with Beam Current

One of the simplest methods to estimate the effective transverse impedance of the total machine is to measure the detuning of the (coherent) betatron oscillation frequencies with current. For a Gaussian bunch of rms length σ

in (transverse) resonator impedances characterised by ω_r and $R_\perp/Q_r$ the tune slope with current, in either the horizontal of the vertical plane, is given by [Brandt92]:

$$\frac{\Delta Q_\beta}{\Delta I_b} = -\frac{e}{2\pi\omega_0\sigma\mathcal{E}}\sum_{i=1}^{N}\langle\beta\rangle_i\left(\frac{R_\perp}{Q_r}\right)_i F(\omega_{ri}\sigma)\ . \tag{11.38}$$

The summation is over all components with non-negligible transverse impedances, where $\langle\beta\rangle_i$ is the beta function averaged over their locations. When quality factors are not too small, $Q_r > 1/2$, the form factor F can be expressed by the *complex error function* [Abramowitz65] $\mathrm{w}(z) = \exp(-z^2)\mathrm{erfc}(-jz)$:

$$F(z) = \frac{\sqrt{\pi}}{Q'_r}\mathrm{Im}\left[z\ \mathrm{w}\left(\frac{jz}{2} + Q'_r z\right)\right]\ , \tag{11.39}$$

where $Q'_r = \sqrt{Q^2 - 1/4}$.

For short bunches, $|z| = \omega_r\sigma \ll 1$, and one needs the first terms of the power series $\mathrm{w}(z) \approx 1 + 2jz/\sqrt{\pi} + ...$ to get $F(z) \approx 2z^2$, while for long bunches it approaches unity. Hence for short bunches in a single machine component the transverse impedance is approximately given by:

$$Z_\perp = \frac{\pi\mathcal{E}}{e\langle\beta\rangle}\frac{\omega_0\sigma}{(\omega_r\sigma)^2}\frac{\Delta Q_\beta}{\Delta I_b} \tag{11.40}$$

At low currents, the bunch length is usually approximately constant, and hence the impedance is directly proportional to the tune slope. However, at higher currents the variation of bunch length with current has to be taken into account.

Such measurements have been performed in *LEP* [Brandt89], and good agreement has been found between computed and measured impedances in the vertical plane. However, the tune shift in the horizontal plane was considerably smaller than expected from a simple broad-band resonator model with different shunt impedances, but equal resonant frequencies in both planes. The influence of the large number of flat *shielded bellows* was found to be quite strong, in particular for short bunches, leading to a lower resonant frequency in the horizontal plane, and hence to a larger ratio of tune shifts in the two planes.

Due to the lack of rotational symmetry of a flat vacuum chamber, a current dependent cross term arises which couples the motion in the horizontal and vertical directions [Wagner96].

11.3.5 Variation of Impedance with Displacement

Another method to obtain the transverse impedance is to measure the variation of the longitudinal impedance with transverse displacement [Nassibian79]. One defines a *generalised impedance* $Z_1(x, x_0)$ as the ratio of the *EM* force acting on a test charge at transverse position x and a beam current at $x_0 = x + \Delta$. Then the longitudinal impedance is simply $Z_{\|} = \lim_{x \to x_0} Z_1(x, x_0)$, while the transverse one is given by:

$$Z_{\perp}(\omega) = \frac{c}{\omega} \frac{\partial^2 Z_1(x, x_0)}{\partial x \, \partial x_0}\bigg|_{x=x_0} . \qquad (11.41)$$

Usually the generalised impedance has a second order variation around the beam axis $Z_1 = Z_{10}[1 + x x_0/a^2]$, and hence

$$Z_{\perp}(\omega) = \frac{c}{\omega} \frac{Z_{10}}{a^2} . \qquad (11.42)$$

For a circular vacuum chamber with radius b, length L, and wall resistivity ρ_c, these quantities can be calculated analytically. For a skin depth δ_s small compared to the wall thickness, and with the complex surface impedance $\zeta_s = (1 + j)\rho_c/\delta_s$, one finds for a small displacement from the axis $Z_1(x, x_0) = \zeta_s L[1 + 2x x_0/b^2]/(2\pi b)$ and thus:

$$Z_{\|} = \frac{\zeta_s L}{2\pi b}\left[1 + 2\frac{x_0^2}{b^2}\right] , \qquad Z_{\perp} = \frac{c}{\omega} \frac{\zeta L}{\pi b^3} , \qquad (11.43)$$

for the resistive components of the impedances, while the reactive ones are obtained from the direct space charge field. The difference of the impedances ΔZ, measured in free space and inside the chamber, with a loop of length L and width w yields:

$$Z_{\perp} = \frac{c}{\omega} \frac{\Delta Z L}{w^2} . \qquad (11.44)$$

Such measurements were performed on several types of vacuum chambers and kicker tanks [Nassibian79], showed good agreement with computations, and permit estimates of the accuracy of the results.

11.3.6 Transfer Function

Some of the best results for determining impedances in existing machines were obtained by measuring the *transfer functions* either in the longitudinal or transverse directions [Moehl71]. This is most easily done with continuous beams, available in proton storage rings, and is more complicated and less accurate with (short) bunches used in electron machines. Basically the method relies on exciting particle beam oscillations either by a continuous harmonic source or - alternatively - by a short pulse, and measuring the phase and amplitude (or real and imaginary parts) of the induced beam oscillation. The measurement is simpler in the transverse direction, which we will discuss here.

In the absence of an impedance, the transverse displacement $y(t)$ of a single particle subjected to continuous excitation is that of a harmonic oscillation with natural *betatron frequency* ω_β due to external focusing, driven by an acceleration $G = \hat{G}\exp(j\omega t)$, with frequency ω and strength $\hat{G}$. In complex notation this can be written [Hofmann77, Borer79]:

$$y(\omega) = \hat{y}\ \mathrm{e}^{j\omega t} = \frac{\hat{G}\mathrm{e}^{j\omega t}}{\omega_\beta^2 - \omega^2}\,. \tag{11.45}$$

Integrating over all particles with different betatron frequencies, described by the normalised distribution function $f(\omega_\beta)$, this can be expressed as:

$$\Big\langle y(\omega) \Big\rangle = \frac{G}{2S} \int_{-\infty}^{\infty} d\omega_\beta \left[\frac{f(\omega_\beta)}{\omega_\beta - \omega} - \frac{f(\omega_\beta)}{\omega_\beta + \omega} \right]\,, \tag{11.46}$$

where $S = \Delta\omega_\beta$ is the spread in betatron frequencies measured at half height. The integrand has two poles, one at $\omega_\beta = \omega$ for the *slow wave*, and the other one at $\omega_\beta = -\omega$ for the *fast wave*. Using the *residue theorem*, the first integral can be written:

$$\Big\langle y(\omega) \Big\rangle = \frac{G}{2S} \left[\pm j\pi f(\omega) + PV \int d\omega_\beta\, \frac{f(\omega_\beta)}{\omega_\beta - \omega} \right]\,, \tag{11.47}$$

where PV stands for the *principal value* of the integral which extends over the *betatron frequency spread* S around the pole. A similar contribution comes from the other pole at $-\omega$. A transverse force causes a change of (transverse) velocity rather than of displacement.

Hence the transfer or *response function* is usually defined as the (complex) ratio of the velocity $\dot{y} = j\omega y$ to the driving acceleration G. It is normalised by multiplication with the frequency spread S to get the *normalised response function*:

$$\mathcal{R}(\omega) = \frac{S\dot{y}}{G} = S \left[\mp\pi f(\omega_\beta) + jPV \int d\omega_\beta\, \frac{f(\omega_\beta)}{\omega_\beta - \omega} \right]\,. \tag{11.48}$$

The amplitude and phase of the *transfer function* can be plotted against frequency, or as a *polar plot* of amplitude against phase, with frequency as running parameter, as illustrated for a Gaussian distribution in Figs. 11.12.

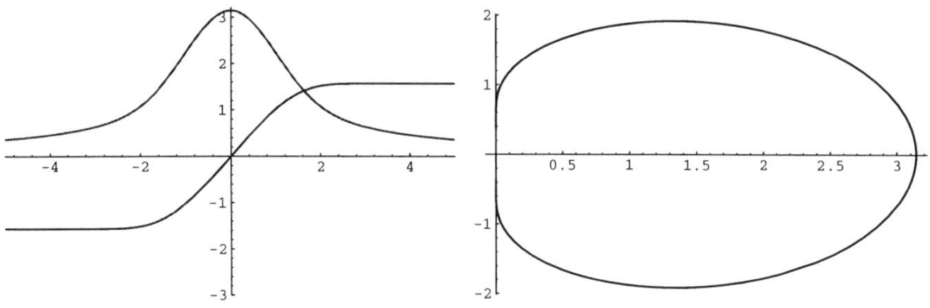

Figure 11.12: Transfer function for a Gaussian distribution: a) amplitude and phase versus normalised frequency and b) polar plot.

The inverse of this response function is called *stability diagram*, in particular when its imaginary part is plotted against its real part. In the presence of a wall impedance, the induced fields create additional forces whose acceleration $G_w = \langle F_\perp \rangle / m_0 \gamma$ has to be added to G.

The average force can be expressed by the impedance $\langle \widetilde{F}_\perp \rangle = jyZ_\perp q^2/2\pi R$, and hence one finds for the (normalised) response $\mathcal{R}_w$ with a wall impedance:

$$\frac{1}{\mathcal{R}_w(\omega)} = \frac{1}{\mathcal{R}(\omega)} + \frac{qI_b Z_\perp}{S\omega m_0 \gamma c} . \tag{11.49}$$

From the complex shift of the inverse response function with beam current one can determine the transverse impedance or the effect of a transverse feedback system at a particular frequency. By *sweeping* the excitation frequency, or exciting the beam with *white noise* and using a *fast Fourier transform* (FFT) analyser, one can determine the impedance over a given frequency range.

The response function can also be measured in the longitudinal direction [Faltens71]. Then one has to replace the distribution function over betatron frequencies by the derivative of the distribution function over synchrotron frequencies, and take the *synchrotron frequency spread* for normalisation.

Appendix 11.A Numerical Methods and Codes

All the analytical and semi-analytical methods, which we have dealt with so far, are important for understanding the nature and the properties of wake fields and coupling impedances. Notwithstanding these advantages, their practical applications are limited to (very) simplified structures. These structures are often quite far from the real ones used in accelerator. Construction details, such as rf windows, connections to wave guides, couplers and plungers, rounded corners, etc. are usually neglected. An additional complication arises from the fact that materials used in real structures are different from the idealised ones, such as perfect conductors or lossless dielectrics, and are often not even uniform over the whole structure.

The only practical way of calculating and studying the *EM* fields in such structures is the application of purely numerical methods. There are dozens of sophisticated and highly developed computer codes based on several computational techniques. We briefly describe a few of them in this Appendix. Our intention is not to provide the reader with a complete guide into this large area of research and applications - a separate book could be written on this subject alone. We will only give a short introduction to the most important computational methods, and an overview of the more frequently used computer codes, without which the subject of our book would be not complete.

The choice of an appropriate code is based on the consideration of several factors. First of all, one should specify the type of problem to be solved, whether it is in frequency domain, such as the calculation of resonant eigenmodes of a cavity, or a problem in time domain, such as the calculation of wake fields excited by a bunch in a cavity. Next, one has to consider the geometry of the problem, whether it can be simplified to a two-dimensional (2-D) structure, usually rotationally symmetric, or whether one needs to include features which make a three-dimensional (3-D) description necessary. The calculational effort, in particular computer time and memory size, depends critically on that choice - 3-D codes need a considerably bigger effort, even if some symmetries can be retained, and thus often limit the number of cases which will be computed.

A further choice is the size of mesh cells needed to obtain a satisfactory result - this often depends on the bunch length, and/or on the importance of details of the surrounding structure. Also one has to specify over what length the wake fields are needed - the computation of long-range wakes is much more cumbersome and needs more accuracy - e.g. smaller mesh sizes - in order to produce correct results, which increases the computer effort even more. Again the desire for more accuracy has to be paid by a vastly larger computer effort. If the properties of the material used for the structure have to be taken into account, this will further complicate the computation. However, in many cases it is sufficient to replace the highly conducting metal walls of most accelerator structures by perfect conductors.

There are two principal methods for solving the Maxwell equations numerically: on one hand the *Finite Difference (FD) Method* or its equivalent, called *Finite Integral Technique (FIT)*, and on the other hand the *Finite Element (FE) Method.* Both methods, in one way or another, reduce the problem of finding the *EM* fields to solving an approximate set of the linear equations of finite size. The actual numerical algorithm to solve such an algebraic system depends on the properties of the matrix defining the normal modes of the problem. In most cases a sparse matrix can be obtained which permits economic inversion.

Below we briefly describe these methods, and a number of programs based on them. A concise compendium of the existing computer programs for accelerator physics is produced by the Los Alamos Accelerator Code Group [Group90].

The capability of solving 2-D or 3-D problems in time domain, i.e. to compute wake potentials, has been demonstrated so far only with the *FD* method, and is called *Finite Difference Time Domain* (*FDTD*) technique. The same method can also be used for obtaining solutions in the frequency domain. Codes based on the *FE* method are exclusively used to find eigenmodes in the frequency domain. However, they permit a much smoother representation of the boundaries, whose *stepped* or *staircase* approximation is a major source of numerical inaccuracies in most *FD* codes.

A large effort has been invested into achieving solutions of high accuracy, and the more popular codes have evolved over a number of years - nevertheless, their utilisation remains difficult and a novice without expert guidance will sometimes obtain wrong results. Further improvements which simplify the usage and improve the attainable accuracy in a given time are desirable. The documentation is often insufficient or too abstract. For 2-D codes in the frequency domains, the precision which can be reached at the present time is sufficient for most practical purposes [Weiland96]. For 3-D codes, however, the computation of wakes of short bunches in long structures remains quite limited even with the fastest computers available.

11.A.1 Finite Difference Methods

The first numerical solutions of the Maxwell equations in time domain were published in Russia [Balakin78, Balakin83]. The program was claimed to use a finite difference method to compute wake potentials of finite bunches in axially symmetric structures. Unfortunately, it did not become available in the West until very recently, and still does not have a user-friendly input structure. Nevertheless, a comparison [Meincke97] with similar codes showed that - at least for a particular structure - it has advantages in converging to the correct solution with a rougher mesh, and hence gets results much faster. The most widely used codes for computing wake potentials in the time domain in 2-D (axially symmetric) structures, such as *TBCI* [Weiland83c, Weiland83d], now incorporated in *MAFIA* [Klatt86a], and *ABCI* [Chin88b], are based on the *Finite Integral Technique (FIT)* [Yee66]. In the frequency domain, *URMEL* [Weiland82, Weiland83a], now also part of *MAFIA*, is based on the *FIT* method and has displaced older codes like *SUPERFISH* [Halbach76]. Recently, the *FDTD* (*Finite Difference Time Domain*) technique, mainly used for antenna design [Taflove75, Umashankar87] has been adapted to accelerators [Juergens92] and led to the development of the code *XWAKE* [Juergens94]. It permits a better description of curved or tapered boundaries, as well as objects with dimensions smaller than a mesh cell.

Mesh and Grid Methods

Consider the homogeneous Maxwell equations for a harmonic *EM* fields:

$$\begin{aligned} \nabla \times \widetilde{\mathbf{H}} &= j\omega\varepsilon_c \widetilde{\mathbf{E}} \,, \\ \nabla \times \widetilde{\mathbf{E}} &= -j\omega\mu_c \widetilde{\mathbf{H}} \,, \\ \nabla \cdot \varepsilon_c \widetilde{\mathbf{E}} &= 0 \,, \\ \nabla \cdot \mu_c \widetilde{\mathbf{H}} &= 0 \,, \end{aligned} \tag{11.50}$$

where ε_c and μ_c are the complex permittivity and permeability as defined in Chapter 1. The problem is to find eigenmodes (i.e. eigenfrequencies and eigenvectors) of this system of equations, with boundary conditions for a given geometry and material distribution.

Several numerical methods have been proposed for solving this system of equations [Albani74, Wilhelm82, Hara82, Tran84]. However, none of these algorithms is free of numerical difficulties and uncertainties. The origin of this difficulties can be traced to the following problem: When the first two Eqs. (11.50) are combined into a single one:

$$\nabla \times \frac{1}{\mu_c} \nabla \times \widetilde{\mathbf{E}} = \omega^2 \varepsilon_c \widetilde{\mathbf{E}} \ , \tag{11.51}$$

this equation has *trivial, static (irrotational) solutions* $\nabla \times \widetilde{\mathbf{E}} = 0$ for $\omega = 0$. This implies that the matrix of the set of approximate, discretised Eqs. (11.50) has vanishing eigenvalues. Hence the solutions for $\omega \neq 0$ are in general not unique, as they may have an undetermined admixture of static fields. For problems which can be solved analytically, the trivial irrotational solutions are excluded by the third Eq. (11.50). However, not every discretisation method is free from "unphysical" solutions even when this is taken into account.

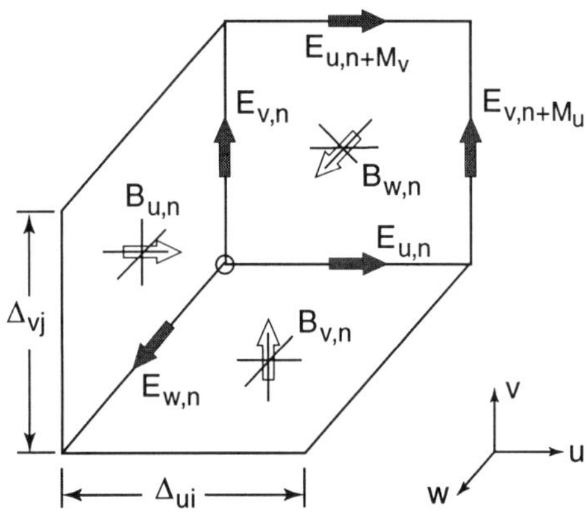

Figure 11.13: Assignment of field components in the *FIT* method

Let us turn now to the *FIT* method [Yee66]. The structure under consideration is approximated by a main 3-D grid G consisting of rectangular blocks, and a dual grid of blocks shifted by half a cell length. In an orthonormal coordinate system $\mathbf{e}_u, \mathbf{e}_v, \mathbf{e}_w$ the main grid nodes are numbered in the following way:

$$n = 1 + (i-1)M_u + (m-1)M_v + (k-1)M_w \ , \tag{11.52}$$

where M_u, M_v, M_w are step increments along the axes $\mathbf{e}_u, \mathbf{e}_v, \mathbf{e}_w$, and $i = 1, \dots \hat{i}, m = 1 \dots \hat{m}, k = 1 \dots \hat{k}$, and hence $n = 1, 2 \dots \hat{i}\hat{m}\hat{k}$. The corresponding coordinates of the nodes are thus limited by the following inequalities:

$$u_1 \leq u_i \leq u_{\hat{i}} \ , v_1 \leq v_m \leq v_{\hat{m}} \ , w_1 \leq w_k \leq w_{\hat{k}} \ . \tag{11.53}$$

The components of the electrical field are calculated at each node. Their assignment is illustrated in Fig. 11.13, which is adapted from ref. [Weiland84]. In a similar way, the components of the magnetic field are assigned to the dual grid $\widetilde{G}$, the nodes of which are shifted by the half a cell length in each coordinate direction.

In first approximation, the fields are evaluated from the integrals along the circumference of each cell surface. For example, from Eq. (11.50) one gets:

$$\oint \widetilde{\mathbf{E}} \cdot d\ell = \frac{\omega}{c} \int \mathbf{B} d\sigma \ , \tag{11.54}$$

where ℓ and σ are the circumference and the area of one of a cell surfaces. For the surface perpendicular to the w axis, this yields:

$$\begin{aligned}(u_{i+1} \ &- \ u_i)(E_{u,n} - E_{u,n+M_v}) + (v_{m+1} - v_m)(E_{v,n+M_u} - E_{v,n}) \\ &= \ \frac{\omega}{c}(u_{i+1} - u_i)(v_{m+1} - v_m)B_{w,n} \ . \end{aligned} \tag{11.55}$$

The key point of the method is that all four Maxwell equations are discretised in a consistent way, i.e. the time harmonics of the field which are source-free in configuration space are also source-free in the space of the grid. A matrix equation is derived for each one of the four equations. Then the matrix equation is derived by combining these equations:

$$\mathcal{B}\widetilde{\mathbf{E}} = (\frac{\omega}{c})^2 \widetilde{\mathbf{E}} \ , \tag{11.56}$$

where the real and symmetric matrix $\mathcal{B}$ is semi-definite (i.e. all its eigenfrequencies are real $\omega^2 \geq 0$), and $\widetilde{\mathbf{E}}$ is the column vector of the electric field normalised by the local density of the stored energy (index T stands for *transpose*):

$$\widetilde{\mathbf{E}}'^T = (E'_{u_1}, E'_{u_2}, \ldots, E'_{u_N}, E'_{v_1}, \ldots, E'_{w_N}) \ , \tag{11.57}$$

The normalised magnetic field is described by the similar column vector $\widetilde{\mathbf{H}}$:

$$\widetilde{\mathbf{H}}'^T = (H'_{u_1}, H'_{u_2}, \ldots, H'_{u_N}, H'_{v_1}, \ldots, H'_{w_N}) \ . \tag{11.58}$$

The prime here denotes the normalised field components.

In the simple case of homogeneous media, the matrix $\mathcal{B}$ reduces to three independent matrices which represent the discrete Laplace operator Δ. The field components then satisfy three independent wave equations, and are coupled only through the boundary conditions. An explicit expression for the system matrix, as well as its detailed derivation, can be found in ref. [Weiland85].

Computer Code TBCI

The program *TBCI* [Weiland83c], [Weiland83d] – **T**ransverse **B**eam **C**avity **I**nteraction - employs the *FIT* method with discretisation of time to calculate wake fields for axially symmetric cavities of an arbitrary shape. It is a development of the original code *BCI* [Weiland80a], which was limited to axially symmetric (m=0) modes. It is now incorporated as a module in the general *MAFIA* [Klatt86a] package, where it is called *T*2, i.e. the 2-D time domain solver.

The basic equations for the field vectors $\tilde{\mathbf{E}}^n$ and $\widetilde{\mathbf{H}}^n$, where the subscript indicates the number of a discrete time steps, can be written:

$$\widetilde{\mathbf{H}}_{n+1} = \widetilde{\mathbf{H}}_n - \frac{1}{Z_0 \Lambda \beta} \mathcal{A}_h \tilde{\mathbf{E}}_{n+1/2} \; ,$$

$$\tilde{\mathbf{E}}_{n+3/2} = \tilde{\mathbf{E}}_{n+1/2} + \frac{Z_0}{\Lambda \beta} \mathcal{A}_e \widetilde{\mathbf{H}}_{n+1} + \tilde{\mathbf{J}}_{n+1} \; , \qquad (11.59)$$

where $Z_0 = \sqrt{\mu_0/\varepsilon_0}$ is the impedance of free space, Λ is the ratio of time to longitudinal mesh steps, and the matrices $\mathcal{A}_e$ and $\mathcal{A}_h$ are the system matrices similar to the matrix $\mathcal{B}$ mentioned above. Finally, $\tilde{\mathbf{J}}$ is the current density associated with a particular azimuthal mode of the beam.

The output of the program contains the fields, the stored energy as function of time and the longitudinal and transverse wake potentials.

Computer Code MAFIA

The acronym *MAFIA* stands for the set of codes designed for solving the **MA**xwell equations by the **F**inite **I**ntegration **A**lgorithm [Klatt86a].

The set consists of:

- a 3-D mesh generator *M3*. It can be used either interactively or in the batch mode. 2-D and 3-D (colour) plots of the input structure, the mesh, and the material distribution can be obtained. A number of different, precalculated shapes are available to facilitate the input and can be repeated any number of times. In addition to the user specified mesh planes and their spacing, additional planes may be inserted to improve the fit to a given structure. Up to 9 different materials may be specified for each cell or half-cell separated by a diagonal plane joining two opposite sides. An *access file* is generated for use by the matrix generator.

- the generator *R3* constructs the 3-D system matrix, whose eigenvalues and eigenvectors yield the frequencies and field components of the normal modes of the *EM* oscillations in the structure. The properties of the materials (conductivity, complex tensor permittivity or permeability) need to be specified. *Material numbers* 0, 1 and 2 are assigned to vacuum, perfectly conducting metal and perfectly magnetic medium, respectively. The finite, but high conductivity of metals is used only in perturbative calculations by the post-processor.

- the eigenvalue solver *E3* (consists actually of two solvers of which *E31* is a modification of the standard routine *EISPACK* [Smith76], while the second one *E32* is still in the process of development).

- the module *T3* [Weiland86] calculates 3-D wake fields in the time domain. This code is an extension of the 2-D code [Weiland83c] to three dimensions.

- the post-processor *P3* for extracting quantities of physical interest, such as shunt impedance, quality factor, loss and kick factors, etc. As an option the post-processor creates (arrow) plots of field vectors in any plane, calculates the line integral of the Lorentz force, etc.

Computer Code URMEL

The computer code *URMEL* [Weiland83b] is designed to calculate the eigenvalues and eigenvectors for the *EM* field oscillations in 2D structures - axially symmetric cavities or waveguides with constant cross section. It is also based on the *FIT* discretisation algorithm. It supersedes the code *SUPERFISH* [Halbach76], which was limited to axially symmetric (m=0) modes, but could handle only closed cavities. Its extension to higher azimuthal modes, called *ULTRAFISH*, had numerical difficulties which are avoided by *URMEL* [Weiland82].

A modification of the code *URMEL-T* [van Rienen86] employs a triangular mesh to improve the approximation of the boundaries. The dual grid is composed of hexagons. Furthermore, this code allows an arbitrary distribution of the permittivity and permeability of materials inside the structure, permitting thus the study of dielectric or magnetic insertions.

The output contains the eigenfrequencies for cavities, or the propagation constants for waveguides, quality factors, shunt impedances, voltages, etc., and a graphical representation of the field distribution for each particular mode.

Computer Code ABCI

The code *ABCI* [Chin88a] is similar to *TBCI*, it is also a 2-D time domain program for computing wake potentials of finite bunches in axially symmetric structures. It was primarily developed to permit unequal mesh sizes in the longitudinal and radial directions - which was originally not possible in *TBCI*, and of which no source code was available for modifications.

In the later versions of *ABCI* many improvements have been added - e.g. the *differential input* which is very useful to describe repetitive structures like disk-loaded wave guides or bellows. More basic improvements are the possibility of describing a structure with *variable radial step sizes* - which permits using a finer mesh near the most important parts of the structure, and a coarse one elsewhere in order to reduce the total number of mesh points, and hence save considerably on computer memory and CPU time.

An even larger saving has been achieved by the *movable mesh* - not to be confused with the windowing option which suppresses only the fields in front and behind the bunch proper. By suppressing also the mesh outside this region, much larger structures or much finer mesh sizes could be tackled without exceeding the limited memory allocations of the computers available.

Finally, the addition of *contour integration* (cf. Section 3.6) allows the computation of wake potentials in structures which protrude into the beam pipes, e.g. collimators, without the necessity of very long entrance or exit tubes which are required when the fields are integrated on the beam axis. With this option it is also possible to find the longitudinal wake function for structures with unequal beam pipe diameters

11.A.2 Finite Element Method

The *Finite Element (FE) Method* was originally developed to analyse the mechanical stability of complicated structures by separating them into simple parts (called finite elements) with known simple behaviour. Later it was realised that the method is a general technique for the numerical solution of partial differential equations, which is based on variational methods.

The discretisation process in the *FE* method begins with reformulation of a differential equation as an equivalent *variational* problem. For example, for finding the equilibrium shape $u(\mathbf{r})$ of an elastic body, its total energy $F(u)$

should be a minimum of the functional $F(v)$ where v is an *admissible function* for the problem. If V_∞ is an infinitely dimensional space of admissible functions v, then the variational problem can be formulated in the following way [Johnson87]:

$$\text{find} \quad u \in V_\infty \quad \text{such that} \quad F(u) \leq F(v) \quad \text{for all} \quad v \in V_\infty \, . \tag{11.60}$$

In general this minimisation procedure can not be solved, but has to be reduced to a finite dimensional problem. The space V_∞ is replaced by a set of subspaces (elements) V_M^i of finite dimensional basis functions. The basis functions are usually chosen to be simple polynomials the (*Galerkin formulation*). Then this problem can be solved, at least numerically, and the solution will be a good approximation to that of the exact problem.

For complicated geometries, general boundary conditions, and variable or nonlinear material properties, the *FE* method has several advantages over the finite difference method. Furthermore, the more solid theoretical foundation of this method gives it added reliability, and in many cases makes possible to estimate errors of the approximate solution.

However, the *FE* method was originally developed only for scalar equations (e.g. Poisson equation, wave equation, diffusion equation, etc.). Its application to the vector equations of electrodynamics are not so well founded. The extension of the method into this area encounters certain difficulties - in particular the appearance of non-physical *spurious modes* [Lynch91]. This may cause bad convergence to the correct solution. At present, the method is undergoing rapid development, and one can expect refinements which will be free from the present limitations [Paulsen91].

The FE Method for the Maxwell Equations

The Helmholtz equation for the electric field in the form

$$\nabla \times \frac{1}{j\omega\mu_c} \nabla \times \widetilde{\mathbf{E}} - \nabla \frac{1}{j\omega\mu_c\varepsilon_c} \nabla \varepsilon_c \widetilde{\mathbf{E}} + j\omega\varepsilon_c \widetilde{\mathbf{E}} = 0 \tag{11.61}$$

is the starting point for the application of the *FE* method to the Maxwell equations [Paulsen91]. Here $\varepsilon_c = \varepsilon - j\sigma/\omega$ is the complex permittivity.

Multiplying this equation by a scalar basis function v_i, integrating over the domain Ω, and using integration by parts produces the *weak formulation* for the *EM* fields:

$$\int_\Omega \frac{d\mathbf{r}}{j\omega\mu_c}(\nabla \times \widetilde{\mathbf{E}}) \times \nabla v_i + \int_\Omega \frac{d\mathbf{r}}{j\omega\mu_c\varepsilon_c}(\nabla\varepsilon_c\widetilde{\mathbf{E}})\nabla v_i + \int_\Omega d\mathbf{r} j\omega\varepsilon_c\widetilde{\mathbf{E}} v_i =$$
$$- \int_\sigma \frac{ds\ v_i\ \mathbf{n}}{j\omega\mu_c} \times (\nabla \times \widetilde{\mathbf{E}}) + \int_\sigma \frac{ds\ v_i\ \mathbf{n}}{j\omega\mu_c\varepsilon_c}(\nabla\varepsilon_c\widetilde{\mathbf{E}}) \ . \qquad (11.62)$$

Here $\mathbf{n}$ is the normal unit vector directed *outward* from the surface σ. This equation takes into account the boundary conditions for both $\nabla \times \widetilde{\mathbf{E}}$ and $\nabla\varepsilon_c\widetilde{\mathbf{E}}$.

An expansion of $\widetilde{\mathbf{E}}$ into *Galerkin basis* functions yields an algebraic set of equations:

$$\mathcal{A}_{ik}\widetilde{\mathbf{E}}_k = \mathcal{R}_{ik}\widetilde{\mathbf{E}}_k \ . \qquad (11.63)$$

The matrices $\mathcal{A}_{ik}$ and $\mathcal{R}_{ik}$ are called *stiffness* and *mass* matrices. The solutions give the eigenmodes of the electric field $\widetilde{\mathbf{E}}_k$, and those of the magnetic field $\widetilde{\mathbf{H}}_k$ can be calculated in a similar way.

Several finite element computer codes for particle accelerator structures exist for 2-dimensional problems, e.g. *YAP* [Nelson93, Zhan96], and *PRIAM/ ANTIGONE*, for which also a 3-D version exists [Meur93].

Computer Code YAP

The code *YAP* [Nelson93] solves the wave equation for axially symmetric structures. All modes $m = 0$ and $m \neq 0$ can be found, but only the *eigenfrequencies* are calculated accurately. The solver for the eigenvectors is yet not refined to produce good results. Both linear, and curved triangular, mesh can be generated. In the latter case, the triangular cells have sides which are described by nonlinear (in most cases, quadratic) polynomials. Both numerical and graphic output are available. However, at present the code is not user friendly and additional work would be required for its general use.

Computer Codes for Parallel *FE* Computation

Another example of the application of the *FE* method is a computer code developed for the calculations of the *EM* fields in *detuned structures* for linear colliders. Such a structure has a large number of cells with small variations of the geometrical dimensions. To get sufficient accuracy, one needs to take into account the rounding of the iris edges. All this prohibits the use of *finite difference* schemes with their structured grids.

On the other hand, the *FE* method allows *unstructured grids* to fit small differences in geometry by the possibility to refine the mesh locally, and, moreover to use quadratic element boundaries. But the memory size and CPU time required for this calculation become quickly prohibitive. These limitation are overcome by way of parallel computation [Zhan96].

The geometry of such a structure is amenable to parallel computation because the *EM* fields in neighbouring cells couple only through small iris openings. This reduces the number of elements with common boundaries between cells, while the vast majority of the stiffness matrix elements can be calculated for each cell independently from all the others.

The accuracy of the solution improves as the mesh is refined. A straightforward subdivision of each element of the whole mesh leads to excessive computer requirements. However, a more efficient method is to do this only in places which are critical for the accuracy of the solution. An *adaptive refinement scheme* can be based e.g. on the local gradient of the stored field energy. The mesh refinement stops when a specified accuracy is reached.

Chapter 12

Accelerator Structures

There are a large number of devices and non-uniformities in the vacuum chamber of an accelerator which may cause a beam to generate *EM* fields when passing through, and which in turn can perturb its own motion, increase its emittance or even lead to instabilities. In previous Chapters we described methods developed and used for the calculation of these fields, and impedances and wakes obtained from them. Here we summarise and complement results for the most important accelerators components, such as *radio frequency* (rf) and *accidental cavities*, *collimators* or *scrapers*, *tapers*, *bellows*, *kickers* or *separators*, *beam position monitors.*

In designing accelerators and storage rings, it has become a reasonable practice to estimate the total *impedance budget* of all the various elements which will be installed. During the construction period, various persons or committees are then charged with the supervision of equipment installed. They are sometimes called quite justly *impedance police* as they have to enforce the adherence to the decisions, or oversee possible adjustments in case of changing requirements.

12.1 Uniform Beam Pipes

The vacuum chamber of an accelerator usually consists, over most of its length, of a smooth, metallic beam pipe of circular or oval cross section. The metals used for the walls are mostly very good electric conductors, such as aluminium or copper, but sometimes steel is preferred for strength reasons, e.g. to make a synchrotron radiation shield (or *liner*) in cryogenically cooled chambers. Since conductivity of steel is smaller then that of copper, copper cladding is often proposed to reduce Ohmic losses and the resistive wall impedance. In some fast-cycling machines, the vacuum chamber is made of insulating, ceramic material to minimise eddy currents. However, in order to avoid static charges accumulating on such walls, a thin conducting layer or at least conducting strips are often applied to its inside.

Here we summarise the expressions needed for calculating the impedances of such chambers, some of which have already been given in Chapter 6.

12.1.1 Space-charge Impedance

The reduced *longitudinal space-charge impedance* of a beam in a perfectly conducting chamber is usually written:

$$\frac{Z_{sc}}{n} = -j\frac{Z_0}{2\beta\gamma^2}\, g_f \ , \tag{12.1}$$

where g_f is called the *g-factor*, which depends on the beam and chamber geometry. For a circular cylindrical beam, with radius a in a concentric chamber of radius b, it is usually given as $g_f = 1 + 2\ln(b/a)$. The constant 1 in this expression should actually be replaced by 1/2, since the average of the electric field over the beam cross section should be used in the impedance calculation rather than its value on the axis. However, such a refinement does not change the result as the logarithmic term usually dominates. Since Z_{sc}/n is frequency independent, the impedance Z_{sc} itself is proportional to frequency $\omega = n\omega_0$. It is thus a *negative inductance*, and *not a capacitance* as is sometimes claimed, since a capacitive impedance is proportional to $1/\omega$.

For the same geometry, the *transverse space-charge impedance* is:

$$Z_\perp{}^{sc} = -j\frac{RZ_0}{\beta^2\gamma^2}\left[\frac{1}{a^2} - \frac{1}{b^2}\right] , \tag{12.2}$$

where R is the circumference of the ring

For other cross sections, it can be evaluated in terms of the electric and magnetic *Laslett coefficients* [Laslett63] $\varepsilon_{1,2}$ and $\xi_{1,2}$:

$$Z_\perp{}^{sc} = j\frac{2RZ_0}{h^2}\begin{cases} \dfrac{1}{\beta^2}(\varepsilon_1 - \xi_1) + (\varepsilon_2 - \xi_2)\dfrac{h^2}{g^2} & \text{for penetrating fields ,} \\ \dfrac{1}{\beta^2\gamma^2}(\varepsilon_1 - \xi_1) & \text{for nonpenetrating fields .} \end{cases} \tag{12.3}$$

Here h is the chamber half-height and g the magnet half-gap. The first expression holds for tunes close to an integer, where the lowest betatron frequency $(n-\nu)\omega_0$ is so small that its skin depth is larger than the wall thickness. The Laslett coefficients have been evaluated for a number of geometries, with expressions in closed form for elliptic cross sections or parallel plates, and as infinite series for rectangular chambers [Laslett63, Zotter72a].

12.1.2 Resistive Wall Impedance

The impedance of a wall with a finite conductivity σ_c can be expressed in terms of the *skin depth* $\delta_s = \sqrt{2/\mu\omega\sigma_c}$, which decreases as the square-root of frequency, or in terms of the dimensionless *surface impedance* $\zeta = \mu_r\omega\delta_s/2c$ which increases as $\sqrt{\omega}$. For a wall thickness large compared to the skin depth, one finds:

$$\frac{Z_{rw}}{n} = Z_0\frac{\beta\mu_r\delta_s}{2b}\,, \qquad Z_\perp{}^{rw} = (1+j)\mu_r R Z_0\frac{\delta_s}{b^3}\,. \tag{12.4}$$

For ultra-relativistic beams the velocity factor $\beta = v/c \approx 1$. The relative permeability $\mu_r \approx 1$ since the vacuum chamber is usually non-magnetic.

The relation $\text{Re}Z_\perp^{(\bar{m}=1)} = (2c/b^2)(Z_\parallel^{(\bar{m}=0)}/\omega)$ holds exactly for the resistive wall impedance, while it is at best a rough approximation for impedances of other structures [Bane80a]. It fails completely for the space-charge impedances given above. On the other hand, the relation $Z_\perp^{(\bar{m})} = (\beta c/b^2)(Z_\parallel^{(\bar{m})}/\omega)$ between impedances with the same azimuthal mode number $\bar{m}$ holds always as it is a consequence of the Maxwell equations, in the form of the Panofsky-Wenzel theorem discussed in Chapter 4.

When the skin depth is larger than the wall thickness t, the EM fields penetrate through the chamber wall and the impedance depends on the structures present outside. If there is a perfectly conducting pipe of larger radius, or if the vacuum extends to infinity, the real part of the impedance vanishes when $t \to 0$ since there is no lossy material. However, if there is perfectly magnetic material, the real part diverges in that limit, and can become quite large for ferro-magnetic materials with finite, but high permeability. Hence magnet poles should preferably not penetrate into the vacuum chamber, although such designs has been chosen for economic reasons [Wilson72].

For thin metal walls $t << \delta_s$, $Z_{rw} = R\rho_c/tb$, the impedance would be independent of frequency, and hence $Z_\parallel/n \sim Z_\perp \sim 1/\omega$, "if there were no alternate paths for the induced current to flow" [Sacherer76]. However, the induced current can actually flow as *displacement current* even in vacuum, and therefore the resistive wall impedance does not increase as $1/\omega$ to very low frequencies, but actually approaches zero at DC [Zotter72b].

For the impedances of thin walls, the important parameter is δ_s^2/tb rather than the skin depth δ_s itself. The real part of the impedance has a maximum when the skin depth equals the geometric mean of the wall thickness and the pipe radius. The rather complicated approximate expressions for this case are given in ref. [Zotter72b]. Similarly, the transverse impedance is changed by the penetration of the EM fields through the vacuum chamber and, to first order, can be found from its relation to Z/n.

12.1.3 Dielectric Wall Impedance

For dielectric walls, the impedance at low frequencies is given by

$$\frac{Z_{\|}}{n} = j\frac{Z_0}{2}\left[1 + \frac{1}{\beta^2\varepsilon'\mu'}\right]\ln\left(1 + \frac{t}{b}\right) , \tag{12.5}$$

where $\varepsilon' = \varepsilon_r(1-j\tan\delta_E)$ is the complex relative permittivity, and $\mu' = \mu_r(1-j\tan\delta_M)$ the complex relative permeability with the *electric* and *magnetic loss tangents* δ_E and δ_M. For a lossless materials, the real part of the impedance is zero.

At higher frequencies, this impedance has resonant peaks whenever an odd multiple of half the *reduced wavelength*, i.e. the wavelength divided by ε, equals the wall thickness. Such large real parts of the resonant impedance limit the effectiveness of ceramic inserts for the purpose to shorten bunches [Burov90], as their spectra extend more and more into the resonant region and drive instabilities leading to bunch lengthening. The resonant impedances can be investigated with the computer program *LAWAL* [Zotter72a].

Similar resonances appear also in the transverse impedance [Zotter69a]. This invalidates the hope that ceramic walls may be used to reduce transverse instabilities in linear accelerators [Simpson86].

12.2 Cavities and Cavity-like objects

Externally powered *radio-frequency (rf) cavities* provide accelerating fields, and are installed in practically all particle accelerators or storage rings. They are required in relatively large numbers in high-energy electron rings, where they are needed to replenish the energy lost to synchrotron radiation. However, they are also present in proton rings, for efficient acceleration and to make up for energy lost in the structures due to wake fields. Additional rf cavities are sometimes installed for *feedback systems*, or to apply *higher-harmonic rf fields* to change the length or the energy spread of a bunch.

Many other objects installed in accelerators form so-called *accidental cavities*, such as kicker or separator tanks, local enlargements of the beam pipe needed to house pickup plates, current transformers, or similar. Before the importance of impedances for beam stability - and hence for the performance of accelerators - was generally appreciated, such enlargements of vacuum chambers were made, often in large numbers, just to fit the size of magnet gaps or to connect vacuum pumps. In some or such machines it was found necessary to provide additional damping to these accidental cavities, by introducing resistive loads into each one of them.

The geometry of rf cavities is usually designed to make their *shunt impedance* large, since for economic reasons it is desirable in order to obtain a maximum accelerating voltage for the available amount of rf power. Such cavities have often small beam tube holes and re-entrant nose cones, and therefore the impedance of most higher-order modes (*HOMs* for short) is then also quite high. In particular in electron machines, the rf cavities are generally the dominant contribution to the total impedance budget. This predicament has become somewhat less severe with the advent of *superconducting cavities*, which have such high quality factors that an optimisation of the *geometric factor* R_s/Q_r is no longer required to obtain a sufficiently high shunt impedance. Therefore these cavities may have larger beam-tube holes, and do not need nose cones. In general, they are constructed as smoothly as possible for a number of reasons. One main reason is to avoid excessive field strength at metallic surfaces which could lead to electric breakdown or discharges such as arcing, sparking, *multipactoring*, or *quenching* of superconducting materials.

12.2.1 Cavity Wake Fields

The calculation of wake fields and impedances in cavities has been discussed already in Chapters 7 to 9, which describe various techniques to solve the wave equations. Perturbation methods are usually limited to small cavities, and we shall discuss this in the section on bellows which can be seen as large numbers of small, identical cavities. In this section we concentrate on larger objects, and discuss first the results for closed pillbox cavities, which can be written in closed form and hence give more insight into the behaviour of the induced fields. However, such idealised cavities without holes for the beam cannot be used in real accelerators, and we therefore have another look at the results of Chapters 8 and 9 for cavities with beam tubes.

Rotationally symmetric cavities with rectangular cross-section and concentric beam tubes can be analysed by the field matching technique. The fact that actual cavities usually have more complicated cross sections can often be neglected, the details of the geometry have only a small effect on the basic behaviour of the solutions, in particular at high frequencies. However, several analytic techniques can be used to find the *EM* fields also for more complicated geometries. For example, an axially symmetric cavity of arbitrary cross section can be *sliced* into a succession of small, simple regions, and the field-matching technique can be applied to each of these slices. By increasing the number of slices, one approaches the results of numerical mesh methods with finite grids approximating the structure.

It is also possible to expand the fields of a complicated geometry into solutions valid in a simpler one. One then obtains integral equations which are related to the *method of moments* [Harrington68] and the *Galerkin method* of variational calculus. However, while such methods are interesting, they often require quite a large computational effort to evaluate the resulting solutions, and it is often more economic to use numerical methods directly as described in the Appendix to Chapter 11. The main strength of analytical methods is to give insight into the physical behaviour, and to derive simple, approximate relations and *scaling laws* which permit to estimate the parameter dependence of a particular problem.

12.2.2 Closed Pillbox Cavities

The wake fields induced in a closed, axially symmetric, circular cylindrical *pillbox cavity* by a point charge (delta-function pulse) moving on its axis are among the few examples of problems which can be solved exactly [Papiernik74, Chao75, Zotter79a]. The main results have been discussed in Section 3.2.3. It was found easier to evaluate the *step function response*, i.e. the integral of the wake function, since it has less discontinuities. It can be expanded into an infinite series of products of Bessel functions and their zeros. This series can be summed explicitly when the cavity radius d tends to infinity, and yields piecewise continuous logarithms, with jumps whenever the argument is a multiple of twice the length g of the cavity:

$$W_s(s) = \frac{2Z_0}{\pi} \ln\left(1 + \frac{1}{s/2g + [s/2g]}\right), \tag{12.6}$$

where the square brackets designate the integer part of the quantity enclosed. The *delta function response*, i.e. the wake function, is the derivative w.r.t. s and thus, in addition to jumps, has delta-function spikes at every multiple of $2g$ including the origin (see Figs. 3.2).

However, the expression in closed form is valid only as long as no reflections can come back to the test charge from the radial wall of the cylindrical cavity, i.e. it actually applies to two parallel plates rather than a cylindrical cavity. However, the reflections from the wall can be taken into account by summing the infinite series over Bessel functions numerically. Although it converges only slowly, with high-speed computers it can be evaluated without much difficulty.

12.2.3 Cylindrical Cavities with Beam Tubes

Actual cavities used in accelerators need to have holes to permit entry and exit of the particle beam. A model consisting of a circular cylindrical cavity with concentric beam tubes is often adequate to study the behaviour of more complicated shapes. Here we have to distinguish between the idealised cases of a single cavity (or a finite number of them), with beam tubes extending to infinity on either side, and the case of periodic cavities repeating indefinitely.

The first case corresponds to that of a cavity in a linear accelerator, the second to a circular accelerator or storage ring where the beam comes to the same cavity on every turn. However, if the circumference is large enough so that the fields have decayed sufficiently before the beam comes again, the second case approaches the first one. The main difference between these two cases is the asymptotic behaviour of the impedance, which tends towards $\omega^{-1/2}$ for the single cavity, but to $\omega^{-3/2}$ for the periodic geometry. The corresponding wake function diverges at the origin in the first case, but remains finite in the second one.

Another distinction has to be made between *short cavities*, with a length comparable or less than the diameter, and *long cavities* where the side walls are so far apart that they can be considered independent steps in radius, without influence upon each other at least over the length of time considered. The case of fields excited in a short cavity is illustrated in Fig. 12.1, and fields excited in a long cavity are shown in Fig. 12.2.

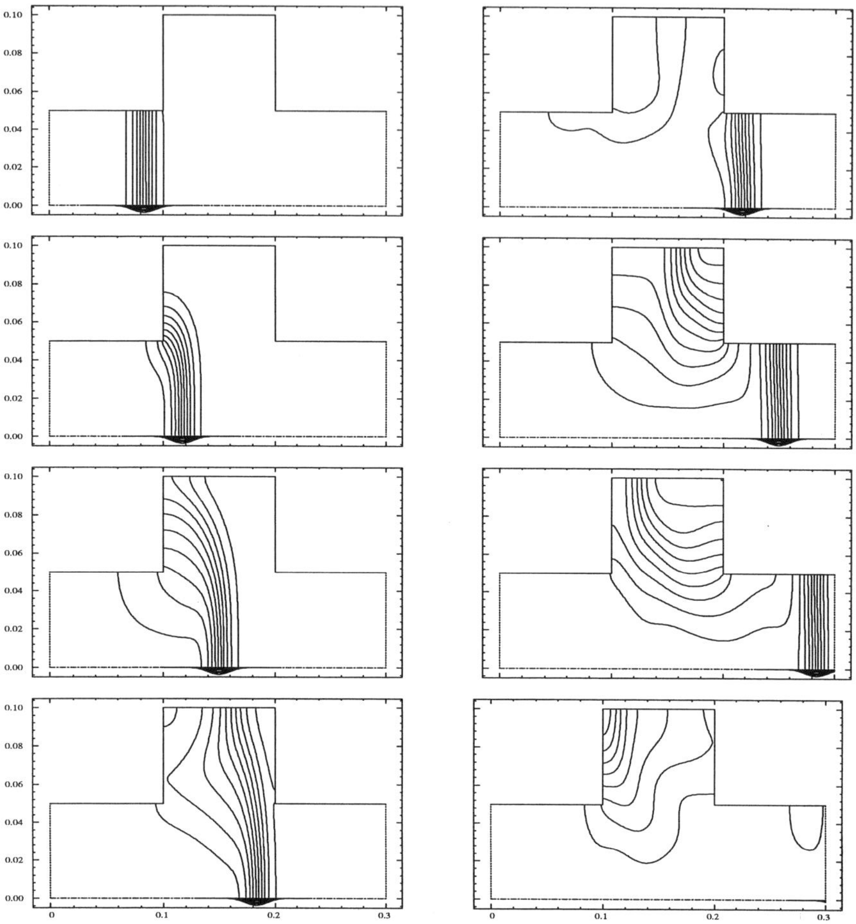

Figure 12.1: Electric field lines of Gaussian bunch passing a short cavity.

The fields drawn in these pictures, and the impedances obtained from them, can be expressed by analytical formulae such as Eq. (8.36), with the coefficients given in Table 8.1, which were obtained by field matching. Simpler formulae are found from the integral equation given in Eq. (9.12). An approximate solution, valid at high frequencies, is given by Eq. (9.23).

The acceleration (or deceleration) of a charge passing a single *step-in* (or *step-out*) corresponds to a positive or negative impedance at zero frequency (*DC*), proportional to $\ln(b_1/b_2)$. This is caused by a change of the *beam*

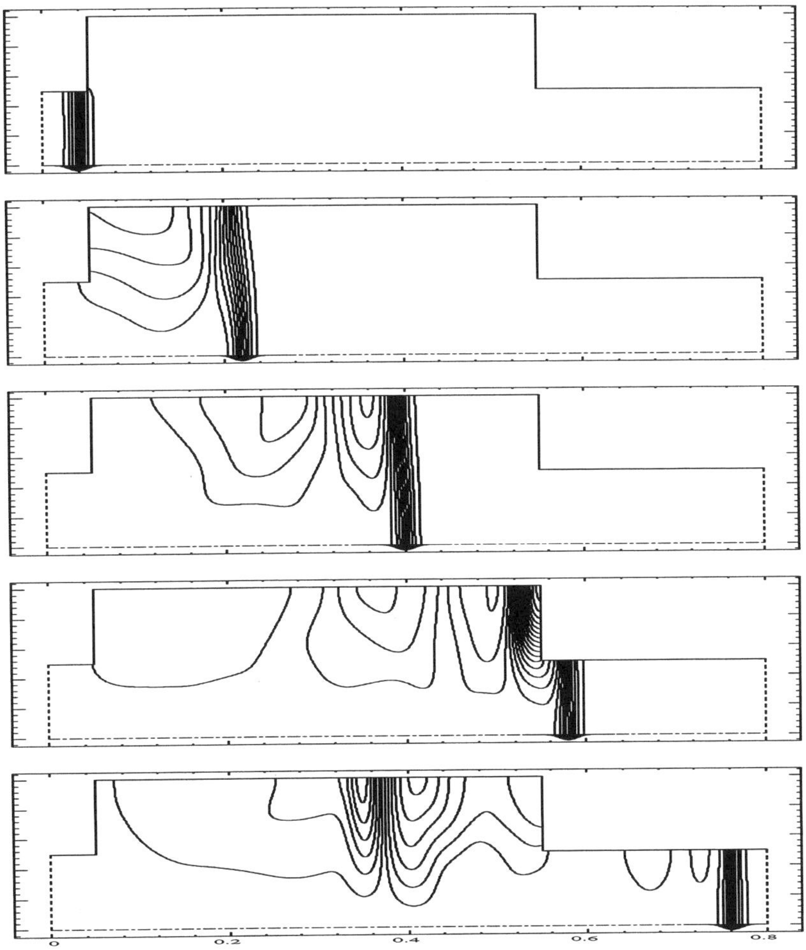

Figure 12.2: Electric field lines of Gaussian bunch passing a long cavity.

potential rather than a proper wake field due to fields induced in the walls. It can be seen that the shape of the field lines in a long cavity approach that of field lines in a single step shown in Fig. 12.3.

The energy gain and loss will cancel exactly when the charge comes back to a beam tube with the same radius. However, the sum of their high-frequency

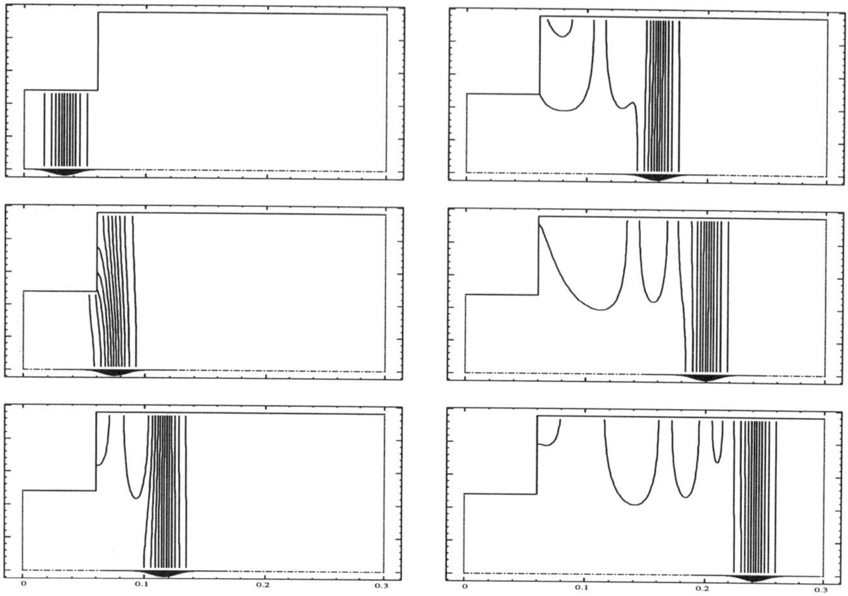

Figure 12.3: Electric field lines of Gaussian bunch passing a single step-out.

impedances will remain different over a quite wide frequency range, since the real part of the step-in approaches zero, while that of the step-out remains constant up to quite high frequencies.

12.3 Bellows

Bellows are sections of corrugated vacuum chamber which are installed mainly to permit longitudinal displacement of the rigid chambers due to temperature variations, in particular if the chambers have to be heated to high temperatures to obtain a better vacuum. Often the bellows should also permit a small transverse displacement of rigid chambers to avoid damage in case of unavoidable motions. A large length of bellows, of the order a few percent of the total machine circumference, is usually needed. This may be achieved by a large number of short bellows with few convolutions, or a smaller number of longer ones.

Usually bellows have circular or oval (elliptic) cross section, and transverse dimensions equal to or slightly larger than the vacuum chambers which they connect. If they extend to the inside, there is not only the danger that they be hit by beam particles or synchrotron radiation, but also they would present an increased impedance to the beam. In order to reduce the impedance, the bellows should have a large diameter, but convolutions of small depth. This gives them little flexibility, therefore they are sometimes optimised mechanically, but *shielded* from induced rf fields by inner sleeves or fingers with sliding contacts. Also inner bellows with smaller convolutions - made of very thin material to reduce their stiffness, but not vacuum tight - have been used for this purpose.

12.3.1 Low-Frequency Impedance of Bellows

The application of field matching techniques, described in Chapter 8, is usually too cumbersome to calculate the impedance of bellows. Several methods have been developed specifically or this purpose. [Krinsky81, Kheifets86b, Kheifets85a, Kurennoy90]. Here we use the method described in Section 7.4 to calculate the low-frequency impedance of the longitudinal $m = 0$ mode in an axially symmetric geometry, shown in Fig. 12.4. Results for the transverse impedance of the dipole mode $m = 1$ can be found in ref.[Kheifets86b].

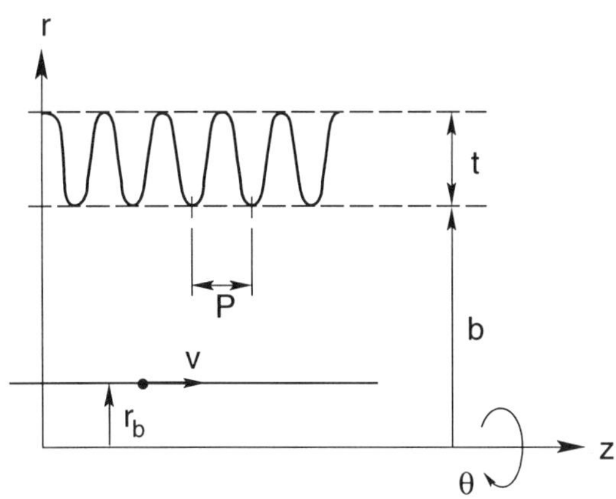

Figure 12.4: Geometry of bellows.

The chamber wall in the plane $\theta = constant$ is described by the periodic function $r(z) = b + t\, s(z)$, with period P. A charged particle beam moves along the axis of the structure ($r_b = 0$). The general solution of the Maxwell equations suitable for this case can be expressed by an infinite series [Krinsky81]. In the limit $\gamma \to \infty$, the longitudinal and radial components of the electric field are given by the expansions:

$$\widetilde{E}_z = \frac{qZ_0 \mathrm{e}^{-jkz}}{2j\pi b} \sum_{p=-\infty}^{\infty} B_p \mathrm{e}^{-2j\pi pz/P} \frac{I_0(\alpha_p r/b)}{I_0(\alpha_p)} ,$$

$$\widetilde{E}_r = \frac{qZ_0 \mathrm{e}^{-jkz}}{2\pi b} \left[\frac{b}{r} + \sum_{p=-\infty}^{\infty} B_p \mathrm{e}^{-2j\pi pz/P} \left(kb + 2\pi \frac{pb}{P} \right) \frac{I_1(\alpha_p r/b)}{\alpha_p I_0(\alpha_p)} \right] , \quad (12.7)$$

where B_p are unknown expansion coefficients, I_0 and I_1 modified Bessel functions of the first kind, and α_p is defined by:

$$\alpha_p^2 = \left(\frac{2\pi bp}{P} \right)^2 + 4\pi p \, \frac{kb^2}{P} . \quad (12.8)$$

The expansion coefficients B_p in this case are given by the boundary condition:

$$\widetilde{E}_z(r(z), z) = -\widetilde{E}_r(r(z), z) dr(z)/dz , \quad (12.9)$$

which leads to the following infinite set of linear algebraic equations:

$$\sum_{p=-\infty}^{\infty} M_{np} B_p = N_n , \qquad n = -\infty \ldots \infty , \quad (12.10)$$

where

$$M_{np} = \begin{cases} \left\langle \left[1 + \frac{j}{2} kr(z) \frac{dr(z)}{dz} \right] \exp \left[\frac{2j\pi nz}{P} \right] \right\rangle & \text{for } p = 0 \\ \left\langle \frac{I_0(\alpha_p r(z))}{\alpha_p^2 I_0(\alpha_p)} \exp \left[\frac{2j\pi (n-p) z}{P} \right] \right\rangle X_{pn} & \text{for } p \neq 0 \end{cases} \quad (12.11)$$

with

$$N_n = \frac{2\pi b}{jP} \left\langle \frac{b}{r(z)} \frac{dr(z)(z)}{dz} \exp \left(\frac{2j\pi nz}{P} \right) \right\rangle ,$$

$$X_{pn} = \left(\frac{2\pi b}{P} \right)^2 pn + \left(\frac{2\pi b}{P} \right) kb(p+n) . \quad (12.12)$$

Here the brackets $\langle f \rangle$ are used to define the value of function $f(u)$ averaged over its period P. The longitudinal impedance of the bellows can be found solving Eq. (12.10) and using Eq. (12.7):

$$Z_{\|}(k) = jZ_0 \frac{PB_0(k)}{2\pi b} \, . \tag{12.13}$$

The system of Eqs. (12.10) must be solved numerically. The computer code *IMPASS* [Kheifets85a] (Impedances of Periodic Axially Symmetric Smooth Structures) calculates the coefficients of the field expansions and finds both the longitudinal and the transverse impedances for $m = 0$ and $m = 1$ modes in the low frequency region. The approach used here is not valid for the high-frequency region, where the impedance has a complicated resonance structure.

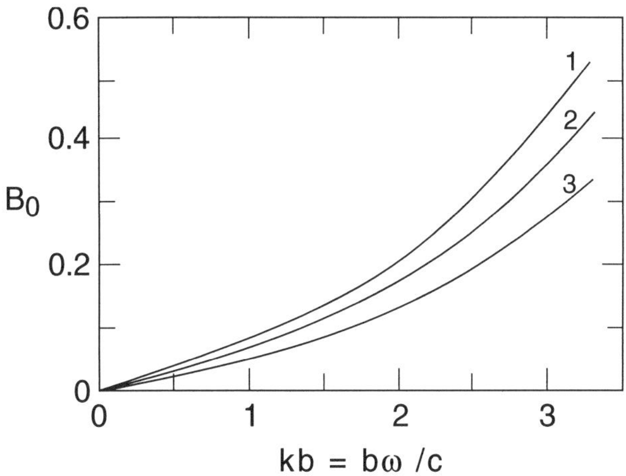

Figure 12.5: Impedance coefficient B_0 of bellows.

Fig. 12.5 [Kheifets85a] shows the impedance coefficient B_0 as function of the normalised frequency kb, for a relative depth of the corrugations $t/2b$ of 9%, and three different values of the bellows parameter $2\pi b/P$. More results can be found in ref. [Kheifets86b].

12.4 Tapers and Collimators

A *taper* is a part of the vacuum chamber which provides a gradual transition from one cross section of the chamber to another. They are used to decrease the total impedance of the vacuum chamber since the energy loss of a beam in a taper is smaller than in a step - the reduction being the bigger the smoother the transition.

A collimator is a restriction of the vacuum chamber which serves the purpose of removing from the beam particles with excessively large amplitudes of the transverse oscillations. The desire to decrease the chamber impedance brings about designs of smooth collimators (that makes them similar to a double sided taper), although step-like collimators are also sometimes employed.

Two types of tapers and collimators are of general interest – axially symmetric and flat ones. The most straightforward approach to compute their effects would seem to use purely numerical codes. However, they are difficult to use because of the need of an excessively fine mesh, which leads to very lengthy calculations for long structures. A number of analytic methods to evaluate the effects of tapers have therefore been developed. A rather slow change from one cross section to another is assumed in all of these. In the present Section we describe such methods.

12.4.1 Axially Symmetric Taper

As an example we consider an axially symmetric taper, i.e. a gradual transition between two beam pipes of circular cross section. In particular, we assume a linear taper, where the slope of the wall is constant. For short bunches, $\sigma \ll b < d$, the energy loss is dominated by the high-frequency modes $kd > kb \gg 1$, where b is the smaller and d the larger tube radius.

We characterise the taper by the angle α between the wall and the axis: $\cot\alpha = g_1/(d-b)$, where g_1 is its length. For a sudden step $g_1 = 0$, and thus $\alpha = \pi/2$. Based on the discussion presented in Section 9.2, we expect that for an ultra-relativistic beam, $\gamma \to \infty$, the main contribution to the impedance comes from the waves with large k diffracted at small angles. The taper will reduce the impedance and the loss factor of the transition effectively if its angle is comparable to or smaller than the Fresnel diffraction angle.

As shown below, for a bunch of r.m.s. length σ, the loss factor decreases with increasing taper length g_1 until it reaches a minimal value at $g_1 \simeq (d-b)^2/\sigma$, and remains constant with any further increases of g_1. A short taper, for which $g_1\sigma/(d-b)^2 \ll 1$, is not effective in reducing the energy loss.

For a circular cylindrical cavity of total length g, with slanted side walls forming two symmetrical tapers of the length g_1 on both sides, its symmetry significantly simplifies calculations. Eq. (9.38) for such a cavity yields:

$$\widetilde{E}_z^{(1)}(r,z) = -j\frac{qZ_0}{8\pi}\int_b^d dr' \int_{-\infty}^{\infty} dp\, \mathrm{e}^{-jpz}\, J_0(\nu r)\, \frac{\partial H_0^{(1)}(\nu r')}{\partial r'}$$
$$\left[\mathrm{e}^{-j(k-p)(r'-b)\cot\alpha} - \mathrm{e}^{-j(k-p)[g-(r'-b)\cot\alpha]}\right]. \qquad (12.14)$$

Here $\nu^2 = k^2 - p^2$ and $\cot\alpha = g_1/(d-b)$. The two terms in this expression describe the waves generated at the two tapers: the taper-out at $z = 0$, and the taper-in at $z = g$. The longitudinal impedance of a taper-out can be obtained by integrating the first term over z in the interval $0 < z < \infty$, and similarly for a taper-in. This gives

$$Z_{\|}(\omega) = -\frac{Z_0}{2\pi}\lim_{P\to\infty}\int_{-\infty}^{\infty}\frac{dp}{k-p}\, J_0(\nu b)\left[1-\exp\{\pm j(k-p)P\}\right]$$
$$\times \int_b^d dr'\, \exp\{\mp j(k-p)(r'-b)\cot\alpha\}\, \frac{\partial H_0^{(1)}(\nu r')}{\partial r'}, \qquad (12.15)$$

where the signs $\pm$ correspond to a taper-out and a taper-in. In the limit $P \to \infty$ one gets:

$$\lim_{P\to\infty}\frac{1}{k-p}\left[1-\mathrm{e}^{\pm j(k-p)P}\right] = \frac{1}{k-p} \pm j\pi\delta(k-p)\,, \qquad (12.16)$$

and the impedance becomes:

$$Z_{\|}(\omega) = \pm\frac{Z_0}{\pi}\ln\frac{d}{b} - \frac{Z_0}{8\pi}\int_{-\infty}^{\infty}\frac{dp}{k-p}\, J_0(\nu b)$$
$$\times \int_b^d dr'\, \exp\{\mp j(k-p)(r'-b)\cot\alpha\}\, \frac{\partial H_0^{(1)}(\nu r')}{\partial r'}\,. \qquad (12.17)$$

The first term corresponds to the difference in the field energy of a particle in beam pipes of different radii and is independent of the angle α. The real part of the second term describes the loss to radiation, which is the same for a taper-out and a taper-in. The sum of the impedances of two tapers gives the impedance of a long tapered cavity; for $\alpha \to 0$ and $g = 0$, it reproduces the result of Eq. (9.46). The difference of the loss factors of two tapers with the same angle α is independent of α:

$$\kappa_{\|}^{out}(\sigma) - \kappa_{\|}^{in}(\sigma) = \frac{1}{2\pi\varepsilon_0\sigma\sqrt{\pi}}\ln\frac{d}{b}\,. \qquad (12.18)$$

This was noticed in numerical simulations [Chan87]. As was shown in ref [Heifets88a], the largest contribution for a step comes from the region of the variable p, where $1/d < \nu < 1/b$. Hence, if

$$\tan\alpha > \frac{d-b}{2kb^2} , \tag{12.19}$$

the exponent in Eq. (12.17) may be replaced by unity, and the impedance of a taper is the same as the impedance of a step. In the opposite case of a small α, the exponent oscillates rapidly unless

$$(k-p)(r'-b)\cot\alpha \ll 1 . \tag{12.20}$$

Restricting the area of integration by this condition, we obtain for the real part of the impedance

$$\text{Re } Z_{\parallel}(\omega) = \text{Re } Z_{\parallel}^{step}(\omega) - \Delta Z_{\parallel}(\omega) , \tag{12.21}$$

where $Z_{\parallel}^{step}$ is the impedance of a step, see Eq. (9.49). With the abbreviation $\xi = \sqrt{1-x^2}$, the correction term is

$$\Delta Z_{\parallel} = \frac{Z_0}{4\pi}\int_{-1}^{x_0} \frac{dx}{1-x}\, J_0(kb\xi)\Big[J_0(kr_\alpha\xi) - J_0(kb\xi)\Big] , \tag{12.22}$$

where $r_\alpha = b + \tan\alpha/(k(d-b))$. To estimate the integral, we notice that $\Delta Z_{\parallel}$ is large only if the argument of the Bessel function in the integrand $kr_\alpha\sqrt{1-x^2}$ is small within the range of integration. This is possible only if $kb\tan\alpha \ll 1$, and then the correction term becomes $\Delta Z_{\parallel} = -(Z_0/4\pi)\ln\zeta$, where

$$\zeta = \begin{cases} b/d & \text{for} \quad kb\tan\alpha < 1 , \\ 2kb\tan\alpha & \text{for} \quad kb\tan\alpha > 1 \end{cases} . \tag{12.23}$$

The loss factor for a taper can be obtained by the convolution of Eq. (12.21) with the bunch distribution of r.m.s. length σ. In the case $d \gg b$, the correction term to the loss factor is:

$$\Delta\kappa_{\parallel} = \begin{cases} \dfrac{\cot\alpha}{4\pi^2\varepsilon_0 b} & \text{for} \quad \tan\alpha \gg \sigma/b , \\ \dfrac{\ln d/b}{4\pi\sqrt{\pi}\varepsilon_0\sigma} & \text{for} \quad \tan\alpha \ll \sigma/b . \end{cases} \tag{12.24}$$

Thus, the energy loss for a taper-out can be smaller than that for a step-out ($\alpha = \pi/2$) at most by a factor of two, even if the angle α is very small [Heifets89a, Yunn89]. For a long cavity, tapered symmetrically from both ends, the correction term given by above equation doubles, reducing the loss factor to zero for a sufficiently small angle $\tan\alpha \ll \sigma/b$.

The dependence of the longitudinal loss factor of a one-sided taper on its angle can be approximated by the formula:

$$\kappa_{\parallel}(\sigma) = \frac{\ln d/b}{2\pi\varepsilon_0\sigma\sqrt{\pi}}\left[1 - \frac{\tilde{\eta}_1}{2}\right] , \tag{12.25}$$

where $\tilde{\eta}_1 = \min(1, \eta_1)$ with $\eta_1 = g_1\sigma/(d-b)^2$. For $\eta_1 > 1$, the loss factor of a taper reaches half the value of the loss factor for a step, and remains constant with any further increase.

A comparison of Eq. (12.25) with the calculations by the code *TBCI* is presented in Fig. 12.6 [Heifets90a] for a cavity of length $g = 1.15\ m$ and radius $b = 0.01\ m$ tapered from one side. The bunch length is assumed to be $\sigma/b = 0.3$. Curves are plotted for the two respective ratios $d/b = 4.0$ and $d/b = 2.0$. The results of the *TBCI* calculations (solid lines) are in reasonably good agreement with those obtained from Eq. (12.25) (dashed lines).

For a symmetric taper, $\tilde{\eta}_1/2$ should be replaced by $\tilde{\eta}_1$:

$$\kappa_{\parallel}(\sigma) = \frac{1}{2\pi\varepsilon_0\sigma\sqrt{\pi}}\,(1-\tilde{\eta}_1)\ln\frac{d}{b} . \tag{12.26}$$

Thus a long, symmetric taper reduces the losses to zero. The independence of the losses on the direction of the particle motion is confirmed by these results.

12.4.2 Axially Symmetric Collimator

Explicit expression for an estimate of the longitudinal impedance of an axially symmetric pipe with smooth slow varying cross section can be obtained from formula Eq. (7.89) by substituting $b(\infty) = b(-\infty)$. This impedance as well as the transverse one Eq. (7.90), are purely imaginary. The impedance acquires a real part only if the next order terms in ϵ are taken into account. We describe here a more general derivation [Warnock93], which reproduces the result [Yokoya90] for low frequencies and, in addition, provides an estimate for the high frequency region.

Let us assume that the tube radius is given by $r_b(z) = b - \epsilon s(z)$, where $s(z)$ is zero for $|z| > g$. The derivative over z of this function $ds(z)/dz$ is

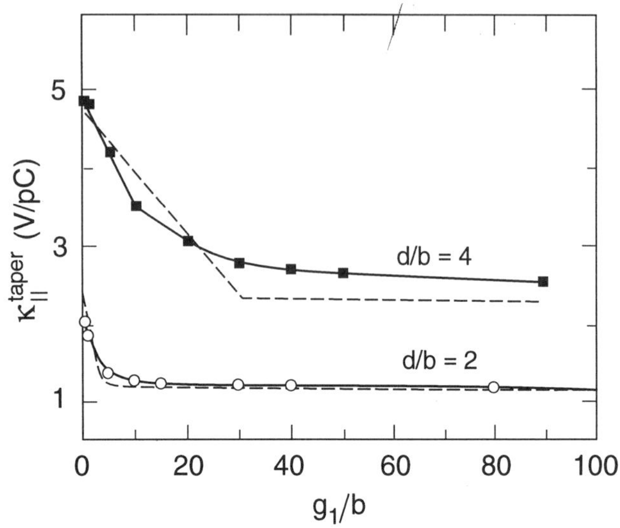

Figure 12.6: Longitudinal loss factor of Gaussian bunch with $\sigma/b = 0.3$ versus ratio of length g_1 to radius b of taper: solid line *TBCI*, dashed Eq. (12.25)

normalised so that $\max|s'| = 1$. Then ϵ measures the effective strength of the wall perturbation.

The charge and current densities are given in Eq. (2.3). In the frequency domain, the charge density becomes $\tilde{\rho} = (q/2\pi)\tilde{\lambda}f(r)\exp(-jkz/\beta)$, and the current density is $\tilde{J}_z = \tilde{\rho}/v$. Only the field components $\tilde{E}_r, \tilde{H}_\theta$ and $\tilde{E}_z$ are non-zero. The Fourier component of the longitudinal bunch density distribution $\lambda(z)$ is $\tilde{\lambda} = \int dt\lambda(z - vt)\exp(-j\omega t)$. The longitudinal and the radial parts of the bunch distribution are normalised $\int dz\lambda(z) = 1, \int rdrf(r) = 1$.

As was shown in Chapter 6 the field components $\tilde{E}_r, \tilde{H}_\theta$ can be expressed in terms of $\tilde{E}_z$, which can be written as

$$\tilde{E}_z(r,z) = \int_C dp \, \mathrm{e}^{-jpz} A(p,k)\frac{I_0(\chi r)}{I_0(\chi b)} + \tilde{E}_z^{(0)}(r,z) \; , \tag{12.27}$$

where $\chi^2 = p^2 - k^2$, the source term is:

$$\tilde{E}_z^{(0)}(r,z) = \frac{jq\tilde{\lambda}f}{2\pi\varepsilon_0 k}\mathrm{e}^{-jkz}\tilde{\lambda}G(r) \; , \tag{12.28}$$

and the contour of integration C is going along the real axis p with indentations below the poles $p = -k_n$ and above $p = k_n$. Such a choice of indentations ensures the absence of incoming waves. The poles are given by $k_n^2 = k^2 - j_{0n}^2/b^2$. Finally, the radial Green function $G(r)$ is regular at $r = 0$:

$$G(r) = I_0(\chi r) \int_b^r udu K_0(\chi u) f(u) - K_0(\chi r) \int_0^r udu I_0(\chi u) f(u) \ . \qquad (12.29)$$

The Fourier component of the radial component of the electric field, see Eq. (6.80), is given by:

$$\widetilde{E}_r(r,z) = \frac{jk}{\chi^2} \frac{\partial \widetilde{E}_z(r,z)}{\partial r} \ . \qquad (12.30)$$

The boundary condition demands that on the wall the tangential component of the vector $\mathbf{E}$ should be zero:

$$\widetilde{E}_z(r_b(z), z) + r_b' \widetilde{E}_r(r_b(z), z) = 0 \ . \qquad (12.31)$$

From this expression an integral equation for the unknown amplitude $A(l,k)$ can be derived:

$$A(l,k) = \int_C dp M(l,p,k) A(p,k) + S(l,k) \ . \qquad (12.32)$$

In the limit $\gamma \to \infty$, the kernel and the free term of this equation are:

$$\begin{aligned} M(l,p,k) &= \frac{1}{2\pi} \int_{-g}^{g} dz \, \frac{e^{-j(p-l)z}}{I_0(\chi b)} \left[I_0(\chi b) - I_0(\chi r_b(z)) - \frac{jp}{\chi} r_b'(z) I_0'(\chi r_b(z)) \right] \\ S(l,k) &= \epsilon Z_0 q c \frac{\tilde{\lambda}}{2\pi} \int dz \frac{s'(z)}{r_b(z)} e^{-j(k-l)z} \ . \end{aligned} \qquad (12.33)$$

To lowest order of the iterative solution of Eq. (12.32), the longitudinal impedance of a taper becomes [Warnock93]:

$$Z_{\parallel} = \frac{\epsilon^2 k Z_0}{2\pi b^3} \sum_{n=1}^{\infty} \frac{1}{k_n} \int_{-g}^{g} dz s'(z) \int_{-g}^{g} du \ s'(u) \exp[-jk_n|z-u| + jk(z-u)] \ . \qquad (12.34)$$

For frequencies below cut-off, $\omega = cj_{01}/b$, integration by parts yields two terms. The first term yields the result Eq. (7.89):

$$Z_Y = j \frac{\epsilon^2 k Z_0}{4\pi} \int_{-g}^{g} dz [s'(z)]^2 \ . \qquad (12.35)$$

However, the neglected terms are not really small. For example, when the wall is described by the function $s(z) = d[1 + \cos(\pi z/g)]/2$ with $|z| < g$, and $\epsilon = \pi d/(2g)$, Eq. (12.33) gives for low frequencies:

$$\begin{aligned} Z(\omega) &= Z_Y \left[1 + \left(1 - 2\frac{I_1(\pi b/g)}{(\pi b/g) I_0(\pi b/g)}\right) + \right. \\ &+ \left. 4\pi^2 (\frac{b}{g})^3 \sum_{n=1}^{\infty} \frac{1 - \exp(-2j_{0n} g/b)}{j_{0n}(j_{0n}^2 + (\pi b/g)^2)^2} \right] . \end{aligned} \tag{12.36}$$

The last two terms are corrections which tend to zero as $(b/g)^2$ and $(b/g)^3$, respectively. Fig. 12.7 [Stupakov95a] gives a comparison of the two results for a shallow collimator with a sinusoidal boundary with small depth of corrugations $\epsilon = 0.078$. As one can see, the agreement is good only for relatively small

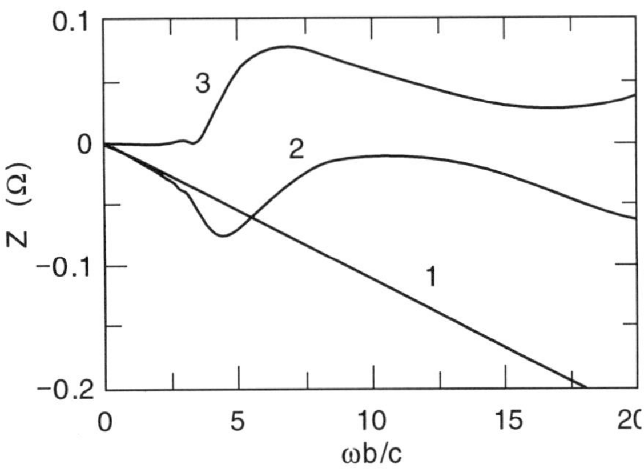

Figure 12.7: Impedance of a shallow collimator.

values of k where the real part of impedance is zero. That indicates that the upper limit of validity of formula Eq. (7.89) is more stringent then $kbb' \ll 1$. Actually it should read $kb^2 \ll g$. Since $b' \approx \Delta b/g$, for large variation of radius $\Delta b \sim b$, both inequalities are equivalent. But for small $\Delta b/b$ condition $kbb' \ll 1$ strongly overestimates the validity region of Eq. (7.89). Furthermore, as shown in ref.[Stupakov95a], the requirement $kb \gg 1$ can be omitted.

12.4.3 High-Frequency Impedance of a Step Collimator

The longitudinal impedance of a step collimator in the high-frequency region (and in the relativistic case $\gamma \gg 1$) can be found analytically using Eq. (8.37). Since asymptotically $\lambda_b \approx k$, only the diffracted field; i.e., the field depending on coefficients $B_s^{(4)}$, contributes to the impedance. Physically that arises from the fact that only the diffracted field radiated forward can reach a relativistic particle. Hence

$$Z^{col}(k) = -2\frac{Z_0}{\pi}\, kd \sum_s B_s^{(4)}\, J_0\left(\frac{j_{0s}}{p}\right) , \tag{12.37}$$

where $p = d/b$ with b is the pipe and d the collimator radius.

The coefficients $B_s^{(4)}$ can be found from the matrix Eq. (8.32). with the expressions for matrix coefficients and the r.h.s. taken from Table 8.1:

$$B_l^{(4)} kdJ_1^2(j_{0l}) = -\frac{J_0(j_{0l}/p)}{j_{0l}^2} + \frac{2kd}{p^2}\sum_m \left[B_m^{(3)} - B_m^{(2)}E_+\right]\phi_{ml}(p^{-1}) , \tag{12.38}$$

where the quantities $\phi_{ml}(p^{-1})$ are defined in Eq. (8.33). Dividing Eq. (12.37) by $J_1^2(j_{0l})$, multiplying by $J_0(j_{0l}/p)$, and summing over l, we obtain:

$$\begin{aligned}\sum_l kdB_l^{(4)} J_0(j_{0l}/p) &= -\sum_l \frac{J_0^2(j_{0l}/p)}{j_{0l}^2 J_1^2(j_{0l})} + \\ 2kd\sum_l \frac{J_0^2(j_{0l}/p)}{J_1^2(j_{0l})}\sum_m (B_m^{(3)} &- B_m^{(2)}E_+)\frac{j_{0m}J_1(j_{0m})}{j_{0m}^2 p^2 - j_{0l}^2} .\end{aligned} \tag{12.39}$$

The summation here can be performed explicitly, using the following form of the *Kneser-Sommerfeld formula* [Erdelyi53]:

$$\sum_l \frac{J_0^2(j_{0l}/p)}{(j_{0l}^2 - x^2)J_1^2(j_{0l})} = \frac{\pi J_0(x/p)}{4J_0(x)}\left[J_0(\frac{x}{p})Y_0(x) - J_0(x)Y_0(\frac{x}{p})\right] , \tag{12.40}$$

where Y_0 is Bessel function of the second kind. From here it follows that the second term of the r.h.s. of Eq. (12.37), containing the coefficients $B_m^{(3)}$ and $B_m^{(2)}$ vanishes, since the sum over l is zero. The first sum in the r.h.s. of Eq. (12.37) according to the same formula is[1]:

[1]There is an apparent discrepancy with Eq. (3.6), but both expressions are nevertheless correct [Erdelyi53].

$$\sum_l \frac{J_0^2(j_{0l}/p)}{j_{0l}^2 J_1^2(j_{0l})} = \frac{\pi}{4} \lim_{x\to 0} \left[Y_0(x) - Y_0(x/p)\right] = \frac{\ln p}{2} . \tag{12.41}$$

Hence, in the high-frequency region, the impedance of a collimator does not depend on frequency and is the following constant:

$$Z_{\parallel}(k) = (Z_0/\pi)\ln(d/b) , \quad \text{for} \quad \gamma > kb >> 1 . \tag{12.42}$$

This equation is not valid for $kb > \gamma$. In this range of frequencies the impedance decreases at least as k^{-2}. The same expression holds for the high-frequency impedance of a pipe with cross section step for the case of a bunch exiting the narrow pipe. In the opposite case the impedance is zero, cf. Section 8.5.

It is interesting to estimate the total energy loss $\Delta\mathcal{E}$ of a charge passing through a collimator. For a Gaussian bunch of r.m.s. length σ, the total energy loss is:

$$\Delta\mathcal{E} = \frac{q^2}{2\pi\sqrt{\pi}\varepsilon_0\sigma} \ln \frac{d}{b} . \tag{12.43}$$

This expression is valid for $\sigma > b/\gamma$, and agrees with the formula for the total energy loss of a charge passing through a sudden change in a pipe cross section obtained in the papers [Balakin83, Kheifets87].

If one assumes that ReZ is constant for $kb < \gamma$ and is zero for $kb > \gamma$, as previously discussed, then the total energy loss of *a point* charge, $\sigma = 0$, is proportional to γ . This conclusion is in agreement with the estimate obtained in ref.[Lawson90] and with numerical calculations [Dnestrovsky59] for a charge passing through a hole in a screen.

12.4.4 Transverse Impedance of a Flat Collimator

In this subsection we describe the results [Stupakov96] obtained for a flat smooth collimator. The collimator of the length g is bounded by perfectly conducting walls and extends in the y direction from $y = -h$ to $y = h$. Its upper and lower walls are given by equations $x = \pm b(z)$, with $b(z)$ being a smooth function for which it is assumed that $b' \ll 1$ and $b \gg h$. The beam propagates parallel to the z axis at a distance x_b from it.

For not too high frequencies such that $kb^2 < g$, where $k = \omega/c \approx 1/\sigma$, the energy loss of the beam due to radiation in the collimator is small. Hence, the real part of the impedance is much smaller than its imaginary part and in the first approximation can be neglected altogether. In this approximation the transverse impedance is purely imaginary and does not depend on frequency. Because of the latter, it is sufficient to consider the impedance in the limit $\omega \to 0$ what greatly simplifies the calculation. In this limit the electric and magnetic fields can be found as solutions of corresponding electro- and magneto-static problems with proper boundary conditions.

Once the fields have been found, the transverse impedance for a unit charge can be calculated from Eq. (4.10):

$$Z_x(x, y, x_b) = \frac{j}{x_b} \int_{-\infty}^{\infty} dz \left[\widetilde{E}_x(x, y, x_b; z) - Z_0 \widetilde{H}_y(x, y, x_b; z) \right] . \tag{12.44}$$

The wake corresponding to this purely imaginary, constant impedance is proportional to a delta function:

$$G_\perp(x, y, x_b, \tau) = Z_x(x, y, x_b)\delta(\tau) . \tag{12.45}$$

As the result of its interaction with the collimator, a particle in the beam experiences a deflection in x direction by the angle

$$\Delta\theta_x = -\frac{N_b e^2 c x_b}{2\mathcal{E}\sqrt{\pi}\sigma} \mathrm{Im} Z_x(x, y, x_b) , \tag{12.46}$$

where N_b is the bunch population.

a) **Electric Field.**

The electric field can be found from the *electrostatic* potential Φ_E which satisfies the Poisson equation with the right hand side representing a linear charge of unit line density:

$$\Delta\Phi_E = -4\pi\delta(x - x_b)\delta(y) , \tag{12.47}$$

with the boundary condition $\Phi_E|_{x=\pm b(z)} = 0$.

The assumption $h \gg b(z)$ permits setting h to infinity. Since the boundary $b(z)$ is a slowly varying function of its argument, this equation can be solved by iteration. In lowest approximation, the variation of the potential Φ_E in the z direction can be neglected. Then the equation reduces to:

$$\left(\frac{\partial^2}{\partial x^2} + \frac{\partial^2}{\partial y^2} \right) \Phi_E^{(0)} = -4\pi\delta(x - x_b)\delta(y) . \tag{12.48}$$

The solution of this equation satisfying the boundary conditions is:

$$\Phi_E^{(0)}(x,y,x_b;b(z)) = \ln \frac{\sinh^2 \pi y/4b(z) + \cos^2 \pi(x+x_b)/4b(z)}{\sinh^2 \pi y/4b(z) + \sin^2 \pi(x-x_b)/4b(z)} . \tag{12.49}$$

In the next approximation, the potential is given by $\Phi_E = \Phi_E^{(0)} + \Phi_E^{(1)}$, where the first order correction $\Phi_E^{(1)}$ satisfies the following equation:

$$\left(\frac{\partial^2}{\partial x^2} + \frac{\partial^2}{\partial y^2}\right)\Phi_E^{(1)} = -\frac{\partial^2}{\partial z^2}\Phi_E^{(0)} \tag{12.50}$$

with the boundary condition $\Phi_E^{(1)}|_{x=\pm b(z)} = 0$.

The solution of this equation in terms of the function $\Phi_E^{(0)}$ is:

$$\Phi_E^{(1)}(x,y,x_b;z) = -\frac{1}{4\pi}\int_{-b(z)}^{b(z)} dx'$$
$$\cdot \int_{-\infty}^{\infty} dy' \frac{\partial^2}{\partial z^2}\Phi_E^{(0)}(x',y',x_b,b(z))\Phi_E^{(0)}(x,y-y',x',b(z)) . \tag{12.51}$$

The main contribution to the integral over y' comes from the region $\approx b$ justifying the assumption $h \to \infty$ made above. The electric field can be obtained from the electrostatic potential: $\tilde{E}_x = -\partial\Phi_E/\partial x$.

b) **Magnetic Field.**

The magnetic field can be found from the *magnetostatic* potential Φ_M which satisfies the Poisson equation with the right hand side representing a linear current with a unit line density:

$$\Delta\Phi_M = 4\pi\delta(x-x_b)\delta(y) , \tag{12.52}$$

with the boundary conditions

$$\Phi_M|_{y=\pm h} = 0 , \quad \frac{\partial \Phi_M}{\partial n}|_{x=\pm b(z)} = 0 , \tag{12.53}$$

where $\mathbf{n} = (\pm 1, 0, -b')$ is the unit vector normal to the wall.

A schematic picture of the currents excited in the walls of the collimator is shown in Fig. 12.8.

Unlike the case of the electric field, the width h must be kept finite, in spite of the fact that the inequality $h \gg b(z)$ does not restrict it. The physical

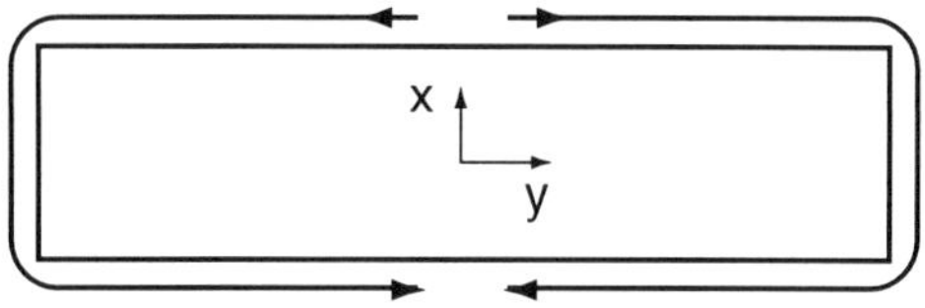

Figure 12.8: Schematic picture of the currents in a collimator.

reason for that are induced currents which flow through the walls at $y = \pm h$, creating a magnetic field proportional to h.

The solution of Eq. (12.52) is again obtained by iteration $\Phi_M = \Phi_M^{(0)} + \Phi_M^{(1)}$ where

$$\Phi_M^{(0)} (x, y, x_b; b(z)) = \ln \Big[16 \left(\sinh^2 \pi y/4b(z) + \cos^2 \pi (x + x_b)/4b(z) \right) \left(\sinh^2 \pi y/4b(z) + \sin^2 \pi (x - x_b)/4b(z) \right) \Big] - \frac{\pi h}{b(z)} , \qquad (12.54)$$

and where

$$\begin{aligned} &\Phi_M^{(1)}(x, y, x_b; z) = \\ &- \frac{1}{4\pi} \int_{-b(z)}^{b(z)} dx' \int_{-h}^{h} dy' \frac{\partial^2}{\partial z^2} \Phi_M^{(0)} \left(x', y', x_b, b(z) \right) \Phi_M^{(0)} \left(x, y - y', x', b(z) \right) \\ &- \frac{b'(z)}{4\pi} \int_{-h}^{h} dy' \frac{\partial}{\partial z} \Phi_M^{(0)} \left(x', y', x_b, b(z) \right) \Phi_M^{(0)} \left(x, y - y', b(z); b(z) \right) \\ &- \frac{b'(z)}{4\pi} \int_{-h}^{h} dy' \frac{\partial}{\partial z} \Phi_M^{(0)} \left(x', y', x_b; b(z) \right) \Phi_M^{(0)} \left(x, y - y', -b(z); b(z) \right) . \qquad (12.55) \end{aligned}$$

The transverse impedance of a smooth collimator can be found now by substituting expressions for the electric and magnetic fields in Eq. (12.44). General rather complicated formula for this quantity can be found in the paper [Stupakov96]. We give here only a particular result for $x, x_b \ll b$ and $y = 0$:

$$Z_x = j \frac{Z_0 h}{2} \int dz \frac{(b')^2}{b^3} . \qquad (12.56)$$

The impedance of a flat collimator turns out to be larger by a factor h/b than the impedance of an axially symmetric collimator of radius d.

12.5 Kickers and Pickups

There is an intimate relation between the properties of the *kickers*, i.e. the devices employed to influence the beam by applying external fields and *pickups*, i.e., the devices used to obtain from the beam a signal, which contains information on its properties [Goldberg92]. Namely, in principle, the same device can be used in both way: as a kicker (longitudinal or transverse) by applying to its electrodes an external voltage, or as a pickup by detecting and analysing the voltage induced on its electrodes by the beam. The mathematical foundation of this property is *the Lorentz Reciprocity Theorem* [Collin50]. However, not every electrode system may be employed in both roles.

Certain type of pickups play especially important role in accelerators. They are known as Beam Position Monitors (*BPM*s). They are employed in accelerators to measure the position of the beam centroid in respect to the vacuum chamber walls. There are a number of non-intercepting devices used for this goal: *split plates* or *split cylinders*, *strip-lines* and more conventional *button BPM*s. The latter represent several (usually four) button pickup electrodes which are connected by coaxial cables to registering electronics and are housed in a special box.

As any other non-uniform element of a vacuum chamber, the pickups create some disturbance of the *EM* field of the beam producing thus wake fields and impedances. An accurate design of these devices is crucial for high performance accelerators since their number in a large accelerator is measured in hundreds and even thousands.

For the precise measurement of the beam position, a *BPM* needs to be calibrated by bench measurements in which the field of the beam is simulated by an rf antenna. However, the *EM* field produced by a wire is somewhat different from that induced by a beam. It is necessary to evaluate the *EM* field induced by a beam in a *BPM* both to find its sensitivity and to be able to estimate its influence on the beam stability.

The electric field of an ultra-relativistic beam is Lorentz contracted in the longitudinal direction and can be considered as produced by a line charge of infinite length. That allows one to treat it as a simple electrostatic problem.

12.5.1 Button BPM

The reciprocity theorem discussed in chapter 2 permits calculating the coupling impedance $Z_{||}$ of a button electrode if the "beam impedance" Z_p of the button, i.e., the ratio of the voltage V_0 induced by the external current I_e on the button capacity, is known [Jacob90].

The beam impedance of the button in turn is related to the impedance of the external circuit Z_e :

$$Z_e = Z_p \frac{I_b}{I_e} , \tag{12.57}$$

where the load resistance R_L and the button capacitance C define the circuit impedance $Z_e = R_L/(1 + j\omega CR_L)$. The corresponding equivalent circuit is shown schematically in Fig. 12.9. In ref. [Jacob90], $Z_p = V_0/I_e$ has been measured with the help of a wire excited with high frequency..

Hence

$$\frac{I_e}{I_b} = \frac{Z_p}{R_L}(1 + j\omega CR_L) \approx j\omega CZ_p . \tag{12.58}$$

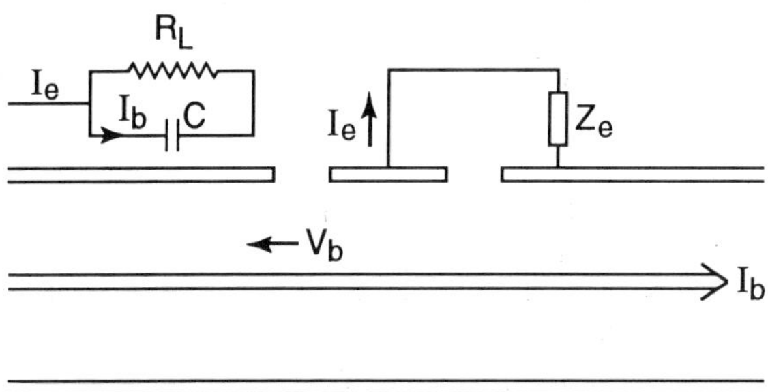

Figure 12.9: Schematic geometry and equivalent circuit for a button *BPM*.

From the definitions of the coupling impedance $Z_{||} = V_b/I_b$, the gap voltage on the button is $V_e = V_b I_e/I_b$, and from the reciprocity theorem Eq. (2.49) follows $Z_{||} = Z_e(I_e/I_b)^2$. Substituting here Eq. (12.58) yields:

$$Z_{||} = j\omega CZ_p^2 . \tag{12.59}$$

Fig. 12.10, taken from the work [Jacob90], reproduces the real and imaginary parts of the longitudinal coupling impedance of a button *BPM* in two frequency regions. Each pickup is a coaxial structure having a 7.6 mm diameter button flush with the wall of the vacuum chamber and connected to a 50 Ω cable. The beam impedance of a button and the button capacitance were measured to be $Z_p = 0.05\Omega$ and $C = 20$ pF, respectively.

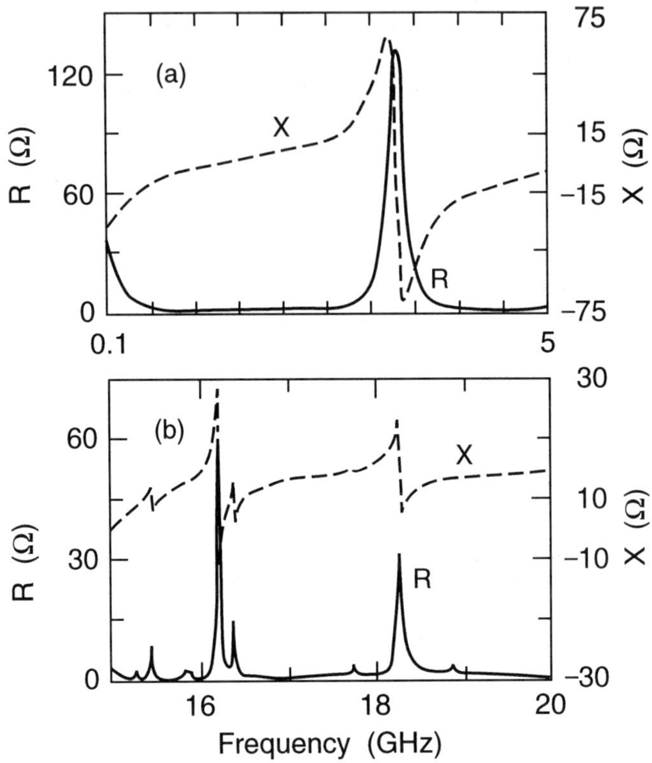

Figure 12.10: Longitudinal coupling impedance of button *BPM*.

12.5.2 Split Cylinder BPM

A common geometry for a beam position monitor at moderately low frequencies is the capacitive pickup consisting of a diagonally-divided cylinder that encloses the beam trajectory. A simplified system is sketched in Fig. 12.11 [Lambertson90]. The Fourier harmonic of the induced current $\tilde{I}$ on each of the electrodes is determined by the rate of change of the induced charge. The

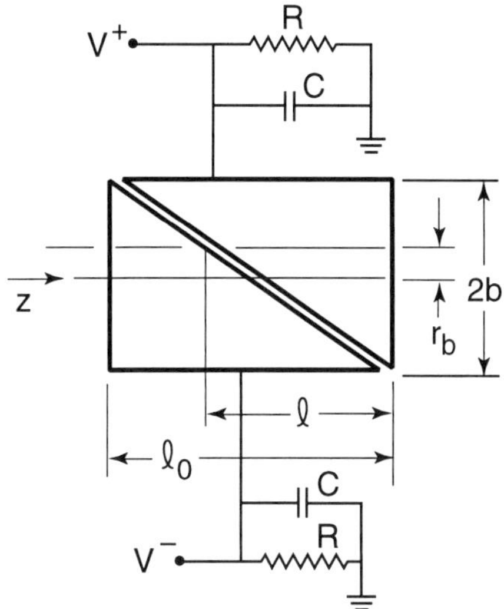

Figure 12.11: Equivalent circuit for a split cylinder.

voltage developed on each half is:

$$\begin{aligned} \widetilde{V}_p^{\pm}(r_b) &= j\frac{\omega l^{\pm}(r_b)}{\beta c}, \frac{R_L\tilde{I}}{1+j\omega CR_L}, \\ l^{\pm}(r_b) &= \frac{l_0}{2}(1 \pm \frac{r_b}{b}) . \end{aligned} \tag{12.60}$$

To find the longitudinal impedance $Z_{||}$, note that for zero beam displacement, the longitudinal field is concentrated at the ends of the electrode, i.e. at $z = \mp l_0/2$. When $\omega l_0/2\beta c \ll 1$ that yields:

$$Z_{||} = \frac{1}{2}\left(\frac{\omega l_0}{\beta c}\right)^2 \frac{R_L}{1+j\omega CR_L} . \tag{12.61}$$

To estimate the transverse impedance $Z_\perp$ in the r-direction, we assume that the transverse electric field averaged over the length l_0 is:

$$\langle \widetilde{E}_\perp \rangle = -\frac{1}{2b}\left[\widetilde{V}_p^{+}(r_b) - \widetilde{V}_p^{-}(-r_b)\right] = -j\frac{r_b}{2b^2}\frac{\omega l_0^2}{\beta c}\frac{R_L\tilde{I}}{1+j\omega CR_L} . \tag{12.62}$$

That gives

$$Z_\perp = \frac{1}{2b^2}\frac{\omega l_0^2}{\beta^2 c}\frac{R_L}{1+j\omega CR_L} . \tag{12.63}$$

An alternative way to evaluate the transverse impedance is to calculate $\widetilde{V}$ as function of r_b and use the Panofsky-Wenzel theorem. The impedance thus obtained coincides with the result Eq. (12.63).

12.5.3 Coaxial Strip-line BPM

A *coaxial strip-line* can be visualised as a section of the length l of a transmission line with a characteristic impedance $Z_T = \sqrt{L/C}$, where L and C are its inductance and capacitance per unit length [Shafer85]. The inner conductor (the pick up electrode) of the strip-line covers a fraction ϕ out of the circumference 2π occupied by the outer conductor (the vacuum chamber wall). The induced current on the strip is reduced by the ratio $\phi/2\pi$.

In reality though, the field between the strip and the outer wall is somewhat different due to its spread at the strip edges. To take this into account, ϕ should be replaced by effective angle $\phi_{eff} = \phi + \Delta\phi$. Consider a strip-line with the outer radius of the inner conductor b and the inner radius of outer conductor $b + h$. The correction angle $\Delta\phi$ is found in work [Papis91]:

$$\Delta\phi = 2\pi \frac{h}{2b + h} \ . \tag{12.64}$$

In the **time domain**, the longitudinal distribution $I(t)$ of the induced signal for relativistic beam is coincident with that of the bunch itself. A signal induced by a bunch on the strip-line will propagate at the relative velocity $\beta_s = 1/\sqrt{\mu\varepsilon}$ where μ and ε are the permeability and permittivity of the media. As the beam pass the upstream gap of impedance $Z_T/2$ between the strip-line and the wall, the equal induced currents flow downstream on the inside and outside of the strip-line. They produce the voltage on the upstream port:

$$V_u(t) = \frac{Z_T}{2}\frac{\phi_{eff}}{2\pi}\left[I(t) - I\left(t - \frac{l}{\beta_s c} - \frac{l}{\beta_b c}\right)\right] \ . \tag{12.65}$$

Similar expression for the downstream gap is:

$$V_d(t) = \frac{Z_T}{2}\frac{\phi_{eff}}{2\pi}\left[I\left(t - \frac{l}{\beta_s c}\right) - I\left(t - \frac{l}{\beta_b c}\right)\right] \ . \tag{12.66}$$

These two signals being different, the strip-line can be used as *directional coupler*.

Note, that if the beam and the signal velocities are the same there is complete cancellation of the voltage at the downstream port. For this reason the signal from the upstream port is used for the beam detection and measurements.

In the **frequency domain** the Fourier component of the bunch current is:

$$\tilde{I}_b(\omega) = I_0 \sin \omega t \ , \tag{12.67}$$

with the phase of the current measured from the midpoint of the strip-line. Then the signal induced on the upstream gap is:

$$\widetilde{V}_u(\omega) = Z_T I_0 \frac{\phi_{eff}}{2\pi} \sin\left[\frac{\omega l}{2c}\left(\frac{1}{\beta_b}+\frac{1}{\beta_s}\right)\right] \sin\left[\omega(t-\tau)+\frac{\pi}{2}\right] . \tag{12.68}$$

Here, τ is the time delay of the signal from the midpoint of the strip-line. The phase shift $\pi/2$ appears since the output signal of the strip-line is essentially a differential of the input signal.

When both the beam and signal velocities are nearly the speed of light, the output power per electrode for a centred beam is

$$P(\omega) = \frac{Z_T}{2} I_0^2 \left(\frac{\phi_{eff}}{2\pi}\right)^2 \sin^2\left(\frac{\omega l}{2}\right) . \tag{12.69}$$

The power produced on a strip-line represents the real part of the longitudinal coupling impedance $Z_{||}$. For a horizontal pair of strip-lines excited by a beam displaced by x in horizontal and y in vertical planes, then [Shafer85]

$$\mathrm{Re} Z_{||}(\omega) = 2Z_T(\frac{\phi_{eff}}{2\pi})^2 G(x,y) \sin^2\left(\frac{\omega l}{2}\right) , \tag{12.70}$$

where the geometric factor $G(x,y)$ is

$$G(x,y) = 1 + \frac{x^2}{b^2}\left[\left(\frac{4}{\phi_{eff}}\right)^2 \sin^2\frac{\phi_{eff}}{2} + \frac{4}{\phi_{eff}} \sin\frac{\phi_{eff}}{2}\right] - \frac{y^2}{b^2}\frac{4}{\phi_{eff}} \sin\frac{\phi_{eff}}{2} . \tag{12.71}$$

The full impedance may be recovered from the real part using the Kramers-Kronig relation:

$$Z_{||}(\omega) = 2Z_T(\frac{\phi_{eff}}{2\pi})^2 G(x,y)\left[\sin^2\frac{\omega l}{2} + j \sin\frac{\omega l}{2}\cos\frac{\omega l}{2}\right] . \tag{12.72}$$

The impedances and wakes of several other accelerator components have also been calculated in the literature, but we have discussed the most important ones here. For structures not discussed here, it will in general be easiest to first compute the wake fields of a simplified geometry - e.g./ a rotationally symmetric approximation to the real structure -with a numerical code, some of which are presented in Appendix 11A. In some cases it may also be justified to estimate the parameter dependence analytically. The authors hope that application of the techniques discussed in this book will be helpful in solving such problems.

Abbreviations for Bibliography

(a) Conference Proceedings

EPAC: European Particle Accelerator Conference:

1-st Rome, Italy 1988
2-nd Nice, France 1990 (editions Frontieres)
3-rd Berlin, Germany 1992 (editions Frontieres)
4-th London, GB 1994 (World Scientific)
5-th Sitges-Barcelona, Spain 1996

PAC: Particle Acclerator Conference (US),
published as IEEE Transactions on Nuclear Science

Washington, D.C. 1965 (NS-12)
Washington, D.C. 1967 (NS-14)
Washington, D.C. 1969 (NS-16)
Chicago, IL 1971 (NS-18)
San Francisco,CA 1973 (NS-20)
Washington, D.C. 1975 (NS-22)
Chicago,IL 1977 (NS-24)
San Francisco,ca 1979 (NS-26)
Washington, D.C. 1981 (NS-28)
Santa Fe, N.M. 1983 (NS-30)
Vancouver, B.C. 1985 (NS-32)
Washington, D.C. 1987 (NS-34)
Chicago, IL 1989 (Proc. IEEE PAC)
San Francisco,CA 1991 (Proc. IEEE PAC)
Washington, D.C. 1993 (Proc. of the 1993 PAC)
Washington, D.C. 1995 (Proc. of the 1995 PAC and ICHEA)

ICHEA: International Conference on High Energy Accelerators:

4-th Dubna, USSR 1963 (USAEC)
5-th Frascati, Italy 1965 (CNEN)
6-th Cambridge, MA, USA 1967 (CEAL 1200)
7-th Erevan, USSR 1969 (ErIP)
8-th Geneva, Switzerland 1971 (CERN),
9-th Stanford, CA USA 1974 (SLAC)
10-th Protvino, USSR 1977 (INP)
11-th Geneva, Switzerland (CERN) 1980 (Birkhaeuser Verlag)
12-th Batavia, IL USA 1983 (FNAL)
13-th Novosibirsk, USSR 1986 (IHEP)
14-th Tsukuba, Japan 1989 (KEK) (Part. Accel. vol 26-33)
15-th Hamburg, Germany 1992 (World Scientific)

(b) Accelerator Schools

CERN Accelerator Schools (CAS)
(published as CERN Yellow Reports):

CERN 77-13 Erice, Sicily 1976 (International)
CERN 85-19,I/I Gif/Yvette, F 1984 (general)
CERN 87-03,I/II Oxford, GB 1985 (advanced)
CERN 87-10 Aarhus, DK 1986 (general)
CERN 89-01 Berlin, D 1987 (advanced)
CERN 90-04 Uppsala, S 1989 (advanced)
CERN 89-05 Salamanca, E 1988 (general)
CERN 91-04 Juelich, D 1990 (general)
CERN 92-0 Noordwijk, N 1991 (advanced)
CERN 94-01,I/II Yvaeskylae, SF 1992 (general)
CERN 95-06 Rhodos, GR 1993 (advanced)

US Particle Accelerator Schools (USPAS)
(published as AIP Conference Proceedings series 28):

AIP 87 (1982) USPAS 1981 FNAL
AIP 105 (1984) USPAS 1982 Stanford
AIP 127 (1985) USPAS 1983 BNL-SUNY
AIP 153 (1987) USPAS 1984 FNAL/1985 SLAC
AIP 184 (1989) USPAS 1987 FNAL/1988 Cornell
AIP 247 (1991) joint 1988 Sardinia
AIP 249 (1992) USPAS 1989 FNAL/1990 Hilton Head, S.C.
AIP 296 (1988) joint 1986 South Padre Island, TX
AIP 343 (1991) joint 1989 Capri

(c) Abbreviations for Journals:

IEEE-AP: IEEE Transactions on Antennas and Propagation
IEEE-MTT: IEEE Transactions on Microwave Techniques
IEEE-NS: IEEE Transactions on Nuclear Science
J.Appl.Phys.: Journal of Applied Physics
NIM-A: Nuclear Instruments and Methods, series A
Part.Accel.: Particle Accelerators (Gordon and Breach)
Phys.Rev.D: Physical Review, series D
Phys.Rev.E: Physical Review, series E
PRL: Physical Review Letters (American Physical Society)

(d) Abbreviations for Reports:

BINP Beijing Institute for Nuclear Physics, Beijing, China
CEBAF Thomas Jefferson National Laboratory, Newport Beach, VA, USA
CERN Centre Europeen de Recherche Nucleaire, Geneva, Switzerland
CESR Cornell Electron Storage Ring, Ithaca, NY, USA
FNAL Fermi National Laboratory, Batatvia, IL, USA
IHEP Institut for High Energy Physics, Novosibirsk, Russia
INP Institut for Nuclear Physics, Serpukhov, Russia
KEK High Energy Physics Laboratory, Tsukuba, Japan
LANL Los Alamos National Laboratory, Los Alamos, NM, USA
LBL Lawrence Berkeley Laboratory, Berkeley, CA, USA
SLAC Stanford Linear Accelerator Center, Stanford, CA, USA

Bibliography

[Abramowitz65] M. Abramowitz and I. Stegun: *Handbook of Mathematical Functions*. Dover, 1965.

[Albani74] H. Albani and M. Bernard: "*A Numerical Method on the Discretization of Maxwell's Equations in Integral Form*". IEEE-MTT 22, pp. 446–449, 1974.

[Ashcroft76] N. Ashcroft and S. Mermin: *Solid State Physics*. Holt, Rinehart and Winston, 1976.

[Baarts94] T. Baarts and W. Chou: "*Computer Code BPERM*". Technical Report 0035, SSC Report SSCL-MAN, 1994.

[Balakin78] V. Balakin, A. Novokhatskii, et al.: "*Beam Dynamics of Colliding Electron-Positron Beams*". In 6-th All-Union Conference on Charged-Particle Accelerators (USSR), 1978.

[Balakin83] V. Balakin and A. Novokhatskii: "*VLEPP: Longitudinal Beam Dynamics*". In 12-th ICHEA, volume 2, pp. 117–8, 1983.

[Balbekov93] V. Balbekov: "*Equivalent Circuit Method for the Calculation of the Longitudinal Impedance of a Circular Accelerator*". Report 93-55, IHEP Serpukhov, 1993.

[Bane80a] K. Bane: "*The Transverse Cavity Impedance in LEP*". Technical report, CERN/ISR-TH/80-48, 1980.

[Bane80b] K. Bane and P. Wilson: "*Longitudinal and Transverse Wake Potentials in SLAC*". In 11-th ICHEA, pp. 592–6, 1980.

[Bane84] K. Bane, P. Wilson, and T. Weiland: "*Wakefields and Wakefield Acceleration*". Technical Report 3528, SLAC-PUB, 1984.

[Bane89] K. Bane: "*Bunch Lengthening in the SLC Damping Rings*". In ANL/APS/TM-5: Impedance and Bunch Instability Workshop, Argonne IL, pp. 191–241, 1989.

[Bane90a] K. Bane: "*Bunch Lengthening in the SLC Damping Rings*". Technical Report 5177, SLAC-PUB, 1990.

[Bane90b] K. Bane and M. Sands: "*Wakefields of Very Short Bunches in an Accelerating Cavity*". Part. Accel. 25, pp. 73–95, 1990.

[Bane91] K. Bane: "*The Short-Range Resistive Wall Effect*". Technical report, SLAC/AP-87, 1991.

[Bane92] K. Bane and R. Gluckstern: "*The Transverse Wakefield of a Detuned X-Band Accelerator Structure*". Technical Report 5783, SLAC-PUB, 1992.

[Bane95] K. Bane and M. Sands: "*The Short-Range Resistive Wall Wake Fields*". In Micro Bunches Workshop at BNL, pp. 131-49, 1995.

[Bassetti86a] M. Bassetti: "*Analytical Formula for the Centrifugal Space Charge Effect*". Report 86-13, CERN/LEP, 1986.

[Bassetti86b] M. Bassetti and D. Brandt: "*Transverse Electro-Magnetic Forces in Circular Trajectory*". Report 86-04, CERN/LEP, 1986.

[Batchelor92] K. Batchelor: "*BNL Accelerator Test Facility*". In 15-th ICHEA, 1992. also Report BNL-44784,1992.

[Bennett53] H. Bennett: "*The Transmission Characteristics of 2-D Lattices*". J. Appl. Phys. 24, pp. 785–810, 1953.

[Berg95] J. S. Berg: "*Observations Involving Broadband Impedance Modelling*". Technical Report 6964, SLAC-PUB, 1995.

[Besnier78] G. Besnier: *Longitudinal Stability.* PhD thesis, Rennes, 1978.

[Bethe44] H. Bethe: "*Theory of Diffraction by Small Holes*". Phys. Rev. 66, pp. 163–182, 1944.

[Birdsall53] J. Birdsall, Whinnery, et al.: "*Theory of Resistive Wall Amplifier*". Proc. IRE 41, p. 865, 1953.

[Bisognano88] J. Bisognano, S. Heifets, and B.Yunn: "*The Loss Parameters for Very Short Bunches*". In 1-st EPAC, Rome, pp. 708–10, 1988. also CEBAF Report PR-88-005.

[Bisognano89] J. Bisognano: "*On the Independence of the Longitudinal Wakefield on Direction of Transit through a Cavity*". Tn-109, CEBAF, 1989.

[Borer79] J. Borer, G. Guignard, et al.: "*Information from Beam Response to Longitudinal and Transverse Excitation*". In IEEE-NS 24, pp. 3405–7, 1979.

[Brandt82] D. Brandt and B. Zotter: "*Calculation of the Wakefield with the Optical Resonator Model*". Technical report, CERN-ISR/TH 82-13, 1982. also CERN-LEP Note 388.

[Brandt85] D. Brandt and B. Zotter: "*SIMTRAC Simulation Tracking Code*". Report 85-44, CERN/ISR-TH, 1985.

[Brandt89] D. Brandt, K. Cornelis, et al.: "*Measurements of the Impedance of LEP*". LEP Commissioning Notes 6 and 21, CERN/SL, 1989.

[Brandt92] D. Brandt, K. Cornelis, et al.: "*LEP Impedance - Summary of Recent Measurements*". LEP Performance Note 75, CERN/SL, 1992.

[Brown55] J. Brown and W. Jackson: "*The Relative Permittivity of Tetragonal Arrays of Perfectly Conducting Thin Discs*". Proc. IEEE-B 102, pp. 37–42, 1955.

[Burov90] A. Burov and A. Novokhatskii: "*Device for Bunch Self-focusing*". Preprint 90-28, INP-Novosibirsk, 1990.

[Caspers85] F. Caspers: "*Beam Impedance Measurement by the Wire Method Using Synthetic Pulse Technique*". IEEE-NS 32, pp. 1914–16, 1985.

[Caspers88] F. Caspers: "*Beam Impedance Measurement by the Wire Method Using the Coaxial Wire Method*". In ESRF Workshop on Impedance and Current Limitations, Section 2.1, 1988.

[Caspers92] F. Caspers et al.: "*Wireless Impedance Measurement and Fault Location*". In 3-rd EPAC, Berlin, pp. 889–91, 1992. CERN Report PS-92-30.

[Caspers95] F. Caspers and T. Scholz: "*Measurement of trapped modes in perforated waveguides*". Part. Accel., 51:251, 1995.

[Chan87] K. Chan and R. Schweinfurth: "*Beam Energy Spread Induced by Beampipe Steps*". Technical Report ATN-87-24, LANL-AT-6, 1987.

[Chao75] A. Chao and P. Morton: "*EM Fields between two Infinite Conducting Plates Produced by a Point Charge*". Technical Report 105, SLAC-PEP-Note, 1975.

[Chao76] A. Chao and J. Gareyte: "*Scaling Law for Bunch Lengthening in SPEAR II*". Technical Report 197, SLAC-SPEAR-Note, 1976.

[Chao82] A. Chao: "*Coherent Instabilities of a Relativistic Bunched Beam*". Technical Report 2946, SLAC-PUB, 1982.

[Chao93] A. Chao: *Physics of Collective Beam Instabilities in High Energy Accelerators.* John Wiley, 1993.

[Chin88a] Y. Chin: "*ABCI - a Comnputer Code*". Report 88-3, CERN-SL, 1988.

[Chin88b] Y. Chin: "*ABCI User's Guide*". Report 88-3, CERN-LEP/TH, 1988.

[Collin50] R. Collin: *Foundations for Microwave Engineering.* McGraw - Hill, N.Y., 1950.

[de Kronig28] R. L. de Kronig and H. A. Kramers: "*Zur Theorie der Absorption und Dispersion in den Roentgen Spektren*". Z. Phys., vol.52, pp.174-9, 1928. Collected Scientific Papers de H.A.Kramers, North Holland, Amsterdam 1956, pp. 382-7.

[Derbenev96] Y. Derbenev and V. Shiltsev: "*Transverse Effects of Microbunch Radiative Interaction*". Technical report, FERMILAB-TN-1974, 1996. also SLAC-PUB-7181.

[Dikansky65] N. Dikansky and A. Skrinski: "*Resistive Wall Effect*". Report, INP-Novosibirsk, 1965.

[Dnestrovsky59] Y. Dnestrovsky and D. Kostamarov: "*Radiation of a Modulated Beam of Charged Particles in Passing through a Round Opening in a Flat Screen*". Soviet Physics 4 (Dokl. Akad. Nauk SSSR 124, pp. 792 and 1026), p. 132 and 158, 1959.

[Dôme85] G. Dôme: "*Wake Potentials of a Relativistic Point Charge Crossing a Beam-Pipe Gap*". PAC 85, IEEE-NS 32, pp. 2531–34, 1985.

[Dôme91] G. Dôme, E. Gianfelice, et al.: "*Longitudinal Coupling Impedance of a Circular Iris*". Il Nuovo Cimento, p. 1241, 1991.

[Dôme92] G. Dôme, L. Palumbo, et al.: "*The Frequency Behaviour of the Longitudinal Coupling Impedance of a Periodic Array of Diaphragms*". In 3-rd EPAC, Berlin, pp. 925–27, 1992.

[Erdelyi53] A. Erdelyi et al.: *Higher Transcendental Functions.* McGraw-Hill, NY, 1953.

[Faltens71] A. Faltens, E. Hartwig, D. Moehl, and A. Sessler: "*Analog Method for Measuring Longitudinal Coupling Impedance*". In 8-th ICHEA, pp. 338–41, 1971.

[Faltens75] A. Faltens and L. J. Laslett: "'*An Estimate of Limits to the Longitudinal Coupling Impedance*". Technical Report 20550, BNL, 1975.

[Friedmann79] M. Friedmann: "*Autoacceleration of an Intense Relativistic Electron Beam*". Technical report, NRL report AD-769-46, 1979.

[Ginzton57] E. Ginzton: *Microwave Measurements.* McGraw-Hill, 1957.

[Gluckstern85] R. Gluckstern and F. Neri: "*Longitudinal Coupling Impedance for a Beam Pipe with a Cavity*". PAC 85, IEEE-NS 32, pp. 2403–4, 1985.

[Gluckstern88] R. Gluckstern and B. Zotter: "*The Coupling Impedance of Asymmetric Cavities*". LEP-Note 613, CERN, 1988.

[Gluckstern89a] R. Gluckstern: "*High-Frequency Behaviour of Longitudinal Impedance for a Cavity of General Shape*". Phys. Rev. D39, pp. 2773–79, 1989.

[Gluckstern89b] R. Gluckstern: "*Longitudinal Impedance of a Periodic Structure at High Frequency*". Phys. Rev. D39, pp. 2780–83, 1989.

[Gluckstern89c] R. Gluckstern and F. Neri: "*Longitudinal Coupling Impedance of a Small Obstacle*". In PAC 89, IEEE-NS 34, pp. 1271–73, 1989.

[Gluckstern91] R.L. Gluckstern and J.A. Diamond: "*Penetration of Fields through a Circular Hole in a Wall of Finite Thickness*". IEEE- MTT 39, pp. 274–9, 1991.

[Gluckstern92a] R. Gluckstern: "*Coupling Impedance of a Many Holes in a Liner within a Beam Pipe*". Technical Report 92-05, CERN-SL, 1992.

[Gluckstern92b] R. Gluckstern: "*Coupling Impedance of a Single Hole in a Thick Wall Beam Pipe*". Technical Report 92-05, CERN-SL, 1992.

[Gluckstern92c] R. Gluckstern: "*Coupling Impedance of Holes in a Liner within a Beam Pipe*". Phys. Rev. A46, pp. 1106–12, 1992.

[Gluckstern93a] R. Gluckstern and B. Radak: "*Penetration of EM Fields Through an Elliptical Hole in a a Wall of Finite Thickness*". Technical report, Univ. of Maryland Report UMDP 93-112, 1993.

[Gluckstern93b] R. Gluckstern, J. van Zeijts, and B. Zotter: "*Coupling Impedance of Beam Pipes of General Cross Section*". Phys. Rev. E47, pp. 656–63, 1993.

[Gluckstern96] R. Gluckstern and B. Zotter: "*Energy Loss to Coaxial Vacuum Chambers in LEP and LHC*". Technical Report 96-56, CERN-SL, 1996.

[Goldberg92] D. Goldberg and G. Lambertson: "*Dynamic Devices: A Primer on Pickups and Kickers*". Technical report, LBL-31664, 1992. also AIP Conf. Proc. 1992.

[Gradshteyn80] I. Gradshteyn and I. Ryzhik: *Table of Integrals, Series, and Products.* Academic Press, 1980.

[Group90] Los Alamos Accel. Code Group: "*Computer Codes for Particle Accelerator Design and Analysis: a Compendium*". Technical Report 90-1766, LANL Report LA-UR, 1990.

[Haeff49] N. Haeff: "*Resistive Wall Amplifier*". Proc. IRE 37, p. 4, 1949.

[Hahn76] H. Hahn, C. Goldstein, and W. Bauer: "*Theory of Iris-Loaded Waveguides*". AEU 30, pp. 297–302, 1976.

[Hahn78] H. Hahn and F. Pedersen: "*Coaxial Wire Measurements of Longitudianl Coupling Impedance*". Technical Report 50870, BNL, 1978.

[Halbach76] K. Halbach and N. Holsinger: "*SUPERFISH - a Computer Program for Evaluation of RF Cavities with Cylindrical Symmetry*". Part. Accel., p. 213, 1976.

[Hara82] M. Hara et al.: "*Calculation of Electromagnetic Field by Finite Element Method*". Sci. Papers of the Institute of Physical and Chemical Research, vol. 75, pp. 143–175, 1982.

[Harrington68] R. Harrington: *Field Compuataion by Moments Method.* Macmillan, 1968.

[Heifets88a] S. Heifets: "*Analytic Study of High Frequency Impedance*". Technical Report 90-014, CEBAF-PR, 1988. Workshop on FEL's and Storage Rings, Dortmund, Germany.

[Heifets88b] S. Heifets and S. Kheifets: "*High-Frequency Limit of the Longitudinal Impedance of an Array of Cavities*". Report 4625, SLAC-PUB, 1988.

[Heifets89a] S. Heifets: "*Diffraction Model of the High-Frequency Impedance*". Phys. Rev. D 40, pp. 3097– 106, 1989.

[Heifets89b] S. Heifets and S. Kheifets: "*High-Frequency Limit of the Longitudinal Impedance of an Array of Cavities*". Phys. Rev. D 39, pp. 960–70, 1989.

[Heifets90a] S. Heifets: "*Determination of the High-Frequency Behavior of the Impedance from the Low-Frequency Data*". Technical Report 90-206, CEBAF-TN, 1990.

[Heifets90b] S. Heifets: "*On Directional Symmetry of Impedance*". Technical report, SLAC/AP-79, 1990.

[Heifets90c] S. Heifets: "*Perturbation Method for Calculation of Narrow-Band Impedance and Trapped Modes*". Part. Accel. 25, pp. 217–228, 1990.

[Heifets90d] S. Heifets and S. Kheifets: "*High-Frequency Limit of the Longitudinal Impedance*". Part. Accel., 25, pp. 61–72, 1990.

[Heifets91a] S. Heifets and S. Kheifets: "*Coupling Impedance in Modern Accelerators*". Review of Modern Physics, vol. 63, pp. 631–73, 1991.

[Heifets91b] S. Heifets and A. Michailichenko: "*On the Impedance due to Synchrotron Radiation*". In PAC 91, San Francisco, Vol.1, pp. 458–60, 1991. also SLAC/AP-83.

[Heifets92] S. Heifets: "*Broad-Band Impedances of the B-Factory*". Technical report, SLAC/AP-93, 1992.

[Heifets94] S. Heifets: "*Narrow-Band Resonator Impedance Model*". Technical report, presented at LBL Seminar, 1994.

[Heifets96] S. Heifets and G. Sabbi: "*Single Bunch Stability in LER of PEP II*". Technical report, SLAC/AP-104, 1996.

[Henke85] H. Henke: "*Point Charge Passing a Resonator with Beam Tubes*". Report 85-41, CERN-LEP-RF, 1985.

[Henke90] H. Henke and O. Napoly: "*Wake Fields between Two Parallel Resistive Plates*". In 2-nd EPAC,, Nice, pp. 1046–8, 1990. also CERN-LEP-RF/89-71, 1989.

[Henke95] H. Henke and S. Vaganian: "*The Panofsky-Wenzel Theorem and General Relations for the Wake Potential*". Part. Accel. 48, pp. 239–42, 1995.

[Henry91] O. Henry and O. Napoly: "*The Resistive-Pipe Wake Potentials for Short Bunches*". Part. Accel. 35, pp. 235–48, 1991. also CLIC Note 142 and SACLAY Report DPhN/STAS/91-R08.

[Hereward75] H. Hereward: "*Coupling Impedance of a Cross -Section Change for High Frequencies*". Report, CERN-ISR/DI/75-47, 1975.

[Hofmann77] A. Hofmann and B. Zotter: "*Measurement of Beam Stability and Coupling Impedance by RF Excitation*". IEEE-NS 24, pp. 1487–9, 1977.

[Hofmann79a] A. Hofmann et al.: "*A Computer Code for the Calculation of Beam Stability (BBI)*". IEEE-NS 26, pp. 3514–17, 1979.

[Hofmann79b] A. Hofmann and J. Maidment: "*Broadband Impedance Model*". Technical Report 168, CERN/LEP-Note, 1979.

[Hofmann88] A. Hofmann and B. Zotter: "*Improved Impedance Models for High-Energy Accelerators and Storage Rings*". Technical Report 88-51, CERN-LEP/TH, 1988.

[Hofmann89] A. Hofmann: "*Characteristics of Synchrotron Radiation*". Technical Report DI/89-55, CERN LEP report, 1989.

[Ishi91] K. Ishi et al.: "*Spectrum of Coherent Synchrotron Radiation in the Far-Infrared Region*". Phys. Rev. A 43, No. 10, pp. 5597–604, 1991.

[Itoh89] T. Itoh: *Numerical Techniques for Microwave and Millimeter-Wave Passive Structures.* John Wiley, 1989.

[Jackson75] J. D. Jackson: *Classical Electrodynamics.* John Wiley, 1975.

[Jacob90] A. Jacob and G. R. Lambertson: "*Impedance Measurements on Button Electrodes*". Technical report, LBL-25955, 1990.

[Johnson87] C. Johnson: *Numerical Solution of Partial Differential Equations by the Finite Element Method.* Cambridge Univ. Press, 1987.

[Juergens92] T. Juergens, A. Taflove, et al.: "*Finite-Difference Time-Domain Modeling of Curved Surfaces*". IEEE- AP 40, pp. 357–66, 1992.

[Juergens94] T. Juergens, F. Harfoush, and G. Saewaert: "*XWAKE*". In 4-th EPAC, pp. 1318–20, 1994.

[Karantzoulis91] E. Karantzoulis: "*An Overview on Impedances and Impedance Measureing Methods for Accelerators*". Note, Sincrotrone Trieste ST-M-91-1, 1991.

[Keil70] E. Keil and B. Zotter: "*Parallel Resonator*". Note, CERN-BEIC, 1970.

[Keil72a] E. Keil: "*Diffraction Radiation of Charged Rings Moving in a Corrugated Cylindrical Pipe*". NIM-A 100, pp. 419– 42, 1972.

[Keil72b] E. Keil and B. Zotter: "*Longitudinal Stability in a Corrugated Resistive Vacuum Chamber*". Part. Accel. 3, pp. 11–20, 1972.

[Kheifets81] S. Kheifets: "*Electromagnetic Fields in an Axial Symmetric Waveguide with Variable Cross Section*". IEEE- MTT-29, pp. 222–9, 1981.

[Kheifets85a] S. Kheifets and P. M. Gygi: "*IMPASS- a Computer Program for Calculations of Impedances of Periodic Axially Symmetric Smooth Structures (Bellows)*". IEEE NS-32, pp. 2338–40, 1985.

[Kheifets85b] S. Kheifets, L. Palumbo, and V. Vaccaro: "*Longitudinal Impedance of a Semi-Infinite Circular Waveguide*". Technical Report 85-23, CERN Report LEP/TH, 1985.

[Kheifets86a] S. Kheifets and S. Heifets: "*Radiation of a Charge in a Perfectly Conducting Cylindrical Pipe with a Jump in its Cross Section*". Technical report, SLAC PUB Report 3965, 1986. also in 1986 Linear Accel. Conference Proc., SLAC-Report-303, Stanford 1986, pp. 493-5.

[Kheifets86b] S. Kheifets and B. Zotter: "*Longitudinal and Transverse Impedances of Bellows in the Low Frequency Range*". NIM A 243, pp. 13–27, 1986. CERN Report LEP/TH 85-27 (1985).

[Kheifets87] S. Kheifets: "*Longitudinal Impedance of Simple Cylindrically Symmetric Structures*". IEEE-MTT-35, pp. 753–760, 1987.

[Kheifets95] S. Kheifets, S. Heifets, and B. Woo: "*Electro magnetic Fields in a Detuned X-Band Accelerating Structure*". IEEE-MTT 43, pp. 1335–41, 1995.

[Klatt86] R. Klatt, T. Weiland, et al.: "*MAFIA – a Three-Dimensional Electromagnetic CAD System for Magnets, RF Structures, and Transient Wake-Field Calculations*". Technical report, Proc. Int. Linear Accel. Conf. SLAC Report SLAC-303, 1986.

[Kogelnik68] H. Kogelnik and R. Li: "*Optical Resonator*". Bell System Techn. Journal, p. 222, 1968.

[Korenev71] I. Korenev and L. Yudin: "*Stability of Electron Ring Beam in an Infinite Cylindrical Pipe with Conductive Walls*". In 7-th ICHEA, pp. 597–601, 1971.

[Kramers27] H. A. Kramers: "*La Diffusion de la Lumiere par les Atomes*". Atti Congr. Internaz. Fisici, Com, vol.2, pp.545-57, 1927.

Collected Scientific Papers de H.A.Kramers, North Holland, Amsterdam 1956, pp. 333-46.

[Krinsky80] S. Krinsky: "*Resonant Impedance of Bellows above Cutoff*". In 11-th ICHEA, pp. 576–80, 1980.

[Krinsky81] S. Krinsky and R. Gluckstern: "*Higher-order Terms in a Perturbation Expansion for the Impedance of a Corrugated Wave Guide*". IEEE-NS 28, p. 2621, 1981.

[Kurennoy90] S. Kurennoy and S. Purtov: "*Method for Calculation of Longitudinal and Transverse Coupling Impedances of Periodic Structures in Accelerators*". Part. Accel. 34, pp. 155–168, 1990. also Report IHEP 89-1.

[Kurennoy92] S. S. Kurennoy: "*Coupling Impedance of Pumping Holes*". Part. Accel. 39, pp. 1–13, 1992. also CERN report SL/91-29.

[Kurennoy95a] S. Kurennoy, R. L. Gluckstern, and G. V. Stupakov: "*A General Approach for Calculating Coupling Impedances of Small Discontinuities*". Phys.Rev E 52, p. 4354 ff, 1995.

[Kurennoy95b] S. S. Kurennoy: "*Trapped modes in waveguides with many small discontinuities*". Phys. Rev., E51:2498, 1995. also Univ. Maryland report PP 95-10.

[Lambertson90] G. Lambertson: "*Beam Impedance of a Split Cylinder*". Technical report, LBL-2914, 1990.

[Landau60] L. D. Landau and E. M. Lifschitz: *Electrodynamics of Continuous Media.* Pergamon, 1960.

[Laslett63] J. Laslett: "*On Intensity Limitations Imposed by Transverse Space-Charge Effects in Circular Accelerators*". In Proceedings of Summer Study on Storage Rings, Accelerators and Experimentation at Super-High Energies, Brookhaven, N.Y., 1963.

[Laslett65] J. Laslett, K. Neil, and A. Sessler: "*Transverse Resistive Instabilities of Intense Coasting Beams in Particle Accelerators*". Rev. Scient.Instr. 36, p.436, 1965.

[Laslett70] L. J. Laslett: *"On the Focusing Effects Arising from the Self-fields of a Toroidal Beam"*. Selected Works of L. Jackson Laslett, v. 2, pp. 3.13–36 and 3.37–54, 1970.

[Lawson68] J. Lawson: *"Radiation from a Ring Charge Passing through a Resonator"*. Rutherford laboratory report, RHEL/M144, 1968.

[Lawson90] J. D. Lawson: *"Radiation from a Ring Charge Passing through a Resonator"*. Part. Accel. 25, pp. 107–12, 1990.

[Lee87] E. P. Lee: *"Cancellation of the Centrifugal Space Charge Force"*. Technical report, LBL-LBID-1376, 1987.

[Leontovich48] L. Leontovich: *"Boundary Conditions"*. Zh. Eksp. Teor. Fiz., 1948.

[Levine49] H. Levine and J. Schwinger: *"On the Radiation of Sound from an Unflanged Circular Pipe"*. Phys. Rev. 73, pp. 383–406, 1949.

[Lynch91] D.R. Lynch and K.D. Paulsen: *"Origin of Vector Parasites in Numerical Maxwell Solutions"*. IEEE-MTT-39, pp. 383–94, 1991.

[Marcuvitz51a] N. Marcuvitz: *Waveguide Handbook.* McGrawHill, 1951.

[Marcuvitz51b] N. Marcuvitz and J. Schwinger: *"On the Representation of the Electric and Magnetic Fields Produced by Currents and Discontinuities in Wave Guides"*. J. Appl. Phys. 22, pp. 806–19, 1951.

[Meincke97] O. Meincke, A. Wagner, and B.Zotter: *"New Wake Field and Bunch Lenghtening Codes"*. Technical report, CERN SL-Note-97-17, 1997.

[Meixner49] R. Meixner: *"Die Kantenbedingungen in der Theory der Beugung von Electro-Magnetischer Wellen an Vollkommen Leitenden Ebenen Schirmen "*. Annalen der Physik, vol.6, pp. 2–9, 1949.

[Meur93] G. Le Meur and F. Touze: "*PRIAM/ ANTIGONE: a 2D/3D package for accelerator design*". In 2-nd EPAC,, pp. 1321–3, 1993.

[Mitra63] R. Mitra: "*Relative Convergence of the Solution of a Doubly Infinite Set of Equations*". Jour. Research NBS 67 D, pp. 243–54, 1963.

[Mitra71] R. Mitra and W. Lee: *Theory of Guided Waves.* Mac Millan, 1971.

[Moehl71] D. Moehl and A. Sessler: "*The Use of RF Knockout for the Determination of the Characteristics of the Transverse Coherent Instabilities of an Intense Beam*". In 8-th ICHEA, pp. 334–7, 1971.

[Morse53] P. Morse and H. Feshbach: *Methods of Theoretical Physics.* McGraw-Hill, NY, Vol. 1, 1953.

[Morton93] P. Morton: "*Introduction to Impedance for Short Relativistic Bunches*". Technical report, SLAC Note AAS 78, 1993.

[Murphy95] J. B. Murphy, S. Krinsky, and R. L. Gluckstern: "*Longitudinal Wakefield for Synchrotron Radiation*". In presented at the 1995 Int. Conf. on High-Energy Accelerator, Dulles, 1995.

[Napoly91] O. Napoly: "*The Wake Potentials from the Fields on the Cavity Boundary*". Part. Accel. 36, pp. 15–23, 1991.

[Napoly93] O. Napoly, Y. Chin, and B. Zotter: "*A Generalized Method for Calculating Wake Potentials*". NIM A 344, p. 255, 1993.

[Nassibian79] G. Nassibian and F. Sacherer: "*A Method for Measuring Transverse Coupling Impedance*". NIM 159, pp. 21–27, 1979. also Report CERN-ISR-TH 77-61.

[Neil65] K. Neil and A. Sessler: "*Longitudinal Resistive Instabilities of Intense Coasting Beams in Particle Accelerators*". Rev. Scient.Instr. 36, p.429-35, 1965.

[Nelson93] E. M. Nelson: "*High Accuracy Electromagnetic Field Solvers for Cylindrical Waveguides and Axisymmetric Structures Using the Finite Element Method*". Technical report, SLAC Report SLAC-431, 1993.

[Nelson96] E. Nelson: "*3-D Computer Code ARGUS*". Report, LANL, 1996.

[Ng89] K. Y. Ng and R. L. Warnock: "*Reactive Impedance of a Smooth Toroidal Chamber below the Resonance Region*". Phys. Rev. D 40, No. 1, pp. 231–43, 1989.

[Ng90] K. Y. Ng: "*Resonant Impedance in a Toroidal Beam Pipe*". Part. Accel. 25, No. 2-4, pp. 153– 81, 1990.

[Nodvick54] J. S. Nodvick and D.S. Saxon: "*Suppression of Coherent Radiation by Electrons in a Synchrotron*". Phys. Rev. 96 , No. 1, pp. 180–4, 1954.

[Novokhatskii88] A. V. Novokhatskii: "*On the Estimation of the Wake Potential for an Ultrarelativistic Charge in an Accelerating Structure*". Preprint 88-39, INP-Novosibirsk, 1988.

[Nys86] V. Nys: "*Computer Simulation of the Longitudinal Motion Including the Wake Field force, Using a Hermite Polynomial Expansion*". Report 86-34, CERN-LEP-TH, 1986.

[Nys87] V. Nys: "*Computer Simulation of a Transverse Single Bunch Instability*". Report 86-35, CERN/LEP-TH, 1987.

[Palmer89] R. Palmer: "*A Diffractional Analysis of Longitudinal Wakes of Short Bunches*". Report 44, SLAC/AAS, 1989.

[Palumbo84] L. Palumbo: "*Radiation of a Point Charge*". Report 84-4, CERN/LEP-TH, 1984.

[Palumbo87] L. Palumbo and V. Vaccaro: "*Wake Fields, Impedances and Green's Function*". Report 87-03, CERN-CAS, 1987.

[Palumbo94] L. Palumbo, V. Vaccaro, and M. Zobov: "*Wake Fields and Impedances*". Report 94-041, INFN-LNF, 1994.

[Panofsky56] W. Panofsky and W. Wenzel: "*Transverse Deflection of Charged Particles in Radiofrequency Fields*". Rev. Sci. Instr., p. 967, 1956.

[Papiernik74] A. Papiernik, M. Chatard-Moulin, and B. Jecko: "*Fast Beam-Cavity Interaction and its Effect on Bunch Shape in Storage Rings*". In 9-th ICHEA, pp. 375–9, 1974.

[Papiernik79] A. Papiernik and M. Chatard-Moulin: "*Boundary Perturbations Method*". PAC 79, IEEE-NS 26, pp. 3523–25, 1979.

[Papis91] J-P. Papis and L. Vos: "*Design Basis for the LHC Beam Position Monitor*". Report 91-6, CERN SL, 1991. LHC Note 14.

[Paulsen91] K.D. Paulsen and D.R. Lynch: "*Elimination of of Vector Parasites in Finite Element Maxwell Solutions*". IEEE-MTT-39, pp. 395–404, 1991.

[Petracca93] S. Petracca: "*Image Coefficients for a Vacuum Chamber with Square Cross Section*". Report 93-13, CERN/SL, 1993.

[Petracca94] S. Petracca: "*Coupling Impedance Computation in Accelerator Pipes with Perturbed Boundaries Based on Reciprocity Theorem*". Report 94-78, CERN/SL, 1994.

[Petracca95a] S. Petracca: "*Coupling Impedances for Complex Multi-Coaxial Accelerator Pipes and Application to LHC*". Report 95-109, CERN/SL, 1995.

[Petracca95b] S. Petracca: "*Reciprocity Theorem for Coupling Impedance Computation in the LHC Liner*". Part.Accel. 50, pp. 211–9, 1995.

[Piwinski77] A. Piwinski: "*Penetration of the Field of a Bunched Beam through a Ceramic Vacuum Chamber with Metallic Coating*". IEEE NS-24, pp. 1364–6, 1977.

[Piwinski85] A. Piwinski: "*On the Transverse Forces Caused by the Curvature*". Technical report, CERN, 1985.

[Puglisi87] M. Puglisi: "*The Radiofrequency Quadrupole Linear Accelerator*". Technical report, CERN 87-03, 1987. CAS Oxford 1985.

[Roos69] B. Roos: *Analytic Functions and Distributions.* Wiley, 1969.

[Ruggiero67] A. Ruggiero: "*Longitudinal Resistive Wall Instabiolities in an Intense Coasting Beam for Elliptic Geometry*". Technical report, CERN/ISR-SRM/67-13, 1967.

[Ruggiero68] A. Ruggiero and V. Vaccaro: "*Solution of the Dispersion Relation of an Intense Coasting Beam in a Circular Accler-ator*". Technical report, CERN/ISR-TH/68-33, 1968.

[Sabbi95] G. Sabbi: "*Simulation of Single Bunch Effects*". Report 95-25, CERN/SL, 1995.

[Sacherer73] F. Sacherer: "*Longitudinal Stability Criterion for Bunched Beams*". In PAC 73, IEEE-NS 20, p. 825, 1973. CERN/MPS-BR/73-3.

[Sacherer74] F. Sacherer: "*Transverse Bunched Beam Instability*". In 9-th ICHEA, pp. 347–51, 1974.

[Sacherer76] F. Sacherer: "*Transverse Stability*". Technical report, CERN-YELLOW REPORT 777-13, 1976. Proc. 1st Accler-ator School, Erice 1976.

[Samarskii83] A. Samarskii: *The Theory of Difference Schemes.* Nauka, Moscow (in Russian), 1983.

[Samarskii89] A. Samarskii and E. Nikolaev: *Numerical Methods of Grid Equations.* Birkhauser, 1989.

[Sands70] M. Sands: "*Physics of Electron Storage Rings*". Technical Report 0121, SLAC, 1970.

[Sands74] M. Sands and J. Rees: "*Bunch Deformation Measurement*". Technical report, SLAC-PEP 95, 1974.

[Sands86] M. Sands: "*Centrifugal Space-Charge Forces in the SLC*". Technical report, SLAC-CN-326, 1986.

[Schiff46] L. Schiff: "*Production of Particle Energies beyond 200 Mev*". Rev. Sci. Instr., 17, No. 1, pp. 6–14, 1946.

[Schnell91] W. Schnell and I. Wilson: "*Microwave Quadrupole Structures for CLIC*". In PAC 91, San Francisco, pp. 3237–9, 1991. also Report CERN/SL-91/20.

[Schott07] G. A. Schott: "*Ueber die Strahlung von Electronengruppen*". Annal. der Physik, vol. 24, pp. 635–60, 1907.

[Schwinger45] J. Schwinger: "*On Radiation by Electrons in a Betatron*". unpublished, 1945. also Report LBL-39088, ed. M. Furman (1996).

[Senior81] T.B.A. Senior: "*Approximate Boundary Condition*". IEEE-AP 29, pp. 826–9, 1981.

[Sessler68] A. Sessler: *Beam Loading in the Accelerating Column of an Electron Ring Accelerator* . unpublished monograph, 1968.

[Shafer85] R. Shafer: "*Characteristics of Directional Coupler Beam Position Monitor*". IEEE NS 32, pp. 1933–7, 1985.

[Siemann83] R. Siemann: "*Simulation Codes*". In PAC 83, Fermilab, pp. 271–5, 1983.

[Simpson86] J. Simpson et al.: "*The Argonne Test Facility*". Technical Report CP-86-46, ANL-HEP, 1986.

[Simpson89] J. Simpson: "*Dielectric Wake Field Accelerators*". Note 109, ANL/WF, 1989.

[Slater52] J. Slater and L. Maier: "*Field Strength Measurements in Resonant Cavities*". J. Appl. Phys. 23, p. 68, 1952.

[Smith76] B. T. Smith et al.: *Matrix Eigensystem Routines– EISPACK Guide*. Springer-Verlag, 1976.

[Sokolov80] A. Sokolov and I. M. Ternov: *Synchrotron Radiation*. Pergamon Pres, 1980.

[Stratton41] J. Stratton: *Electromagnetic Theory*. McGraw-Hill, 1941.

[Stupakov94] G. Stupakov and S. Kurennoy: "*Trapped Electromagnetic Modes in a Waveguide with a Small Discontinuity*". Phys. Rev. E 49, p. 794, 1994.

[Stupakov95a] G. Stupakov: "*Geometrical Wake of a Smooth Taper*". Technical report, SLAC-PUB-7086, 1995.

[Stupakov95b] G. V. Stupakov: "*Coupling Impedance in a Periodic Array of Diaphragms*". In PAC 95, 1995. also ICHEA95 and SLAC-PUB-6802.

[Stupakov96] G. Stupakov: "*Geometrical Wake of a Smooth Flat Collimator*". Technical report, SLAC-PUB-7167, 1996.

[Taflove75] A. Taflove and M. Brodwin: "*Numerical Solution of Steady-state Electromagnnetic Scattering Problems Using Time-dependent Maxwell's Equations*". IEEE-MTT-23, p. 262, 1975.

[Talman86] R. Talman: "*A Novel Relativistic Effect Important in Accelerators*". Phys. Rev. Letters 56, pp. 1429–32, 1986.

[Teichmann75] T. Teichmann and E. Wigner: "*Electromagnetic Field Expansions in Loss-Free Cavities Excited through Holes*". J. ppl.Phys., pp. 623–30, 1975.

[Teng63] L. Teng: "*Space-charge impedance*". Technical report, MURA Report, 1963.

[Thomson94] K.A. Thomson et al.: "*Design and Simulation of Accelerating Structures for Future Linear Colliders*". Part. Accel. 47, pp. 65–109, 1994.

[Tran84] D. T. Tran, A. Cailleaux, and G. Berthevas: "*Computer-Aided Design of RF Structires*". In Proc. of the 1984 Linear Accel. Conf, Darmstadt, GSI-84-11, pp. 414–416, 1984.

[Umashankar87] K. Umashankar, A. Taflove, and B. Beker: "*Calculation and Experimental Validation of Induced Currents on Coupled Wires in an Arbitrarily Shaped Cavity*". IEEE-AP-35, pp. 1248–57, 1987.

[Vainshtein63] L. Vainshtein: "*Open Resonator for Lasers*". Sov. Phys. JETP 17, pp. 709–19, 1963. Zh. Eksp. Teor. Fiz. 44, pp. 1415-18.

[Vainshtein69] L.A. Vainshtein: *Theory of Diffraction*. Golden Press, Boulder, CO, 1969.

[van Rienen85] U. van Rienen and T. Weiland: "*Triangular Discretization Method for the Evaluation of RF Fields in Cylindrically Symmetric Cavities*". In Proc. COMPUMAG, Fort Collins, pp. 287–290, 1985.

[van Rienen86] U. van Rienen and T. Weiland: "*Cavity and Waveguide Design by Triangular Mesh Code URMEL-T*". Technical report, Proc. Int. Linear Accel. Conf. SLAC Report SLAC-303, 1986.

[Von Bladel85] J. Von Bladel: *Electromagnetic Fields*. Hemisphere, Wash., 2nd edition, 1985.

[Vos86] L. Vos: "*Computer Calculation of the Longitudinal Impedance of Cylindrically Symmetric Structures and its Application to the SPS*". In PAC 87, IEEE-NS 32, pp. 1237–39, 1986. also Report CERN/SPS 86-21.

[Wagner96] A. Wagner and B. Zotter: "*A 3-dimensional Simulation of Collective Effects in Particle Accelerators*". In 5-th EPAC, Barcelona, 1996. also report CERN-SL-96-29-AP.

[Walkinshaw49] W. Walkinshaw: "*Wave Guides for Slow Waves*". J. Appl. Phys. 20, pp. 634–5, 1949.

[Wang89] T. S. Wang and B. Zotter: "*An Expansion Method for Calculating Wake Potentials*". Report 89-75, CERN/LEP-TH, 1989.

[Warnock82] R. Warnock, G. Bart, and S. Fenster: "*Longitudinal Stability of a Particle Beam in the Presence of Resonant Cavities, Part 1*". Part. Accel. 12, pp. 179–203, 1982. also Report LBL 13180 (1981).

[Warnock90] R. L. Warnock and P. Morton: "*Fields Excited by a Beam in a Smooth Toroidal Chamber. Part 1: Longitudinal Coupling Impedance*". Part. Accel. 25, pp. 113–51, 1990. also SLAC-PUB-4562.

[Warnock91] R. L. Warnock: "*'Shielded Coherent Synchrotron Radiation and its Possible Effect in the Next Linear Collider*". Technical report, SLAC-PUB-5523, 1991.

[Warnock93] R. Warnock: "*An Integro-Algebraic Equation for High Frequency Wake Fields in a Tube with Smoothly Varying Radius*". Technical report, SLAC-PUB-6038, 1993.

[Watson66] G. Watson: *A Treatise on the Theory of Bessel Functions.* Dover, 1966.

[Weiland80a] T. Weiland: "*Transient EM Fields of Bunched Beam in Accelerating Structures with Infinitely Long Tubes*". Technical Report 80-46, CERN-ISR/TH, 1980.

[Weiland80b] T. Weiland and B. Zotter: "*Wake Field of a Relativistic Current in a Cavity*". Part. Accel. 11, p. 143, 1980.

[Weiland82] T. Weiland: "*URMEL User Guide (Modal Analysis of Cylindrically Symmetric Cavities)*". Technical Report M-82-24, DESY, 1982.

[Weiland83a] T. Weiland: "*Comment on Wakefield Computation in the Time Domain (TBCI)*". NIM-A 216, pp. 31–34, 1983.

[Weiland83b] T. Weiland: "*On the Computation of Resonant Modes in Cylindrically Symmetric Cavities (URMEL)*". NIM-A 216, pp. 329–48, 1983.

[Weiland83c] T. Weiland: "*Transverse Beam Cavity Interaction. Part I: Short Range Forces*". NIM-A 212, pp. 13–21, 1983.

[Weiland83d] T. Weiland: "*Transverse Beam Cavity Interaction. Part II: Long Range Forces*". Technical Report M-83-005, DESY, 1983.

[Weiland84] T. Weiland: "*On the Numerical Solution of Maxwell's Equations and Applications in the Field of Accelerator Physics*". Part. Accel. 15, pp. 245–292, 1984.

[Weiland85] T. Weiland: "*On the Unique Numerical Solution of Maxwellian Eigenvalue Problems in Three Dimensions*". Part. Accel. 17, pp. 227–42, 1985.

[Weiland86] T. Weiland and R. Klatt: "*'Wake Field Calculations with Three-Dimensional BCI Code*". Technical report, Proc. Int. Linear Accel. Conf. SLAC Report SLAC-303, 1986.

[Weiland88] T. Weiland: "*Solving Maxwell's Equations by Means of the MAFIA CAD System*". presented at Internat. Symp. on EM Fields in Electrical Engineering, Beijing, China, 1988. also Report DESY-M-88-11, 1988.

[Weiland96] T. Weiland: "*High Precision Eigenmode Computation*". submitted to Part. Accel., 1996.

[Whittaker69] E. Whittaker and G. Watson: *Theory of Diffraction.* Golem Press, 1969.

[Wiener31] N. Wiener and E. Hopf: "*Ueber eine Klasse Singulaerer Integralgleichungen*". Sitzungsber. Deutsch. Akad. Wiss., Berlin, Math. Phys. Tech., pp. 696–706, 1931. also N. Wiener, Collected Works, MIT Press 1966.

[Wilhelm82] W. Wilhelm: "*CAVIT and CAV3D – Computer Programs for RF-Cavities with Constant Cross Section or any Three Dimensional Form*". Part. Accel. 12, p. 139, 1982.

[Wilson72] R. Wilson: "*Synchrotron Radiation in the 20-TeV Proton Accelerator*". In Possibilities and Limitations of Accelerators and Detectors, 1972.

[Wilson77] P. Wilson, K. Bane, et al.: "*Bunch Lengthening and Related Effects in SPEAR II*". In PAC 77, Chicago, IEEE-NS 24, pp. 1211–14, 1977.

[Wilson87] P. Wilson: "*The Fundamental Theorem of Beam Loading*". PUB 4441, SLAC, 1987.

[Wilson89] P. Wilson: "*Introduction to Wake Fields and Wake Potentials*". PUB 4547, SLAC, 1989. also AIP 184, 525.

[Wilson91] P. Wilson: "*High Energy Electron Linacs: Application to Storage Ring RF Systems and Linear Colliders*". Technical Report 2284 (Rev), SLAC, 1991. revised version of paper in AIP Conf. Proc. 87.

[Winick80] H. Winick: *Synchrotron Radiation.* Pergamon Pres, 1980.

[Wolfram89] S. Wolfram: *MATHEMATICA.* Addison-Wesley, 1989.

[Yamamoto91] M. Yamamoto et al.: "*Simulation of Long Range Wake Field in a Detuned Structure Based on Equivalent Circuit Model*". Technical Report 91-153, KEK-Preprint, 1991.

[Yee66] K. S. Yee: "*Numerical Solution of Initial Boundary Value Problems Involving Maxwell's Equations in Isotropid Media*". IEEE-AP-14, pp. 302–7, 1966.

[Yokoya90] K. Yokoya: "*Impedance of Slowly Tapered Structures*". Technical Report 90-88, CERN-SL, 1990.

[Yunn89] B. Yunn: "*A Note on Tapering a Step*". Technical Report 89-134, CEBAF-TN, 1989.

[Zhan96] X. Zhan and K. Ko: "*Parallel Computation of Transverse Wakes in Linear Colliders*". In presented at the Conf. on Computational Physiscs CAP96, Williamsburg, Virginia, 1996.

[Zisman86] M. Zisman, S. Chattopadhyay, and J. Bisognano: "*ZAP - User's Manual*". Report, LBL-21270, 1986.

[Zotter69a] B. Zotter: "*Transverse Oscilaltions of a Relativistic Particle Beam in a Laminated Vacuum Chamber*". Technical report, CERN 69-15, 1969.

[Zotter69b] B. Zotter: "*Transverse Oscillations in a Laminated Vacuum Chamber*". Yellow Report 69-16, CERN, 1969.

[Zotter72a] B. Zotter: "*Coherent Q-Shift of a Relativistic Particle Beam in a Metallic Vacuum Chamber*". Technical report, CERN-ISR/TH/72-8, 1972.

[Zotter72b] B. Zotter: "*Longitudinal Instability inside Walls of Finite Thickness*". Part. Accel. 1, pp. 311–26, 1972.

[Zotter78a] B. Zotter: "*The Effective Coupling Impedance for Bunched Beam Instabilities*". Report 78-16, CERN/ISR-TH, 1978.

[Zotter78b] B. Zotter: "*On the Summation of Infinite Algebraic and Fourier Series*". Report 78-9, CERN/ISR-TH, 1978.

[Zotter79a] B. Zotter: "*Wakefields in a Pillbox Cavity*". Note 310, SLAC/PEP, 1979.

[Zotter79b] B. Zotter and K. Bane: "*Transverse Resonances of Periodically Widened Cylindrical Tubes with Circular Cross Section*". Note 308, SLAC/PEP, 1979.

[Zotter80] B. Zotter: "*The effective Coupling Impedance for Instabilities of Gaussian Bunches*". Report 80-03, CERN/ISR-TH, 1980.

Index